# ESSENTIALS OF
# SOCIAL STATISTICS
## FOR A
# DIVERSE SOCIETY

# ESSENTIALS OF
# SOCIAL STATISTICS
## FOR A
# DIVERSE SOCIETY

## Anna Leon-Guerrero
*Pacific Lutheran University*

## Chava Frankfort-Nachmias
*University of Wisconsin–Milwaukee*

Los Angeles | London | New Delhi
Singapore | Washington DC

Los Angeles | London | New Delhi
Singapore | Washington DC

FOR INFORMATION:

SAGE Publications, Inc.
2455 Teller Road
Thousand Oaks, California 91320
E-mail: order@sagepub.com

SAGE Publications Ltd.
1 Oliver's Yard
55 City Road
London EC1Y 1SP
United Kingdom

SAGE Publications India Pvt. Ltd.
B 1/I 1 Mohan Cooperative Industrial Area
Mathura Road, New Delhi 110 044
India

SAGE Publications Asia-Pacific Pte. Ltd.
33 Pekin Street #02-01
Far East Square
Singapore 048763

Acquisitions Editor:   Jerry Westby
Editorial Assistant:   Erim Sarbuland
Production Editor:   Melanie Birdsall
Typesetter:   C&M Digitals (P) Ltd.
Proofreader:   Cheryl Rivard
Indexer:   Sheila Bodell
Cover Designer:   Gail Buschman
Marketing Manager:   Erica DeLuca
Permissions Editor:   Karen Ehrmann

Printed in the United States of America

Library of Congress Cataloging-in-Publication Data

Leon-Guerrero, Anna.
Essentials of social statistics for a diverse society/Anna Leon-Guerrero, Chava Frankfort-Nachmias.

p. cm.
Rev. ed. of: Social statistics for a diverse society by Chava Frankfort-Nachmias and Anna Leon-Guerrero.
Includes bibliographical references and index.

ISBN 978-1-4522-0583-0 (pbk. : acid-free paper)

1. Social sciences—Statistical methods. 2. Statistics.
I. Frankfort-Nachmias, Chava. II. Frankfort-Nachmias, Chava. Social statistics for a diverse society. III. Title.

HA29.L364 2012
519.5—dc23          2011031172

This book is printed on acid-free paper.

11 12 13 14 15 10 9 8 7 6 5 4 3 2 1

# BRIEF CONTENTS

# Detailed Contents

# ABOUT THE AUTHORS

**Anna Leon-Guerrero** is Professor of Sociology at Pacific Lutheran University in Washington. She received her PhD in sociology from the University of California, Los Angeles. She teaches courses in statistics, social theory, and social problems. Her areas of research and publications include family business, social welfare policy, and social service program evaluation. She is also the author of *Social Problems: Community, Policy, and Social Action* and coeditor of *Contemporary Readings in Social Problems*.

**Chava Frankfort-Nachmias** is now an Emeritus Professor of Sociology at the University of Wisconsin–Milwaukee, where until recently she had been Director of the Center for Jewish Studies and teaching courses in research methods, statistics, and gender. She is the coauthor of *Research Methods in the Social Sciences* (with David Nachmias) and coeditor of *Sappho in the Holy Land* (with Erella Shadmi) and numerous publications on ethnicity and development, urban revitalization, science and gender, and women in Israel. She was the recipient of the University of Wisconsin System teaching improvement grant on integrating race, ethnicity, and gender into the social statistics and research methods curriculum.

# PREFACE

You may be reading this introduction on the first day or sometime during the first week of your statistics class. You probably have some questions about statistics and concerns about what your course will be like. Math, formulas, calculations? Yes, those will be part of your learning experience. However, there is more.

Throughout our text, we emphasize the relevance of statistics in our daily and professional lives. In fact, statistics is such a part of our lives that its importance and uses are often overlooked. How Americans feel about a variety of political and social topics—safety in schools, the environment, the economy, abortion, health care reform, or our president—is measured by surveys and polls and reported daily by the news media. The latest from a health care study on women was just reported on a morning talk show. And that outfit you just purchased—it didn't go unnoticed. The study of consumer trends, specifically focusing on teens and young adults, helps determine commercial programming, product advertising and placement, and, ultimately, consumer spending.

Statistics is not just a part of our lives in the form of news bits or information. And it isn't just numbers either. Throughout this book, we encourage you to move beyond being just a consumer of statistics and begin to recognize and use the many ways that statistics can increase our understanding of our world. As social scientists, we know that statistics can be a valuable set of tools to help us analyze and understand the differences in our American society and the world. We use statistics to track demographic trends, to assess differences among groups in society, and to make an impact on social policy and social change. Statistics can help us gain insight into real-life problems that affect our lives.

## ◙ TEACHING AND LEARNING GOALS

Three teaching and learning goals are the guiding principles of our book.

The first goal is to introduce you to social statistics and demonstrate its value. Although most of you will not use statistics in your own student research, you will be expected to read and interpret statistical information presented by others in professional and scholarly publications, in the workplace, and in the popular media. This book will help you understand the concepts behind the statistics so that you will be able to assess the circumstances in which certain statistics should and should not be used.

Our second goal is to demonstrate that substance and statistical techniques are truly related in social science research. A special quality of this book is its integration of statistical techniques with substantive issues of particular relevance in the social sciences. Your learning will not be limited to statistical calculations and formulas. Rather, you will become proficient in statistical techniques while learning about social differences and inequality through numerous substantive examples and real-world data applications. Because the world we live in is characterized by a growing diversity—where personal and social realities are increasingly shaped by race, class, gender, and other categories of experience—this book teaches you basic statistics while incorporating social science research related to the dynamic interplay of social variables.

Many of you may lack substantial math background, and some of you may suffer from the "math anxiety syndrome." This anxiety often leads to a less-than-optimal learning environment, with students trying to memorize every detail of a statistical procedure rather than attempting to understand the general concept involved. Hence, our third goal is to address math anxiety by using straightforward prose to explain statistical concepts and by emphasizing intuition, logic, and common sense over rote memorization and derivation of formulas.

## ▣ DISTINCTIVE FEATURES OF OUR BOOK

The three learning goals we emphasize are accomplished through a variety of specific and distinctive features throughout this book.

*A Close Link Between the Practice of Statistics, Important Social Issues, and Real-World Examples.* A special quality of this book is its integration of statistical technique with pressing social issues of particular concern to society and social science. We emphasize how the conduct of social science is the constant interplay between social concerns and methods of inquiry. In addition, the examples throughout the book—mostly taken from news stories, government reports, scholarly research, the National Opinion Research Center General Social Survey, and the Monitoring the Future survey—are formulated to emphasize to students like you that we live in a world in which statistical arguments are common. Statistical concepts and procedures are illustrated with real data and research, providing a clear sense of how questions about important social issues can be studied with various statistical techniques.

*A Focus on Diversity: U.S. and International.* A strong emphasis on race, class, and gender as central substantive concepts is mindful of a trend in the social sciences toward integrating issues of diversity in the curriculum. This focus on the richness of social differences within our society and our global neighbors is manifested in the application of statistical tools to examine how race, class, gender, and other categories of experience shape our social world and explain social behavior. Data examples from the International Social Survey Programme data set help expand our statistical focus beyond the United States to other nations.

*Reading the Research Literature.* In your student career and in the workplace, you may be expected to read and interpret statistical information presented by others in professional and scholarly publications. The statistical analyses presented in these publications are a good deal more complex than most class and textbook presentations. To guide you in reading and interpreting research reports written by social scientists, most chapters include a section presenting excerpts of published research reports using the statistical concepts under discussion.

*Tools to Promote Effective Study.* Each chapter concludes with a list of main points and key terms discussed in that chapter. Boxed definitions of the key terms also appear in the body of the chapter, as do learning checks keyed to the most important points. Key terms are also clearly defined and explained in the glossary, another special feature in our book. Answers to all the odd-numbered exercises and Learning Checks in the text are included at the end of the book, as well as on the study site at www.sagepub.com/ssdsessentials. Complete step-by-step solutions are in the manual for instructors, available from the publisher on adoption of the text.

*Emphasis on Computing.* SPSS for Windows is used throughout the book, although the use of computers is not required to learn from the text. Real data are used to motivate and make concrete the coverage of statistical topics. These data, from the General Social Survey (GSS), Health Information National Trends Survey (HINTS), and the Monitoring the Future (MTF) survey, are available on the study site at www .sagepub.com/ssdsessentials. At the end of each chapter, we feature a demonstration of a related SPSS procedure along with a set of exercises.

## ▣ HIGHLIGHTS

- *Real-world examples and exercises:* A hallmark of this edition is the extensive use of real data from a variety of sources for chapter illustrations and exercises.
- *SPSS version 18.0:* Packaged with this text, on an optional basis, is IBM® SPSS®Student version 18.0. SPSS demonstrations and exercises have been updated, using version 18.0 format. Please contact the publisher at (805) 499-4224 or access the publisher's website at www.sagepub.com to learn how to order the book packaged with the student version of SPSS version 18.0.
- *GSS 2008, HINTS 2007, and MTF 2008:* As a companion to this edition's SPSS demonstrations and exercises, we have created four data sets. Those of you with the student version of SPSS 18.0 can work with four separate files: GSS Module A, GSS Module B, HINTS2007, and MTF2008. The GSS08PFP.SAV contains an expanded selection of variables and cases from the 2008 GSS. The HINTS2007.SAV contains 50 variables from the 2007 Health Information National Trends Survey, administered by the National Cancer Institute. HINTS, a nationally representative survey collected in both English and Spanish, aims to monitor changes in the rapidly evolving field of health communication. The MTF2008.SAV contains a selection of variables and cases from the Monitoring the Future 2008 survey conducted by the University of Michigan Survey Research Center. MTF is a survey of 12th-grade students, and it explores drug use and criminal behavior. SPSS exercises, available on the text's website, use certain variables from all data modules. There is ample opportunity for instructors to develop their own SPSS exercises using these data.

## ▣ ACKNOWLEDGMENTS

We are both grateful to Jerry Westby, Series Editor for SAGE Publications, for his commitment to our book and for his invaluable assistance through the production process.

Many manuscript reviewers recruited by SAGE provided invaluable feedback. For their comments to this edition, we thank

Robert Abbey, American University, Washington, DC

John Alexander, Texas Wesleyan University

Melanie Arthur, University of Alaska, Fairbanks

Robert Bickel, Marshall University, West Virginia

Joyce Clapp, University of North Carolina at Greensboro

Paul Croll, Augustana College, Illinois

Todd Daniel, Drury University, Missouri

Regina Davis-Sowers, Santa Clara University, California

Tina Eyraud, Northern Arizona University

Julie Ford, State University of New York, Brockport

Ginny Garcia, University of Texas at San Antonio

Claudia Geist, University of Utah

Tony Giordano, Bloomfield College, New Jersey

Kathy Green, University of Denver

Meredith Greif, Georgia State University

Michael M. Harrod, Central Washington University

Robert Kunovich, University of Texas at Arlington

Bonnie Lewis, Southeastern Louisiana University

Thomas J. Linneman, College of William and Mary

Kyle Longest, Furman University, South Carolina

Ginger Macheski, Valdosta State University, Georgia

Maryhelen MacInnes, Michigan State University

Patricia Nishimoto, Hawaii Pacific University

Mary Noonan, University of Iowa

Ezekiel Olagoke, Waynesburg University, Pennsylvania

Kate Parks, Loras College, Iowa

Mitchell Peck, University of Oklahoma

David Pettinicchio, University of Washington

Heather Rodriguez, Marian University, Indiana

Christabel Rogalin, Purdue University North Central, Indiana

Sara Skiles, University of Notre Dame

Carrie Lee Smith, Millersville University of Pennsylvania

Carrie E. Spearin, Brown University

Pamela Stone, Hunter College of The City University of New York

Nancy Whittier, Smith College, Massachusetts

Loretta Winters, California State University, Northridge

We are grateful to Melanie Birdsall for guiding the book through the production process.

Both of us extend our deepest appreciation to Caleb Schaffner. Our work on this edition would not have been possible without his contribution. We are grateful to Caleb for his editing attention and thoroughness.

Anna Leon-Guerrero expresses her thanks to the following: I wish to thank my statistics students. My passion for and understanding of teaching statistics grow with each semester and class experience. I am grateful for the teaching and learning opportunities that we have shared.

I would like to express my gratitude to friends and colleagues for their encouragement and support throughout this project. My love and thanks to my husband, Brian Sullivan.

Chava Frankfort-Nachmias would like to thank and acknowledge the following: My profound gratitude goes to friends and colleagues who have stood by me, cheered me on, and understood when I was unavailable for long periods due to the demands of this project.

I am grateful to my students at the University of Wisconsin–Milwaukee, who taught me that even the most complex statistical ideas can be simplified. The ideas presented in this book are the products of many years of classroom testing. I thank my students for their patience and contributions.

Finally, I thank my partner, Marlene Stern, and my daughters, Anat and Talia, for their love, support, and faith in me.

Anna Leon-Guerrero
*Pacific Lutheran University*

Chava Frankfort-Nachmias
*University of Wisconsin–Milwaukee*

# Chapter 1

# The What and the Why of Statistics

## Chapter Learning Objectives

❖ Understanding the research process
❖ Identifying and distinguishing between independent and dependent variables
❖ Identifying and distinguishing between three levels of measurement
❖ Understanding descriptive versus inferential statistical procedures

Are you taking statistics because it is required in your major—not because you find it interesting? In this book, we will show you that statistics can be a lot more interesting and easy to understand than you may have been led to believe. As we draw on your previous knowledge and experience and relate materials to interesting and important social issues, you'll begin to see that statistics is not just a course you have to take but a useful tool as well.

There are two major reasons why learning statistics may be of value to you. First, you are constantly exposed to statistics every day of your life. Marketing surveys, voting polls, and the findings of social research appear daily in newspapers and popular magazines. By learning statistics, you will become a sharper consumer of statistical material. Second, as a major in the social sciences, even if conducting research is not a part of your job, you may still be expected to understand and learn from other people's research or to be able to write reports based on statistical analyses.

Just what *is* statistics anyway? You may associate the word with numbers that indicate birthrates, conviction rates, per capita income, marriage and divorce rates, and so on. But the word **statistics** also refers to a set of procedures used by social scientists. They use these procedures to organize, summarize, and communicate information. Only information represented by numbers can be the subject of statistical analysis. Such information is called **data**; researchers use statistical procedures to analyze data to answer research questions and test theories. It is the latter usage—answering research questions and testing theories—that this textbook explores.

*Statistics*   A set of procedures used by social scientists to organize, summarize, and communicate information.

*Data*   Information represented by numbers, which can be the subject of statistical analysis.

## ◉ THE RESEARCH PROCESS

To give you a better idea of the role of statistics in social research, let's start by looking at the **research process**. We can think of the research process as a set of activities in which social scientists engage so that they can answer questions, examine ideas, or test theories.

As illustrated in Figure 1.1, the research process consists of five stages:

1. Asking the research question

2. Formulating the hypotheses

3. Collecting data

4. Analyzing data

5. Evaluating the hypotheses

**Figure 1.1**   The Research Process

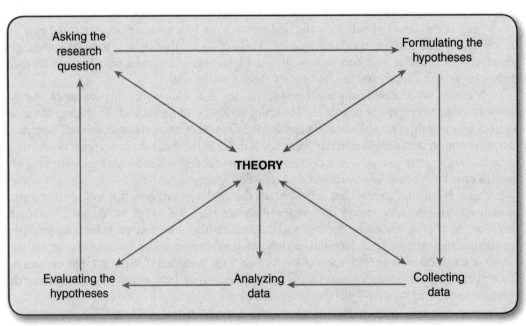

Each stage affects the *theory* and is affected by it as well. Statistics is most closely tied to the data analysis stage of the research process. As we will see in later chapters, statistical analysis of the data helps researchers test the validity and accuracy of their hypotheses.

---

*Research process*   A set of activities in which social scientists engage to answer questions, examine ideas, or test theories.

---

## ▣ ASKING RESEARCH QUESTIONS

The starting point for most research is asking a *research question.* Consider the following research questions taken from a number of social science journals:

Does managed health care influence the quality of health care?

Does social class influence voting behavior?

What factors influence the economic mobility of female workers?

These are all questions that can be answered by conducting **empirical research**—research based on information that can be verified by using our direct experience. To answer research questions, we cannot rely on reasoning, speculation, moral judgment, or subjective preference. For example, the questions, "Is racial equality good for society?" and "Is an urban lifestyle better than a rural lifestyle?" cannot be answered empirically because the terms *good* and *better* are concerned with values, beliefs, or subjective preference and, therefore, cannot be independently verified. One way to study these questions is by defining *good* and *better* in terms that can be verified empirically. For example, we can define *good* in terms of economic growth and *better* in terms of psychological well-being. These questions could then be answered by conducting empirical research.

---

*Empirical research*   Research based on evidence that can be verified by using our direct experience.

---

You may wonder how to come up with a research question. The first step is to pick a question that interests you. Ideas for research problems are all around you, from media sources to personal experience or your own intuition. Talk to other people, write down your own observations and ideas, or learn what other social scientists have written about.

Take, for instance, the issue of gender and work. In 2008, women who were employed full-time earned about $682 per week on average; men who were employed full-time earned $876 per week on average.[1] Women's and men's work are also very different. Women continue to be the minority in many of the higher-ranking and higher-salaried positions in professional and managerial occupations. For example, in 2008 women made up 10.4% of civil engineers, 30.5% of physicians, 27.2% of dentists, and 24.8% of architects. In comparison, among all those employed as preschool and kindergarten teachers, 97.6% were women. Among all administrative assistants in 2008, 96.1% were women.[2] Another noteworthy development in the history of labor in the United States took place in January 2010: Women outnumbered men for the first time by holding 50.3% of the nonfarm payroll jobs.[3] These observations may prompt us to ask research questions such as the following: Are women paid, on an average, less than men for the same type of work? How much change has there been in women's work over time?

✓ Learning Check

> *Identify one or two social science questions amenable to empirical research.*

## ▣ THE ROLE OF THEORY

You may have noticed that each preceding research question was expressed in terms of a *relationship*. This relationship may be between two or more attributes of individuals or groups, such as gender segregation in the workplace and income disparity. The relationship between attributes or characteristics of individuals and groups lies at the heart of social scientific inquiry.

Most of us use the term *theory* quite casually to explain events and experiences in our daily life. We may have a "theory" about why our boss has been so nice to us lately or why we didn't do so well on our last history test. In a somewhat similar manner, social scientists attempt to explain the nature of social reality. To the social scientist, a theory is a more precise explanation that is frequently tested by conducting research.

A **theory** is an explanation of the relationship between two or more observable attributes of individuals or groups. The theory attempts to establish a link between what we observe (the data) and our conceptual understanding of why certain phenomena are related to each other in a particular way. Suppose we wanted to understand the reasons for the income disparity between men and women; we may wonder whether the types of jobs men and women have and the organizations in which they work have something to do with their wages.

---

*Theory*   An elaborate explanation of the relationship between two or more observable attributes of individuals or groups.

---

One explanation for gender inequality in wages is *gender segregation in the workplace*—the fact that American men and women are concentrated in different kinds of jobs and occupations. For example, in 2008, of the approximately 69 million women in the labor force, more than one third (34%) worked in only 1 of the 14 industries listed by the census.[4]

The jobs in which women and men are segregated are not only different but also unequal. Although the proportion of women in the labor force has markedly increased, women are still concentrated in occupations with low pay, low prestige, and few opportunities for promotion. In particular, women's segregation into different jobs and occupations from those of men is the most immediate cause of the pay gap. Women receive lower pay than men do even when they have the same level of education, skills, and experience as men in comparable occupations.

## ▣ FORMULATING THE HYPOTHESES

So far, we have come up with a number of research questions about the income disparity between men and women in the workplace. We have also discussed a possible explanation—a theory—that helps us make sense of gender inequality in wages. Is that enough? Where do we go from here?

Our next step is to test some of the ideas suggested by the gender segregation theory. Theories suggest specific concrete predictions about the way that observable attributes of people or groups are interrelated in real life. These predictions, called **hypotheses**, are tentative answers to research problems. Hypotheses are tentative because they can be verified only after they have been tested empirically.[5] For example, one hypothesis we can derive from the gender segregation theory is that wages in occupations in which the majority of workers are female are lower than the wages in occupations in which the majority of workers are male.

---

*Hypothesis*    A tentative answer to a research problem.

---

Not all hypotheses are derived directly from theories. We can generate hypotheses in many ways—from theories, directly from observations, or from intuition. Probably, the greatest source of hypotheses is the professional literature. A critical review of the professional literature will familiarize you with the current state of knowledge and with hypotheses that others have studied.

Let's restate our hypothesis:

Wages in occupations in which the majority of workers are female are lower than the wages in occupations in which the majority of workers are male.

Note that this hypothesis is a statement of a relationship between two characteristics that vary: *wages* and *gender composition* of occupations. Such characteristics are called variables. A **variable** is a property of people or objects that takes on two or more values. For example, people can be classified into a number of *social class* categories, such as upper class, middle class, or working class. Similarly, people have different levels of education; therefore, *education* is a variable. *Family income* is a variable; it can take on values from zero to hundreds of thousands of dollars or more. *Wages* is a variable, with values from zero to thousands of dollars or more. Similarly, *gender composition* is a variable. The percentage of females (or males) in an occupation can vary from 0 to 100. (See Figure 1.2 for examples of some variables and their possible values.)

---

*Variable*    A property of people or objects that takes on two or more values.

---

**Figure 1.2**   Variables and Value Categories

| Variable | Categories |
|---|---|
| Social class | Upper class<br>Middle class<br>Working class |
| Religion | Christian<br>Jewish<br>Muslim |
| Monthly income | $1,000<br>$2,500<br>$10,000<br>$15,000 |
| Gender | Male<br>Female |

Each variable must include categories that are both *exhaustive* and *mutually exclusive*. Exhaustiveness means that there should be enough categories composing the variables to classify every observation. For example, the common classification of the variable *marital status* into the categories "married," "single," "divorced," and "widowed" violates the requirement of exhaustiveness. As defined, it does not allow us to classify same-sex couples or heterosexual couples who are not legally married. (We can make every variable exhaustive by adding the category "other" to the list of categories. However, this practice is not recommended if it leads to the exclusion of categories that have theoretical significance or a substantial number of observations.)

Mutual exclusiveness means that there is only one category suitable for each observation. For example, we need to define *religion* in such a way that no one would be classified into more than one category. For instance, the categories "Protestant" and "Methodist" are not mutually exclusive because Methodists are also considered Protestant and, therefore, could be classified into both categories.

✓ *Learning Check*

*Review the definitions of exhaustive and mutually exclusive. Now look at Figure 1.2. What other categories could be added to the variable* religion *to be exhaustive and mutually exclusive? What other categories could be added to* social class? *To* income?

Social scientists can choose which level of social life to focus their research on. They can focus on individuals or on groups of people such as families, organizations, and nations. These distinctions are referred to as **units of analysis.** A variable is a property of whatever the unit of analysis is for the study. Variables can be properties of individuals, of groups (e.g., the family or a social group), of organizations (e.g., a hospital or university), or of societies (e.g., a country or a nation). For example, in a study that looks at the relationship between individuals' level of education and their income, the variable *income* refers to the income level of an individual. On the other hand, a study that compares how differences in corporations' revenues relate to differences in the fringe benefits they provide to their employees uses the variable *revenue* as a characteristic of an organization (the corporation). The variables *wages* and *gender composition* in our example are characteristics of occupations. Figure 1.3 illustrates different units of analysis frequently employed by social scientists.

---

*Unit of analysis*   The level of social life on which social scientists focus. Examples of different levels are individuals and groups.

---

✓ *Learning Check*

*Remember that research question you came up with? Can you formulate a hypothesis you could test? Remember that the variables must take on two or more values and you must determine the unit of analysis.*

## Independent and Dependent Variables: Causality

Hypotheses are usually stated in terms of a relationship between an *independent* and a *dependent variable*. The distinction between an independent and a dependent variable is important in the language of research. Social theories often intend to provide an explanation for social patterns or causal relations between variables. For example, according to the gender segregation theory, gender segregation in the workplace is the primary explanation (although certainly not the only one) of the male-female earning gap. Why should jobs where the majority of workers are women pay less than jobs that employ mostly men? One explanation is that

societies undervalue the work women do, regardless of what those tasks are, because women do them. . . . For example, our culture tends to devalue caring or nurturant work at least partly because it is done by women. This tendency accounts for child care workers' low rank in the pay hierarchy.[6]

In the language of research, the variable the researcher wants to explain (the "effect") is called the **dependent variable**. The variable that is expected to "cause" or account for the dependent variable is called the **independent variable**. Therefore, in our example, *gender composition of occupations* is the independent variable, and *wages* is the dependent variable.

---

*Dependent variable*   The variable to be explained (the "effect").

*Independent variable*   The variable expected to account for (the "cause" of) the dependent variable.

---

Cause-and-effect relationships between variables are *not* easy to infer in the social sciences. To establish that two variables are causally related, you need to meet three conditions: (1) The cause has

**Figure 1.3**   Examples of Units of Analysis

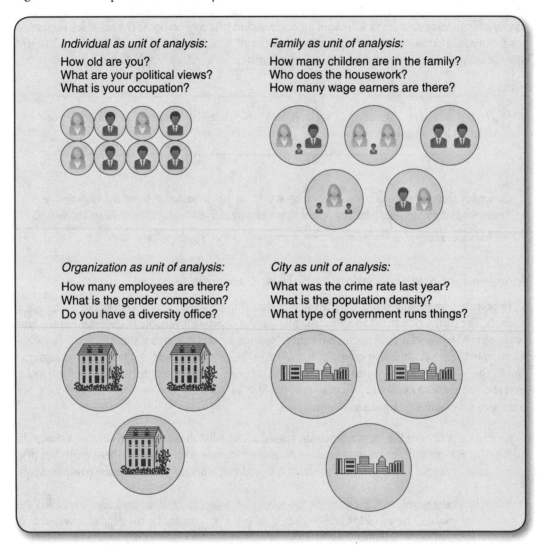

Individual as unit of analysis:

How old are you?
What are your political views?
What is your occupation?

Family as unit of analysis:

How many children are in the family?
Who does the housework?
How many wage earners are there?

Organization as unit of analysis:

How many employees are there?
What is the gender composition?
Do you have a diversity office?

City as unit of analysis:

What was the crime rate last year?
What is the population density?
What type of government runs things?

to precede the effect in time, (2) there has to be an empirical relationship between the cause and the effect, and (3) this relationship cannot be explained by other factors.

## Independent and Dependent Variables: Guidelines

Because of the limitations in inferring cause-and-effect relationships in the social sciences, be cautious about using the terms *cause* and *effect* when examining relationships between variables. However, using the terms *independent variable* and *dependent variable* is still appropriate even when this relationship is not articulated in terms of direct cause and effect. Here are a few guidelines that may help you identify the independent and dependent variables:

1. The dependent variable is always the property that you are trying to explain; it is always the object of the research.

2. The independent variable usually occurs earlier in time than the dependent variable.

3. The independent variable is often seen as influencing, directly or indirectly, the dependent variable.

The purpose of the research should help determine which is the independent variable and which is the dependent variable. In the real world, variables are neither dependent nor independent; they can be switched around depending on the research problem. A variable defined as independent in one research investigation may be a dependent variable in another.[7] For instance, *educational attainment* may be an independent variable in a study attempting to explain how education influences political attitudes. However, in an investigation of whether a person's level of education is influenced by the social status of his or her family of origin, *educational attainment* is the dependent variable. Some variables, such as race, age, and ethnicity, because they are primordial characteristics that cannot be explained by social scientists, are never considered dependent variables in a social science analysis.

---

*Identify the independent and dependent variables in the following hypotheses:*

- *Younger Americans are more likely to support health care reform than older Americans.*
- *People who attend church regularly are more likely to oppose abortion than people who do not attend church regularly.*
- *Elderly women are more likely to live alone than elderly men.*
- *Individuals with postgraduate education are likely to have fewer children than those with less education.*

*What are the independent and dependent variables in your hypothesis?*

✓ *Learning Check*

---

## ▣ COLLECTING DATA

Once we have decided on the research question, the hypothesis, and the variables to be included in the study, we proceed to the next stage in the research cycle. This step includes measuring our variables and collecting the data. As researchers, we must decide how to measure the variables of interest to us, how to select the cases for our research, and what kind of data collection techniques we will be using. A wide variety of data collection techniques are available to us, from direct observations to survey research, experiments, or secondary sources. Similarly, we can construct numerous measuring instruments. These instruments can be as simple as a single question included in a questionnaire or as complex as a composite measure constructed through the combination of two or more questionnaire items. The choice of a particular data collection method or instrument to measure our variables depends on the study objective. For instance, suppose we decide to study how social class position is related to attitudes about abortion. Since attitudes about abortion are not directly observable, we need to collect data by asking a group of people questions about their attitudes and opinions. A suitable method of data collection for this project would be a *survey* that uses some kind of questionnaire or interview guide to elicit verbal reports from respondents. The questionnaire could include numerous questions designed to measure attitudes toward abortion, social class, and other variables relevant to the study.

How would we go about collecting data to test the hypothesis relating the gender composition of occupations to wages? We want to gather information on the proportion of men and women in different occupations and the average earnings for these occupations. This kind of information is routinely collected by the government and published in sources such as bulletins distributed by the U.S. Department of Labor's Bureau of Labor Statistics and the *Statistical Abstract of the United States*. The data obtained from these sources could then be analyzed and used to test our hypothesis.

## Levels of Measurement

The statistical analysis of data involves many mathematical operations, from simple counting to addition and multiplication. However, not every operation can be used with every variable. The type of statistical operations we employ depends on how our variables are measured. For example, for the variable *gender*, we can use the number 1 to represent females and the number 2 to represent males. Similarly, 1 can also be used as a numerical code for the category "one child" in the variable *number of children*. Clearly, in the first example, the number is an arbitrary symbol that does not correspond to the property "female," whereas in the second example the number 1 has a distinct numerical meaning that does correspond to the property "one child." The correspondence between the properties we measure and the numbers representing these properties determines the type of statistical operations we can use. The degree of correspondence also leads to different ways of measuring—that is, to distinct *levels of measurement*. In this section, we will discuss three levels of measurement: *nominal*, *ordinal*, and *interval ratio*.

### Nominal Level of Measurement

At the **nominal** level of measurement, numbers or other symbols are assigned a set of categories for the purpose of naming, labeling, or classifying the observations. *Gender* is an example of a nominal-level variable. Using the numbers 1 and 2, for instance, we can classify our observations into the categories "females" and "males," with 1 representing females and 2 representing males. We could use any of a variety of symbols to represent the different categories of a nominal variable; however, when numbers are used to represent the different categories, we do not imply anything about the magnitude or quantitative difference between the categories. Because the different categories (e.g., males vs. females) vary in the quality inherent in each but not in quantity, nominal variables are often called *qualitative*. Other examples of nominal-level variables are political party, religion, and race.

---

*Nominal measurement*   Numbers or other symbols are assigned to a set of categories for the purpose of naming, labeling, or classifying the observations.

---

### Ordinal Level of Measurement

Whenever we assign numbers to rank-ordered categories ranging from low to high, we have an **ordinal** level of measurement. *Social class* is an example of an ordinal variable. We might classify individuals with respect to their social class status as "upper class," "middle class," or "working class." We can say that a person in the category "upper class" has a higher class position than a person in a "middle-class" category (or that a "middle-class" position is higher than a "working-class" position), but we do not know how much higher "upper class" is compared with the "middle class."

Many attitudes that we measure in the social sciences are ordinal-level variables. Take, for instance, the following statement used to measure attitudes toward same-sex marriages: "Same-sex partners should have the right to marry each other." Respondents are asked to mark the number representing their degree of agreement or disagreement with this statement. One form in which a number might be made to correspond with the answers can be seen in Table 1.1. Although the differences between these numbers represent higher or lower degrees of agreement with same-sex marriage, the distance between any two of those numbers does not have a precise numerical meaning.

---

**Ordinal measurement**   Numbers are assigned to rank-ordered categories ranging from low to high.

---

**Table 1.1**   Ordinal Ranking Scale

| Rank | Value |
|------|-------|
| 1 | Strongly agree |
| 2 | Agree |
| 3 | Neither agree nor disagree |
| 4 | Disagree |
| 5 | Strongly disagree |

### Interval-Ratio Level of Measurement

If the categories (or values) of a variable can be rank ordered, and if the measurements for all the cases are expressed in the same units, then an **interval-ratio** level of measurement has been achieved. Examples of variables measured at the interval-ratio level are *age, income,* and *SAT scores.* With all these variables, we can compare values not only in terms of which is larger or smaller but also in terms of *how much* larger or smaller one is compared with another. In some discussions of levels of measurement, you will see a distinction made between interval-ratio variables that have a natural zero point (where zero means the absence of the property) and those variables that have zero as an arbitrary point. For example, weight and length have a natural zero point, whereas temperature has an arbitrary zero point. Variables with a natural zero point are also called *ratio variables.* In statistical practice, however,

**Table 1.2**   Gender Composition of Four Major Occupational Groups

| *Occupational Group* | *Women in Occupation (%)* |
|---|---|
| Management, professional, and related occupations | 50.8 |
| Service occupations | 57.2 |
| Production, transportation, and materials occupations | 22.4 |
| Natural resources, construction, and maintenance occupations | 4.2 |

*Source:* U.S. Bureau of the Census, *Statistical Abstract of the United States*, 2010, Table 603.

ratio variables are subjected to operations that treat them as interval and ignore their ratio properties. Therefore, no distinction between these two types is made in this text.

---

*Interval-ratio measurement*   Measurements for all cases are expressed in the same units.

---

### Cumulative Property of Levels of Measurement

Variables that can be measured at the interval-ratio level of measurement can also be measured at the ordinal and nominal levels. As a rule, properties that can be measured at a higher level (interval-ratio is the highest) can also be measured at lower levels, but not vice versa. Let's take, for example, *gender composition of occupations,* the independent variable in our research example. Table 1.2 shows the percentage of women in four major occupational groups as reported in the 2010 *Statistical Abstract of the United States.*

The variable *gender composition* (measured as the percentage of women in the occupational group) is an interval-ratio variable and, therefore, has the properties of nominal, ordinal, and interval-ratio measures. For example, we can say that the management group differs from the natural resources group (a nominal comparison), that service occupations have more women than the other occupational categories (an ordinal comparison), and that service occupations have 34.8 percentage points more women (57.2 − 22.4) than production occupations (an interval-ratio comparison).

The types of comparisons possible at each level of measurement are summarized in Table 1.3 and Figure 1.4. Note that differences can be established at each of the three levels, but only at the interval-ratio level can we establish the magnitude of the difference.

### Levels of Measurement of Dichotomous Variables

A variable that has only two values is called a **dichotomous variable.** Several key social factors, such as gender, employment status, and marital status, are dichotomies—that is, you are male or female, employed or unemployed, married or not married. Such variables may seem to be measured at the nominal level: You fit in either one category or the other. No category is naturally higher or lower than the other, so they can't be ordered.

Table 1.3    Levels of Measurement and Possible Comparisons

| Level | Different or Equivalent | Higher or Lower | How Much Higher |
|---|---|---|---|
| Nominal | Yes | No | No |
| Ordinal | Yes | Yes | No |
| Interval-ratio | Yes | Yes | Yes |

Figure 1.4    Levels of Measurement and Possible Comparisons: Education Measured on Nominal, Ordinal, and Interval-Ratio Levels

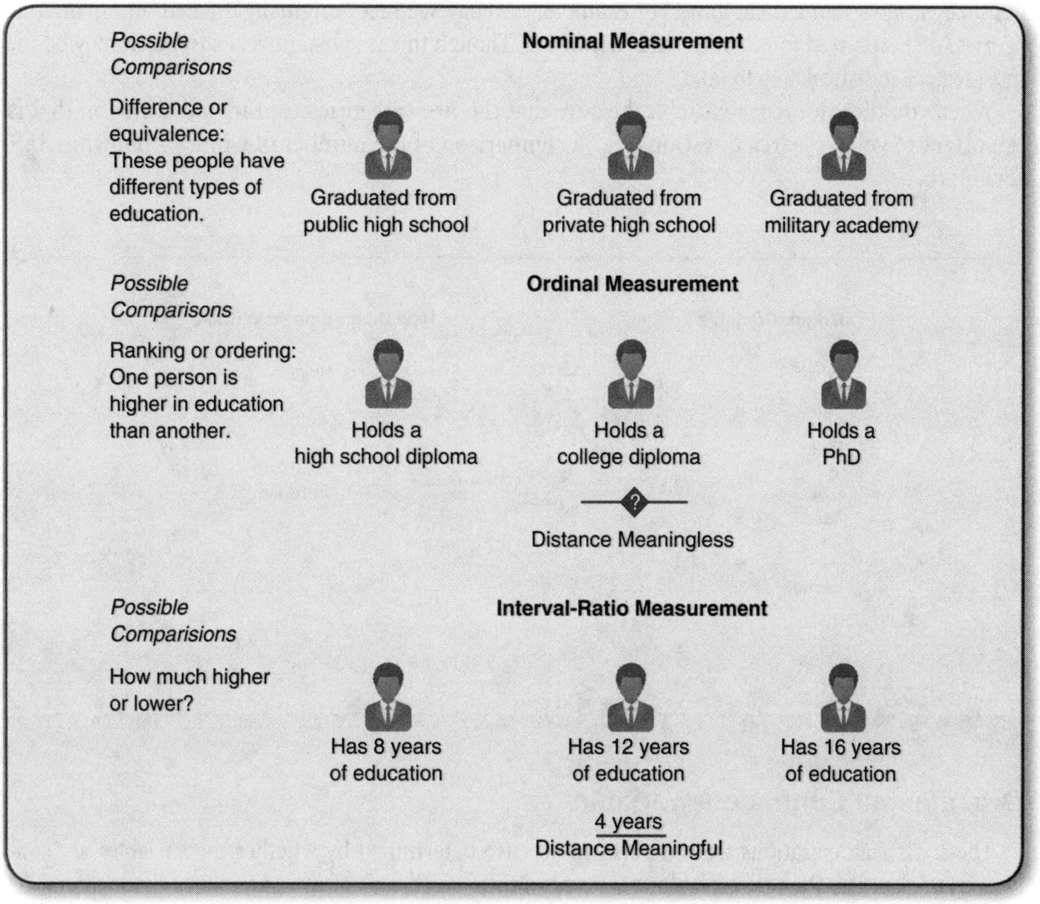

*Possible Comparisons*

**Nominal Measurement**

Difference or equivalence: These people have different types of education.

Graduated from public high school      Graduated from private high school      Graduated from military academy

*Possible Comparisons*

**Ordinal Measurement**

Ranking or ordering: One person is higher in education than another.

Holds a high school diploma      Holds a college diploma      Holds a PhD

Distance Meaningless

*Possible Comparisions*

**Interval-Ratio Measurement**

How much higher or lower?

Has 8 years of education      Has 12 years of education      Has 16 years of education

4 years
Distance Meaningful

---

*Dichotomous variable*    A variable that has only two values.

---

However, because there are only two possible values for a dichotomy, we can measure it at the ordinal or the interval-ratio level. For example, we can think of "femaleness" as the ordering principle for gender, so that "female" is higher and "male" is lower. Using "maleness" as the ordering principle, "female" is lower and "male" is higher. In either case, with only two classes, there is no way to get them out of order; therefore, gender could be considered at the ordinal level.

Dichotomous variables can also be considered to be interval-ratio level. Why is this? In measuring interval-ratio data, the size of the interval between the categories is *meaningful:* The distance between 4 and 7, for example, is the same as the distance between 11 and 14. But with a dichotomy, there is only one interval. Therefore, there is really no other distance to which we can compare it.

Mathematically, this gives the dichotomy more power than other nominal-level variables (as you will notice later in the text).

For this reason, researchers often dichotomize some of their variables, turning a multicategory nominal variable into a dichotomy. For example, you may see race (originally divided into many categories) dichotomized into "white" and "nonwhite." Though this is substantively suspect, it may be the most logical statistical step to take.

When you dichotomize a variable, be sure that the two categories capture a distinction that is important to your research question (e.g., a comparison of the number of white vs. nonwhite U.S. senators).

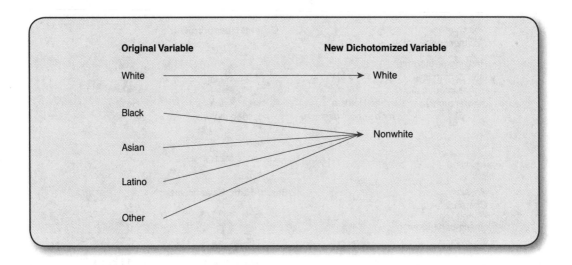

## Discrete and Continuous Variables

The statistical operations we can perform are also determined by whether the variables are continuous or discrete. *Discrete* variables have a minimum-sized unit of measurement, which cannot be subdivided. The number of children per family is an example of a discrete variable because the minimum unit is one child. A family may have two or three children, but not 2.5 children. The variable *wages* in our research example is a discrete variable because currency has a minimum unit (1 cent), which cannot be subdivided. One can have $101.21 or $101.22 but not $101.21843. Wages cannot differ by less than 1 cent—the minimum-sized unit.

Unlike discrete variables, continuous variables do not have a minimum-sized unit of measurement; their range of values can be subdivided into increasingly smaller fractional values. *Length* is an example of a continuous variable because there is no minimum unit of length. A particular object may be 12 in. long, it may be 12.5 in. long, or it may be 12.532011 in. long. Although we cannot always measure all possible length values with absolute accuracy, it is possible for objects to exist at an infinite number of lengths.[8] In principle, we can speak of a tenth of an inch, a ten thousandth of an inch, or a ten trillionth of an inch. The variable *gender composition of occupations* is a continuous variable because it is measured in proportions or percentages (e.g., the percentage of women in medicine), which can be subdivided into smaller and smaller fractions.

This attribute of variables—whether they are continuous or discrete—affects subsequent research operations, particularly measurement procedures, data analysis, and methods of inference and generalization. However, keep in mind that, in practice, some discrete variables can be treated as if they were continuous, and vice versa.

*Name three continuous and three discrete variables. Determine whether each of the variables in your hypothesis is continuous or discrete.*

✓ *Learning Check*

## ▣ A CAUTIONARY NOTE: MEASUREMENT ERROR

Social scientists attempt to ensure that the research process is as error free as possible, beginning with how we construct our measurements. We pay attention to two characteristics of measurement: *reliability* and *validity*.

Reliability means that the measurement yields consistent results each time it is used. For example, asking a sample of individuals "Do you approve or disapprove of President Obama's job performance?" is more reliable than asking "What do you think of President Obama's job performance?" While responses to the second question are meaningful, the answers might be vague and could be subject to different interpretation. Researchers look for the consistency of measurement over time, in relationship with other related measures, or in measurements or observations made by two or more researchers. Reliability is a prerequisite for validity: We cannot measure a phenomenon if the measure we are using gives us inconsistent results.

Validity refers to the extent to which measures indicate what they are intended to measure. While standardized IQ tests are reliable, it is still debated whether such tests measure intelligence or one's test-taking ability. A measure may not be valid due to individual error (individuals may want to provide socially desirable responses) or method error (questions may be unclear or poorly written).

Specific techniques and practices for determining and improving measurement reliability and validity are the subject of research methods courses.

## ▣ ANALYZING DATA AND EVALUATING THE HYPOTHESES

Following the data collection stage, researchers analyze their data and evaluate the hypotheses of the study. The data consist of codes and numbers used to represent our observations. In our example, each occupational group would be represented by two scores: (1) the percentage of women and (2) the average wage. If we had collected information on 100 occupations, we would end up with 200 scores, 2 per occupational group. However, the typical research project includes more variables; therefore, the amount of data the researcher confronts is considerably larger. We now must find a systematic way to organize these data, analyze them, and use some set of procedures to decide what they mean. These last steps make up the *statistical analysis* stage, which is the main topic of this textbook. It is also at this point in the research cycle that statistical procedures will help us *evaluate* our research hypothesis and assess the theory from which the hypothesis was derived.

### Descriptive and Inferential Statistics

Statistical procedures can be divided into two major categories: *descriptive statistics* and *inferential statistics*. Before we can discuss the difference between these two types of statistics, we need to understand the terms *population* and *sample*. A **population** is the total set of individuals, objects, groups, or events in which the researcher is interested. For example, if we were interested in looking at voting behavior in the last presidential election, we would probably define our population as all citizens who voted in the election. If we wanted to understand the employment patterns of Latinas in our state, we would include in our population all Latinas in our state who are in the labor force.

---

*Population*   The total set of individuals, objects, groups, or events in which the researcher is interested.

---

Although we are usually interested in a population, quite often, because of limited time and resources, it is impossible to study the entire population. Imagine interviewing all the citizens of the United States who voted in the last election or even all the Latinas who are in the labor force in our state. Not only would that be very expensive and time-consuming, but we would also probably have a very hard time locating everyone! Fortunately, we can learn a lot about a population if we carefully select a subset from that population. A subset selected from a population is called a **sample**. Researchers usually collect their data from a sample and then generalize their observations to the larger population.

---

*Sample*   A relatively small subset selected from a population.

---

**Descriptive statistics** includes procedures that help us organize and describe data collected from either a sample or a population. Occasionally data are collected on an entire population, as in a census.

**Inferential statistics**, on the other hand, is concerned with making predictions or inferences about a population from observations and analyses of a sample. For instance, the General Social Survey (GSS), from which numerous examples presented in this book are drawn, is conducted every other year by the National Opinion Research Center (NORC) on a representative sample of several thousands of respondents (e.g., the sample size in 2006 was 4,510, and in 2008 it was 2,023). The survey, which includes several hundred questions, is designed to provide social science researchers with a readily accessible database of socially relevant attitudes, behaviors, and attributes of a cross section of the U.S. adult population. NORC has verified that the composition of the GSS samples closely resembles census data. But because the data are based on a sample rather than on the entire population, the average of the sample does not equal the average of the population as a whole. The tools of statistical inference help determine the accuracy of the sample average obtained by the researchers.

---

*Descriptive statistics*   Procedures that help us organize and describe data collected from either a sample or a population.

*Inferential statistics*   The logic and procedures concerned with making predictions or inferences about a population from observations and analyses of a sample.

---

## Evaluating the Hypotheses

At the completion of these descriptive and inferential procedures, we can move to the next stage of the research process: the assessment and evaluation of our hypotheses and theories in light of the analyzed data. At this next stage, new questions might be raised about unexpected trends in the data and about other variables that may have to be considered in addition to our original variables. For example, we may have found that the relationship between gender composition of occupations and earnings can be observed with respect to some groups of occupations but not others. Similarly, the relationship between these variables may apply for some racial/ethnic groups but not for others.

These findings provide evidence to help us decide how our data relate to the theoretical framework that guided our research. We may decide to revise our theory and hypothesis to take account of these later findings. Recent studies are modifying what we know about gender segregation in the workplace. These studies suggest that race as well as gender shapes the occupational structure in the United States and helps explain disparities in income. This reformulation of the theory calls for a modified hypothesis and new research, which starts the circular process of research all over again.

Statistics provides an important link between theory and research. As our example on gender segregation demonstrates, the application of statistical techniques is an indispensable part of the research process. The results of statistical analyses help us evaluate our hypotheses and theories, discover unanticipated patterns and trends, and provide the impetus for shaping and reformulating our theories. Nevertheless, the importance of statistics should not diminish the significance of the preceding phases of the research process. Nor does the use of statistics lessen the importance of our own judgment in the entire process. Statistical analysis is a relatively small part of the research process, and even the most rigorous statistical procedures cannot speak for themselves. If our research questions are poorly conceived or our data are flawed due to errors in our design and measurement procedures, our results will be useless.

## ▣ LOOKING AT SOCIAL DIFFERENCES

By the middle of this century, if current trends continue unchanged, the United States will no longer be a predominantly European society. Due mostly to renewed immigration and higher birthrates, the United States is being transformed into a "global society" in which nearly half the population will be of African, Asian, Latino, or Native American ancestry.

Is the increasing diversity of American society relevant to social scientists? What impact will such diversity have on the research methodologies we employ?

In a diverse society stratified by race, ethnicity, class, and gender, less partial and distorted explanations of social relations tend to result when researchers, research participants, and the research process itself reflect that diversity. Such diversity shapes the research questions we ask, how we observe and interpret our findings, and the conclusions we draw.

A statistical approach that focuses on social differences uses statistical tools to examine how variables such as race, class, and gender as well as other demographic categories such as age, religion, and sexual orientation shape our social world and explain our social behavior. Numerous statistical procedures can be applied to describe these processes, and we will begin to look at some of those options in the next chapter.

Whichever model of social research you use—whether you follow a traditional one or integrate your analysis with qualitative data, whether you focus on social differences or any other aspect of social behavior—remember that any application of statistical procedures requires a basic understanding of the statistical concepts and techniques. This introductory text is intended to familiarize you with the range of descriptive and inferential statistics widely applied in the social sciences. Our emphasis on statistical techniques should not diminish the importance of human judgment and your awareness of the person-made quality of statistics. Only with this awareness can statistics become a useful tool for viewing social life.

### MAIN POINTS

• Statistics are procedures used by social scientists to organize, summarize, and communicate information. Only information represented by numbers can be the subject of statistical analysis.

• The research process is a set of activities in which social scientists engage to answer questions, examine ideas, or test theories. It consists of the following stages: asking the research question, formulating the hypotheses, collecting data, analyzing data, and evaluating the hypotheses.

• A theory is an elaborate explanation of the relationship between two or more observable attributes of individuals or groups.

• Theories offer specific concrete predictions about the way observable attributes of people or groups would be interrelated in real life. These predictions, called hypotheses, are tentative answers to research problems.

• A variable is a property of people or objects that takes on two or more values. The variable that the researcher wants to explain (the "effect") is called the dependent variable. The variable that is expected to "cause" or account for the dependent variable is called the independent variable.

• Three conditions are required to establish causal relations: (1) The cause has to precede the effect in time, (2) there has to be an empirical relationship between the cause and the effect, and (3) this relationship cannot be explained by other factors.

• At the nominal level of measurement, numbers or other symbols are assigned to a set of categories to name, label, or classify the observations. At the ordinal level of measurement, categories can be rank ordered from low to high (or vice versa). At the interval-ratio level of measurement, measurements for all cases are expressed in the same unit.

• A population is the total set of individuals, objects, groups, or events in which the researcher is interested. A sample is a relatively small subset selected from a population.

• Descriptive statistics includes procedures that help us organize and describe data collected from either a sample or a population. Inferential statistics is concerned with making predictions or inferences about a population from observations and analyses of a sample.

## KEY TERMS

| | | |
|---|---|---|
| data | independent variable | research process |
| dependent variable | inferential statistics | sample |
| descriptive statistics | interval-ratio measurement | statistics |
| dichotomous variable | nominal measurement | theory |
| empirical research | ordinal measurement | unit of analysis |
| hypothesis | population | variable |

## ON YOUR OWN

 Log on to the web-based student study site at **www.sagepub.com/ssdsessentials** for additional study questions, web quizzes, web resources, flashcards, codebooks and datasets, web exercises, appendices, and links to social science journal articles reflecting the statistics used in this chapter.

## CHAPTER EXERCISES

1. In your own words, explain the relationship of data (collecting and analyzing) to the research process. (Refer to Figure 1.1.)

2. Construct potential hypotheses or research questions to relate the variables in each of the following examples. Also, write a brief statement explaining why you believe there is a relationship between the variables as specified in your hypotheses.
   a. Gender and educational level
   b. Income and race
   c. The crime rate and the number of police in a city
   d. Life satisfaction and age

    e. A nation's military expenditures as a percentage of its gross domestic product (GDP) and that nation's overall level of security

    f. Care of elderly parents and ethnicity

3. Determine the level of measurement for each of the following variables:
   a. The number of people in your family
   b. Place of residence classified as urban, suburban, or rural
   c. The percentage of university students who attended public high school
   d. The rating of the overall quality of a textbook, on a scale from "Excellent" to "Poor"
   e. The type of transportation a person takes to work (e.g., bus, walk, car)
   f. Your annual income
   g. The U.S. unemployment rate
   h. The presidential candidate that the respondent voted for in 2008

4. For each of the variables in Exercise 3 that you classified as interval ratio, identify whether it is discrete or continuous.

5. Why do you think men and women, on average, do not earn the same amount of money? Develop your own theory to explain the difference. Use three independent variables in your theory, with annual income as your dependent variable. Construct hypotheses to link each independent variable with your dependent variable.

6. For each of the following examples, indicate whether it involves the use of descriptive or inferential statistics. Justify your answer.
   a. The number of unemployed people in the United States
   b. Determining students' opinion about the quality of food at the cafeteria based on a sample of 100 students
   c. The national incidence of breast cancer among Asian women
   d. Conducting a study to determine the rating of the quality of a new automobile, gathered from 1,000 new buyers
   e. The average GPA of various majors (e.g., sociology, psychology, English) at your university
   f. The change in the number of immigrants coming to the United States from Southeast Asian countries between 2005 and 2010

7. Identify three social problems or issues that can be investigated with statistics. (One example of a social problem is hate crimes.) Which one of the three issues would be the most difficult to study? Which would be the easiest? Why?

8. Construct measures of political participation at the nominal, ordinal, and interval-ratio levels. (*Hint:* You can use behaviors such as voting frequency or political party membership.) Discuss the advantages and disadvantages of each.

9. Variables can be measured according to more than one level of measurement. For the following variables, identify at least two levels of measurement. Is one level of measurement better than another? Explain.
   a. Individual age
   b. Annual income
   c. Religiosity
   d. Student performance
   e. Social class
   f. Attitude toward affirmative action

Exercises

# Chapter 2

# The Organization and Graphic Presentation of Data

## Chapter Learning Objectives

❖ Understanding how to construct and analyze frequency, percentage, and cumulative distributions

❖ Understanding how to calculate proportions and percentages

❖ Recognizing the differences in frequency distributions for nominal, ordinal, and interval-ratio variables

❖ Reading statistical tables in research literature

❖ Constructing and interpreting a pie chart, bar graph, histogram, line graph, and time-series chart

❖ Analyzing and interpreting charts and graphs in literature

As social researchers, we often have to deal with very large amounts of data. For example, in a typical survey, by the completion of your data collection phase you will have accumulated thousands of individual responses represented by a jumble of numbers. To make sense out of these data, you will have to organize and summarize them in some systematic fashion. The most basic method for organizing data is to classify the observations into a frequency distribution. A **frequency distribution** is a table that reports the number of observations that fall into each category of the variable we are analyzing. Constructing a frequency distribution is usually the first step in the statistical analysis of data.

*Frequency distribution*    A table reporting the number of observations falling into each category of the variable.

# ▣ FREQUENCY DISTRIBUTIONS

Sweeping immigration proposals by the Bush administration and the U.S. Congress in 2006 and 2007 sparked nationwide debate about the status of immigrants in the United States. Globalization has fueled labor migration, particularly since the beginning of the 21st century. Workers migrate because of the promise of employment and higher standards of living than their home countries. Data reveal that the United States is the destination for many migrants.[1] The U.S. Census Bureau uses the term *foreign born* to refer to those who are not U.S. citizens at birth. The U.S. Census estimates that nearly 12% of the U.S. population or 37 million people are foreign born.[2] Today, one in eight U.S. residents is foreign born.[3] Immigrants are not one homogeneous group but are many diverse groups. Table 2.1 shows the frequency distribution of the world region of birth for the foreign-born population.

Table 2.1   Frequency Distribution for Categories of World Region of Birth for Foreign-Born Population, 2008

| World Region of Birth | Frequency (f) |
| --- | --- |
| Europe | 4,644,000 |
| Asia | 9,978,000 |
| Latin America | 20,034,000 |
| Other areas | 2,602,000 |
| Total ($N$) | 37,258,000 |

*Source:* U.S. Census Bureau, *Statistical Abstract of the United States*, 2010, Table 43.

Note that the frequency distribution is organized in a table, which has a number (2.1) and a descriptive title. The title indicates the kind of data presented: "Categories of World Region of Birth for Foreign-Born Population." The table consists of two columns. The first column identifies the variable (world region of birth) and its categories. The second column, headed "Frequency (f)," tells the number of cases in each category as well as the total number of cases ($N = 37,258,000$). Note also that the source of the table is clearly identified in a source note. It tells us that the data are from a U.S. census report and the data come from a 2010 report. In general, the source of data for a table should appear as a source note unless it is clear from the general discussion of the data.

What can you learn from the information presented in Table 2.1? The table shows that as of 2008, approximately 37 million people were classified as foreign born. Out of this group, the majority, about 20 million people, were from Latin America, 10 million were from Asia, followed by 4.6 million from Europe.

# ▣ PROPORTIONS AND PERCENTAGES

Frequency distributions are helpful in presenting information in a compact form. However, when the number of cases is large, the frequencies may be difficult to grasp. To standardize these raw frequencies, we can translate them into relative frequencies—that is, proportions or percentages.

A **proportion** is a relative frequency obtained by dividing the frequency in each category by the total number of cases. To find a proportion ($p$), divide the frequency ($f$) in each category by the total number of cases ($N$):

$$p = \frac{f}{N} \tag{2.1}$$

where

$f$ = frequency

$N$ = total number of cases

Thus, the proportion of foreign born originally from Latin America is

$$\frac{20,034,000}{37,258,000} = 0.54$$

The proportion of foreign born who were originally from Asia is

$$\frac{9,978,000}{37,258,000} = 0.27$$

The proportion of foreign born who were originally from Europe is

$$\frac{4,644,000}{37,258,000} = 0.12$$

And finally, the proportion of foreign born who were originally from other areas is

$$\frac{2,602,000}{37,258,000} = 0.07$$

Proportions should always sum to 1.00 (allowing for some rounding errors). Thus, in our example the sum of the four proportions is

$$0.54 + 0.27 + 0.12 + 0.07 = 1.00$$

To determine a frequency from a proportion, we simply multiply the proportion by the total $N$:

$$f = p(N) \tag{2.2}$$

Thus, the frequency of foreign born from Asia can be calculated as

$$0.27(37,258,000) = 10,059,660$$

Note that the obtained frequency differs somewhat from the actual frequency of 9,978,000. This difference is due to rounding off of the proportion. If we use the actual proportion instead of the rounded proportion, we obtain the correct frequency:

$$0.267808255945032(37,258,000) = 9,978,000$$

> *Proportion*  A relative frequency obtained by dividing the frequency in each category by the total number of cases.

We can also express frequencies as percentages. A **percentage** is a relative frequency obtained by dividing the frequency in each category by the total number of cases and multiplying by 100. In most statistical reports, frequencies are presented as percentages rather than proportions. Percentages express the size of the frequencies as if there were a total of 100 cases.

To calculate a percentage, simply multiply the proportion by 100:

$$\text{Percentage } (\%) = \frac{f}{N}(100) \tag{2.3}$$

or

$$\text{Percentage } (\%) = p(100) \tag{2.4}$$

Thus, the percentage of respondents who were originally from Asia is

$$0.27(100) = 27\%$$

The percentage of respondents who were originally from Latin America is

$$0.54(100) = 54\%$$

> *Percentage*  A relative frequency obtained by dividing the frequency in each category by the total number of cases and multiplying by 100.

✓ *Learning Check*  *Calculate the proportion of males and females in your statistics class. What proportion is female?*

## ▣ PERCENTAGE DISTRIBUTIONS

Percentages are usually displayed as percentage distributions. A **percentage distribution** is a table showing the percentage of observations falling into each category of the variable. For example, Table 2.2 presents the frequency distribution of categories of places of origin (Table 2.1) along with the corresponding percentage distribution. Percentage distributions (or proportions) should always show the base ($N$) on which they were computed. Thus, in Table 2.2 the base on which the percentages were computed is $N = 37,258,000$.

> *Percentage distribution*  A table showing the percentage of observations falling into each category of the variable.

Table 2.2   Frequency and Percentage Distributions for Categories of
World Region of Birth for Foreign Born, 2008

| World Region of Birth | Frequency (f) | Percentage |
|---|---|---|
| Europe | 4,644,000 | 12.5 |
| Asia | 9,978,000 | 26.8 |
| Latin America | 20,034,000 | 53.8 |
| Other areas | 2,602,000 | 7.0 |
| Total (*N*) | 37,258,000 | 100.1 |

*Source:* U.S. Census Bureau, *Statistical Abstract of the United States*, 2010, Table 43.

## ▣ COMPARISONS

In Table 2.2, we illustrated that there are four primary places of origin for foreign born in the United States. These distinctions help us understand the specific characteristics and backgrounds of each group. We can resist the temptation to group all foreign born in one category and ask, for instance, is one group more educated than another? Is one group younger than the other groups?

The decision to consider these groups separately or to pool them depends to a large extent on our research question. For instance, we know that in 2008, 16.5% of the foreign-born population were living in poverty.[4] Among the foreign born, poverty rates were highest among those from Latin America (21%) and lowest among those from Europe (9%). What do these figures tell us about the demographic characteristics of foreign borns? Of Latin American foreign borns? Of European foreign borns? To answer these questions and determine whether the two categories of region of birth have markedly different social characteristics, we need to *compare* them.

As students, as social scientists, and even as consumers, we are frequently faced with problems that call for some way to make a clear and valid comparison. For example, in 2008, 22.8% of children lived with only their mothers.[5] Is this figure high or low? In 2008, 26% of those 18 years of age or older had never been married.[6] Does this reflect a change in the American family? In each of these cases, comparative information is required to answer the question and reach a conclusion.

## ▣ STATISTICS IN PRACTICE: LABOR FORCE PARTICIPATION AMONG LATINOS

Very often, we are interested in comparing two or more groups that differ in size. Percentages are especially useful for making such comparisons. For example, we know that differences in socioeconomic status mark divisions between populations, indicating differential access to economic opportunities. Labor participation (either employed or seeking employment) is an important indicator of access to economic opportunities and is strongly associated with socioeconomic status. Table 2.3 shows the raw frequency distributions for the variable *labor force participation* for three categories of Latino (or Hispanic) identity.

Table 2.3   Employment Status of the Civilian Population,
Selected Hispanic Categories, 2008

| Employment Status | Mexican | Puerto Rican | Cuban |
|---|---|---|---|
| Employed | 12,931,000 | 1,634,000 | 841,000 |
| Unemployed | 1,078,000 | 188,000 | 57,000 |
| Not in labor force | 6,465,000 | 1,032,000 | 525,000 |
| Total (N) | 20,474,000 | 2,854,000 | 1,423,000 |

Source: U.S. Census Bureau, Statistical Abstract of the United States, 2010, Table 39.

Which group has the highest relative number of persons who are not in the labor force? Because of the differences in the population sizes of the three groups, this is a difficult question to answer based on only the raw frequencies. To make a valid comparison, we have to compare the percentage distributions for all three groups. These are presented in Table 2.4. Note that the percentage distributions make it easier to identify differences between the groups. Compared with Puerto Ricans and Cubans, Mexicans have the highest percentage employed in the labor force (63% vs. 57% and 59%). Among the three groups, Cubans have the lowest percentage (4% vs. 5.3% and 6.6%) of persons who are unemployed.

Table 2.4   Employment Status of the Civilian Population,
Selected Hispanic Categories, 2008

| Employment Status | Mexican (%) | Puerto Rican (%) | Cuban (%) |
|---|---|---|---|
| Employed | 63.2 | 57.2 | 59.1 |
| Unemployed | 5.3 | 6.6 | 4.0 |
| Not in labor force | 31.6 | 36.2 | 36.9 |
| Total | 100.1 | 100.0 | 100.0 |
| (N) | 20,474,000 | 2,854,000 | 1,423,000 |

Source: U.S. Census Bureau, Statistical Abstract of the United States, 2010, Table 39.

✓ Learning
Check

*Examine Table 2.4 and answer the following questions: What is the percentage of Mexicans who are employed? What is the base (N) for this percentage? What is the percentage of Cubans who are not in the labor force? What is the base (N) for this percentage?*

Whenever one group is compared with another, the most meaningful conclusions can usually be drawn based on comparison of the relative frequency distributions. In fact, we are seldom interested in a single distribution. Most interesting questions in the social sciences are about differences between two or more groups.[7] The finding that the labor force participation patterns vary among different

Latino groups raises doubt about whether Latinos can be legitimately regarded as a single, relatively homogeneous ethnic group. Further analyses could examine *why* differences in Latino identity are associated with differences in labor force participation patterns. Other variables that explain these differences could be identified. These kinds of questions can be answered using more complex multivariate statistical techniques that involve more than two variables. The comparison of percentage distributions is an important foundation for these more complex techniques.

## ▣ THE CONSTRUCTION OF FREQUENCY DISTRIBUTIONS

In this section, you will learn how to construct frequency distributions. Most often, this can be done by your computer, but it is important to go through the process to understand how frequency distributions are actually put together.

For nominal and ordinal variables, constructing a frequency distribution is quite simple. Count and report the number of cases that fall into each category of the variable along with the total number of cases (*N*). For the purpose of illustration, let's take a small random sample of 40 cases from a General Social Survey (GSS) sample and record their scores on the following variables: gender, a nominal-level variable; degree, an ordinal measurement of education; and age and number of children, both interval-ratio variables. The use of "male" and "female" in parts of this book is in keeping with the GSS categories for the variable "sex" (respondent's sex).

You can see that it is going to be difficult to make sense of these data just by eyeballing Table 2.5. How many of these 40 respondents are males? How many said that they had a graduate degree? How many were older than 50 years of age? To answer these questions, we construct the frequency distributions for all four variables.

### Frequency Distributions for Nominal Variables

Let's begin with the nominal variable, gender. First, we tally the number of males, then the number of females (the column of tallies has been included in Table 2.6 for the purpose of illustration). The tally results are then used to construct the frequency distribution presented in Table 2.6. The table has a title describing its content ("Frequency Distribution of the Variable Gender: GSS Subsample"). Its categories (male and female) and their associated frequencies are clearly listed; in addition, the total number of cases (*N*) is also reported. The Percentage column is the percentage distribution for this variable. To convert the Frequency column to percentages, simply divide each frequency by the total number of cases and multiply by 100. Percentage distributions are routinely added to almost any frequency table and are especially important if comparisons with other groups are to be considered. Immediately, we can see that it is easier to read the information. There are 25 females and 15 males in this sample. Based on this frequency distribution, we can also conclude that the majority of sample respondents are female.

*Construct a frequency and percentage distribution for males and females in your statistics class.*

✓ *Learning Check*

**Table 2.5**  A GSS Subsample of 40 Respondents

| Gender of Respondent | Degree | Number of Children | Age |
|---|---|---|---|
| M | Bachelor | 1 | 43 |
| F | High school | 2 | 71 |
| F | High school | 0 | 71 |
| M | High school | 0 | 37 |
| M | High school | 0 | 28 |
| F | High school | 6 | 34 |
| F | High school | 4 | 69 |
| F | Graduate | 0 | 51 |
| F | Bachelor | 0 | 76 |
| M | Graduate | 2 | 48 |
| M | Graduate | 0 | 49 |
| M | Less than high school | 3 | 62 |
| F | Less than high school | 8 | 71 |
| F | High school | 1 | 32 |
| F | High school | 1 | 59 |
| F | High school | 1 | 71 |
| M | High school | 0 | 34 |
| M | Bachelor | 0 | 39 |
| F | Bachelor | 2 | 50 |
| M | High school | 3 | 82 |
| F | High school | 1 | 45 |
| M | High school | 0 | 22 |
| M | High school | 2 | 40 |
| F | High school | 2 | 46 |
| M | High school | 0 | 29 |
| F | High school | 1 | 75 |
| F | High school | 0 | 23 |
| M | Bachelor | 2 | 35 |
| M | Bachelor | 3 | 44 |
| F | High school | 3 | 47 |
| M | High school | 1 | 84 |
| F | Graduate | 1 | 45 |
| F | Less than high school | 3 | 24 |
| F | Graduate | 0 | 47 |
| F | Less than high school | 5 | 67 |
| F | High school | 1 | 21 |
| F | High school | 0 | 24 |
| F | High school | 3 | 49 |
| F | High school | 3 | 45 |
| F | Graduate | 3 | 37 |

*Note:* M, male; F, female.

Table 2.6    Frequency Distribution of the Variable Gender: GSS
             Subsample

| Gender | Tallies | Frequency (f) | Percentage |
|--------|---------|---------------|------------|
| Male | ||||| ||||| ||||| | 15 | 37.5 |
| Female | ||||| ||||| ||||| ||||| ||||| | 25 | 62.5 |
| Total (N) | | 40 | 100.0 |

## Frequency Distributions for Ordinal Variables

To construct a frequency distribution for ordinal-level variables, follow the same procedures outlined for nominal-level variables. Table 2.7 presents the frequency distribution for the variable degree. The table shows that 60.0%, a majority, indicated that their highest degree was a high school degree.

Table 2.7    Frequency Distribution of the Variable Degree: GSS
             Subsample

| Degree | Tallies | Frequency (f) | Percentage |
|--------|---------|---------------|------------|
| Less than high school | |||| | 4 | 10.0 |
| High school | ||||| ||||| ||||| ||||| |||| | 24 | 60.0 |
| Bachelor | ||||| | | 6 | 15.0 |
| Graduate | ||||| | | 6 | 15.0 |
| Total (N) | | 40 | 100.0 |

The major difference between frequency distributions for nominal and ordinal variables is the order in which the categories are listed. The categories for nominal-level variables do not have to be listed in any particular order. For example, we could list females first and males second without changing the nature of the distribution. Because the categories or values of ordinal variables are rank ordered, however, they must be listed in a way that reflects their rank—from the lowest to the highest or from the highest to the lowest. Thus, the data on degree in Table 2.7 are presented in inclining order from "less than high school" (the lowest educational category) to "graduate" (the highest educational category).

✓ Learning
Check

Figures 2.1, 2.2, and 2.3 illustrate the gender and degree data in stages as presented in Tables 2.5, 2.6, and 2.7. To convince yourself that classifying the respondents by gender (Figure 2.2) and by degree (Figure 2.3) makes the job of counting much easier, turn to Figure 2.1 and answer these questions: How many men are in the group? How many women? How many said that they completed a bachelor's degree? Now turn to Figure 2.2: How many men are in the group? How many women? Finally, examine Figure 2.3: How many said that they completed a bachelor's degree?

**Figure 2.1** Forty Respondents From the GSS Subsample, Their Gender and Their Level of Degree (see Table 2.5)

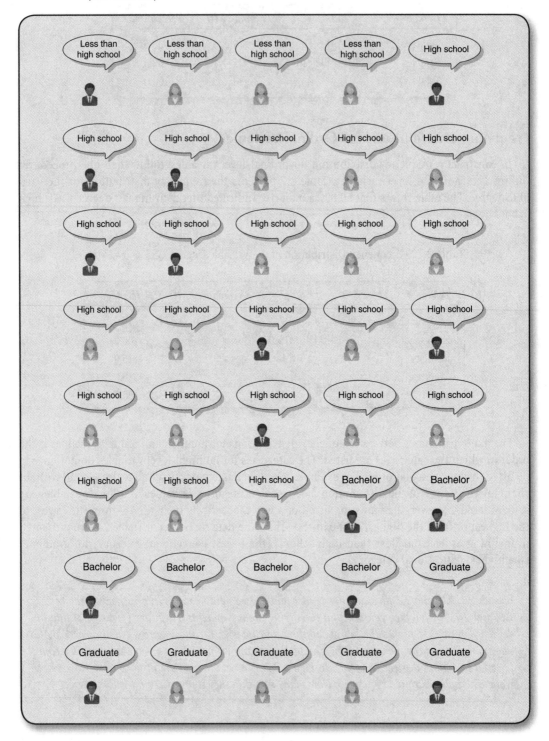

**Figure 2.2** Forty Respondents From the GSS Subsample, Classified by Gender (see Table 2.6)

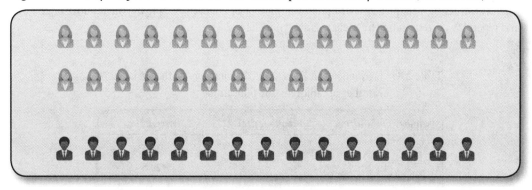

**Figure 2.3** Forty Respondents From the GSS Subsample, Classified by Gender and Degree

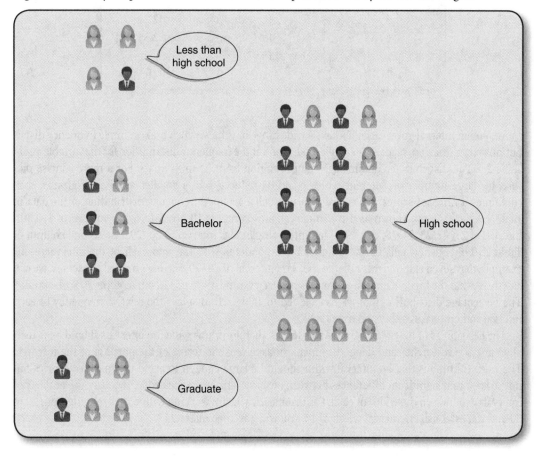

## Frequency Distributions for Interval-Ratio Variables

Constructing frequency distributions for nominal- and ordinal-level variables is rather straightforward. Simply list the categories and count the number of observations that fall into each category. Building a frequency distribution for interval-ratio variables with relatively few values is also easy. For example, when constructing a frequency distribution for number of children, simply list the number of children and report the corresponding frequency, as shown in Table 2.8.

**Table 2.8** Frequency and Percentage Distribution for the Variable Number of Children: GSS Subsample

| Number of Children | Frequency (f) | Percentage |
|---|---|---|
| 0 | 13 | 32.5 |
| 1 | 9 | 22.5 |
| 2 | 6 | 15.0 |
| 3 | 8 | 20.0 |
| 4 | 1 | 2.5 |
| 5 | 1 | 2.5 |
| 6 | 1 | 2.5 |
| 7+ | 1 | 2.5 |
| Total (N) | 40 | 100.0 |

Very often interval-ratio variables have a wide range of values, which makes simple frequency distributions very difficult to read. For example, take a look at the frequency distribution for the variable *age* in Table 2.9. The distribution contains age values ranging from 21 to 84 years. For a more concise picture, the large number of different scores could be reduced into a smaller number of groups, each containing a range of scores. Table 2.10 displays such a grouped frequency distribution of the data in Table 2.9. Each group, known as a *class interval*, now contains 10 possible scores instead of 1. Thus, the ages of 21, 22, 23, 24, 28, and 29 all fall into a single class interval of 20–29. The second column of Table 2.10, Frequency, tells us the number of respondents who fall into each of the intervals—for example, that seven respondents fall into the class interval of 20–29. Having grouped the scores, we can clearly see that the biggest single age group is between 40 and 49 years (12 out of 40, or 30% of sample). The percentage distribution that we have added to Table 2.10 displays the relative frequency of each interval and emphasizes this pattern as well.

The decision as to how many groups to use and, therefore, how wide the intervals should be is usually up to the researcher and depends on what makes sense in terms of the purpose of the research. The rule of thumb is that an interval width should be large enough to avoid too many categories but not so large that significant differences between observations are concealed.[8] Obviously, the number of intervals depends on the width of each. For instance, if you are working with scores ranging from 10 to 60 and you establish an interval width of 10, you will have five intervals.

✓ *Learning Check*

*Can you verify that Table 2.10 was constructed correctly? Use Table 2.9 to determine the frequency of cases that fall into the categories of Table 2.10.*

**Table 2.9**    Frequency Distribution of the Variable Age: GSS Subsample

| Age of Respondent | Frequency (f) | Age of Respondent | Frequency (f) |
|---|---|---|---|
| 21 | 1 | 59 | 1 |
| 22 | 1 | 62 | 1 |
| 23 | 1 | 67 | 1 |
| 24 | 2 | 69 | 1 |
| 28 | 1 | 71 | 4 |
| 29 | 1 | 75 | 1 |
| 32 | 1 | 76 | 1 |
| 34 | 2 | 82 | 1 |
| 35 | 1 | 84 | 1 |
| 37 | 2 | | |
| 39 | 1 | | |
| 40 | 1 | | |
| 43 | 1 | | |
| 44 | 1 | | |
| 45 | 3 | | |
| 46 | 1 | | |
| 47 | 2 | | |
| 48 | 1 | | |
| 49 | 2 | | |
| 50 | 1 | | |
| 51 | 1 | | |

**Table 2.10**    Grouped Frequency, Cumulative Frequency, and Percentage
Distribution for the Variable Age: GSS Subsample

| Age Category | Frequency (f) | Cf | Percentage |
|---|---|---|---|
| 20–29 | 7 | 7 | 17.5 |
| 30–39 | 7 | 14 | 17.5 |
| 40–49 | 12 | 26 | 30.0 |
| 50–59 | 3 | 29 | 7.5 |
| 60–69 | 3 | 32 | 7.5 |
| 70–79 | 6 | 38 | 15.0 |
| 80–89 | 2 | 40 | 5.0 |
| Total (*N*) | 40 | | 100.0 |

✓ *Learning Check* *If you are having trouble distinguishing between nominal, ordinal, and interval-ratio variables, go back to Chapter 1 and review the section on levels of measurement. The distinction between these three levels of measurement is important throughout the book.*

## ▣ CUMULATIVE DISTRIBUTIONS

Sometimes, we may be interested in locating the relative position of a given score in a distribution. For example, we may be interested in finding out how many or what percentage of our sample was younger than 40 or older than 60. Frequency distributions can be presented in a cumulative fashion to answer such questions. A **cumulative frequency distribution** shows the frequencies at or below each category of the variable.

---

*Cumulative frequency distribution*    A distribution showing the frequency at or below each category (class interval or score) of the variable.

---

Cumulative frequencies are appropriate only for variables that are measured at an ordinal level or higher. They are obtained by adding to the frequency in each category the frequencies of all the categories below it.

Let's look at the cumulative frequencies in Table 2.10. The cumulative frequency column, denoted by $Cf$, shows the number of persons at or below each interval.

To construct a cumulative frequency distribution, start with the frequency in the lowest class interval (or with the lowest score, if the data are ungrouped), and add to it the frequencies in the next highest class interval. Continue adding the frequencies until you reach the last class interval. The cumulative frequency in the last class interval will be equal to the total number of cases ($N$). In Table 2.10, the frequency associated with the first class interval (20–29) is 7. The cumulative frequency associated with this interval is also 7, since there are no cases below this class interval. The frequency for the second class interval is 7. The cumulative frequency for this interval is 7 + 7 = 14. To obtain the cumulative frequency of 26 for the third interval, we add its frequency (12) to the cumulative frequency associated with the second class interval (14). Continue this process until you reach the last class interval. Therefore, the cumulative frequency for the last interval is equal to 40, the total number of cases ($N$).

We can also construct a cumulative percentage distribution ($C\%$), which has wider applications than the cumulative frequency distribution ($Cf$). A **cumulative percentage distribution** shows the percentage at or below each category (class interval or score) of the variable. A cumulative percentage distribution is constructed using the same procedure as for a cumulative frequency distribution except that the percentages—rather than the raw frequencies—for each category are added to the total percentages for all the previous categories.

---

*Cumulative percentage distribution*    A distribution showing the percentage at or below each category (class interval or score) of the variable.

---

In Table 2.11, we have added the cumulative percentage distribution to the frequency and percentage distributions shown in Table 2.10. The cumulative percentage distribution shows, for example, that 35% of the sample was younger than 40 years of age—that is, 39 years or younger.

Like the percentage distributions described earlier, cumulative percentage distributions are especially useful when you want to compare differences between groups. For an example of how cumulative percentages are used in a comparison, we have used the 2008 GSS data to contrast the opinions of white women and white men about their standard of living compared with other American families. Respondents were asked the following question: "The way things are in America, people like me and my family have a good chance of improving our standard of living—do you agree or disagree?"

The percentage distribution and the cumulative percentage distribution for white women and white men are shown in Table 2.12. The cumulative percentage distributions suggest that slightly

**Table 2.11**   Grouped Frequency, Percentage, and Cumulative Percentage Distribution for the Variable Age: GSS Subsample

| Age Category | Frequency (f) | Percentage | C% |
|---|---|---|---|
| 20–29 | 7 | 17.5 | 17.5 |
| 30–39 | 7 | 17.5 | 35.0 |
| 40–49 | 12 | 30.0 | 65.0 |
| 50–59 | 3 | 7.5 | 72.5 |
| 60–69 | 3 | 7.5 | 80.0 |
| 70–79 | 6 | 15.0 | 95.0 |
| 80–89 | 2 | 5.0 | 100.0 |
| Total (N) | 40 | 100.0 | |

**Table 2.12**   Will One's Standard of Living Improve: White Males Versus White Females

| | White Males | | White Females | |
|---|---|---|---|---|
| | % | C% | % | C% |
| Strongly agree | 18.6 | 18.6 | 14.4 | 14.4 |
| Agree | 42.3 | 60.9 | 39.9 | 54.3 |
| Neither | 13.0 | 73.9 | 16.5 | 70.8 |
| Disagree | 24.2 | 98.1 | 24.8 | 95.6 |
| Strongly disagree | 2.0 | 100.1 | 4.4 | 100.0 |
| Total | 100.1 | | 100.0 | |
| (N) | 355 | | 411 | |

Source: General Social Survey, 2008.

more white men agree to the statement that their standard of living will improve. The two groups are separated by 6.6 percentage points—54.3% of the white women indicated that they either strongly agreed or agreed to the statement, while 60.9% of white males said the same. (Note that a higher percentage of women disagree with the statement than men.) What might explain these differences? Gender and wage discrimination might play a role in the gap between white women and white men.

## ▣ RATES

Terms such as *birthrate*, *unemployment rate*, and *marriage rate* are often used by social scientists and demographers and then quoted in the popular media to describe population trends. But what exactly are rates, and how are they constructed? A **rate** is a number obtained by dividing the number of actual occurrences in a given time period by the number of possible occurrences. For example, to determine the poverty rate for 2008, the U.S. Census Bureau took the number of men and women in poverty in 2008 (actual occurrences) and divided it by the total population in 2008 (possible occurrences). The rate for 2008 can be expressed as

$$\text{Poverty rate, } 2008 = \frac{\text{Number of people in poverty in 2008}}{\text{Total population in 2008}}$$

Since 40,136,000 people were poor in 2008 and the number for the total population was 304,060,000, the poverty rate for 2008 can be expressed as

$$\text{Poverty rate, } 2008 = \frac{40,136,000}{304,060,000} = 0.132$$

The poverty rate in 2008 as reported by the U.S. Census Bureau was 13.2% ($0.132 \times 100$). Rates are often expressed as rates per thousand or hundred thousand to eliminate decimal points and make the number easier to interpret. For example, to express the poverty rate per thousand we multiply it by 1,000: $0.132 \times 1,000 = 132$. This means that for every 1,000 people, 132 were poor according to the U.S. Census Bureau definition.

The preceding poverty rate can be referred to as a *crude rate* because it is based on the total population. Rates can be calculated on the general population or on a more narrowly defined select group. For instance, poverty rates are often given for the number of people who are under 18 years—highlighting how our young are vulnerable to poverty. The poverty rate for those under 18 years is as follows:

$$\text{Poverty rate (under 18 years, 2008)} = \frac{13,454,000}{73,922,000}$$
$$= 0.182 \times 1,000$$
$$= 182$$

We could even take a look at the poverty rate for older Americans:

$$\text{Poverty rate (65 years or older, 2008)} = \frac{3,842,000}{38,812,000}$$
$$= 0.099 \times 1,000$$
$$= 99$$

---

*Rate*   A number obtained by dividing the number of actual occurrences in a given time period by the number of possible occurrences.

---

*Law enforcement agencies routinely record crime rates (the number of crimes committed relative to the size of a population), arrest rates (the number of arrests made relative to the number of crimes reported), and conviction rates (the number of convictions relative to the number of cases tried). Can you think of some other variables that could be expressed as rates?*

✓ *Learning Check*

## 回 READING THE RESEARCH LITERATURE: STATISTICAL TABLES[9]

In this section, we present some guidelines for how to read and interpret statistical tables displaying frequency distributions. The purpose is to help you see that some of the techniques described in this chapter are actually used in a meaningful way. Remember that it takes time and practice to develop the skill of reading tables. Even experienced researchers sometimes make mistakes when interpreting tables. So take the time to study the tables presented here, do the chapter exercises, and you will find that reading, interpreting, and understanding tables will become easier in time.

### Basic Principles

The first step in reading any statistical table is to understand what the researcher is trying to tell you. Begin your inspection of the table by reading its title. It usually describes the central contents of the table. Check for any source notes to the table. These tell the source of the data or the table and any additional information that the author considers important. Next, examine the column and row headings and subheadings. These identify the variables, their categories, and the kind of statistics presented, such as raw frequencies or percentages. The main body of the table includes the appropriate statistics (frequencies, percentages, rates, etc.) for each variable or group as defined by each heading and subheading.

Table 2.13 was taken from an article written by Professors Eric Fong and Kumiko Shibuya about the residential patterns of different racial and ethnic groups. In their study, the researchers attempted to compare different home ownership patterns and suburbanization patterns among whites, blacks, five major Asian groups, and three major Hispanic groups. Data from the 1990 1% Public Use Microdata

Sample (PUMS) based on 15 Primary Metropolitan Statistical Areas (PMSAs) and Metropolitan Statistical Areas (MSAs) are used for their analysis.

In Table 2.13, the researchers display the percentages for housing status by location for each racial/ethnic group for 1990. Note that the frequency (*f*) for each category is not reported. Although the table is quite simple, it is important to examine it carefully, including its title and headings, to make sure that you understand what the information means.

**Table 2.13**    Percentage Distribution of Housing Tenure Status by Residential Location of Major Racial and Ethnic Groups, 1990

|  | Suburban Owner | Central City Owner | Suburban Renter | Central City Renter | Total |
|---|---|---|---|---|---|
| Whites | 50.4 | 16.8 | 17.1 | 15.7 | 100.0 |
| Blacks | 14.6 | 27.2 | 13.9 | 44.3 | 100.0 |
| Asians | 34.2 | 20.6 | 14.7 | 30.5 | 100.0 |
| Chinese | 33.7 | 26.1 | 9.9 | 30.3 | 100.0 |
| Japanese | 39.2 | 20.8 | 17.8 | 22.2 | 100.0 |
| Koreans | 29.8 | 13.8 | 17.1 | 39.3 | 100.0 |
| Vietnamese | 31.0 | 11.8 | 24.3 | 32.9 | 100.0 |
| Filipino | 38.1 | 21.5 | 15.8 | 24.6 | 100.0 |
| Hispanics | 22.9 | 15.3 | 20.8 | 40.9 | 99.9 |
| Cubans | 36.3 | 16.6 | 16.2 | 30.9 | 100.0 |
| Mexicans | 26.0 | 15.5 | 26.8 | 31.8 | 100.1 |
| Puerto Ricans | 7.1 | 14.2 | 8.4 | 70.3 | 100.0 |
| Total | 41.7 | 18.5 | 16.8 | 23.0 | 100.0 |

*Source:* 1990 U.S. 1% PUMS Data. Adapted from Eric Fong and Kumiko Shibuya, "Suburbanization and Home Ownership: The Spatial Assimilation Process in U.S. Metropolitan Cities," *Sociological Perspectives* 43, no. 1 (2000): 143.

✓ *Learning Check*

*Inspect Table 2.13 and answer the following questions:*

- *What is the source of this table?*
- *How many variables are presented? What are their names?*
- *What is represented by the numbers presented in the first column? In the third column?*

What do the authors tell us about the table?

Table 2.13 [Table 1], which shows tenure status cross-classified by housing location, reveals four distinctive patterns. Column 1, which presents the proportions of suburban home owners, shows remarkable variations among groups. Although 42 percent of the total population of these 15 metropolitan areas have been able to own a house in the suburbs, the figures suggest that it is atypical for some groups. Whites have the highest share: about 50 percent are suburban home owners. Blacks and Puerto Ricans, on the other hand, show an extremely low level of suburban home ownership. Asians in general have higher suburban home ownership rates (34%) than Hispanics (23%). There are, however, substantial variations within these two groups, ranging from 39 percent for Japanese to only 7 percent for Puerto Ricans.

Home ownership rates in the central city (Column 2) show a different picture. Blacks have the highest rate: about 27 percent of blacks own homes in central city areas. In fact, two-thirds of black home owners are found in those areas. Their higher home ownership rates in the central city may be explained by the fact that the cost of owning a house in the suburbs is higher. It is also possible that blacks are barred from the suburban areas because of discrimination in the housing market. Asian ethnic groups vary substantially in home ownership rates in the central city. The rate ranges from 12 percent for Vietnamese to 26 percent for Chinese. A higher rate of Chinese home owners in central cities may reflect the fact that Chinatowns have traditionally been located there. However, the rates among Hispanic ethnic groups are similar to one another. They range from 14 percent for Cubans and 17 percent for Mexicans.

The results in column 3 suggest a pattern for suburban renters that is distinct from home ownership patterns in either the suburbs or the central city. The results indicate that whites and Hispanics have the highest proportions of suburban renters. Although the percentage of Asian suburban renters is low on average, some specific Asian ethnic groups, such as Vietnamese, have higher rates than whites. The relatively high number of Vietnamese renters may be related to the results of the dispersion by refugee settlement programs. The higher rate of suburban renters among these Asian groups may be related to the absence of their ethnic communities in central cities....

The results in column 4 reveal another unique pattern for central city renters. Among all groups, blacks and Puerto Ricans have the highest rates in this category. About 70 percent of Puerto Ricans and 44 percent of blacks are central city renters. Asians have a moderate percentage of central city renters, ranging from a low of 22 percent for Japanese to a high of 39 percent for Koreans. Whites have the lowest percentage of central city renters: about 16 percent.

Overall, these data show the complicated nature of the residential distribution patterns of racial/ethnic groups in contemporary cities. The results suggest four distinctive pictures of tenure status in suburbs and central cities, which would be undetected if suburbanization or home ownership were studied separately. These distinctive patterns, however, may simply reflect the differences in the socioeconomic resources and acculturation levels of each group in both locations.[10]

For a more detailed analysis of the relationships between these variables, you need to consider some of the more complex techniques of bivariate analysis and statistical inference. We consider these more advanced techniques beginning with Chapter 7.

## ▣ GRAPHIC PRESENTATION OF DATA

You have probably heard that "a picture is worth a thousand words." The same can be said about statistical graphs because they summarize hundreds or thousands of numbers. Many people are intimidated by statistical information presented in frequency distributions or in other tabular forms, but they find the same information to be readable and understandable when presented graphically. Graphs tell a story in "pictures" rather than in words or numbers. They are supposed to make us think about the substance rather than the technical details of the presentation.

As we introduce the various graphical techniques, we also show you how to use graphs to tell a "story." The particular story we tell in this section is that of the elderly in the United States. The different types of graphs introduced in this section demonstrate the many facets of the aging of society over the next four decades. People have tended to talk about seniors as if they were a homogeneous group, but the different graphical techniques we illustrate here dramatize the wide variations in economic characteristics, living arrangements, and family status among people aged 65 and older. Most of the statistical information presented in this chapter is based on reports prepared by statisticians from the U.S. Census Bureau and other government agencies that gather information about the elderly in the United States and internationally.

Numerous graphing techniques are available to you, but here we focus on just a few of the most widely used ones in the social sciences. The first two, the pie chart and bar graph, are appropriate for nominal and ordinal variables. The next two, histograms and line graphs, are used with interval-ratio variables. We also discuss time-series charts. Time-series charts are used to show how some variables change over time.

## ▣ THE PIE CHART: RACE AND ETHNICITY OF THE ELDERLY

The elderly population of the United States is racially heterogeneous. As the data in Table 2.14 show, of the total elderly population (defined as persons 65 years and older) in 2008, about 33.7 million were white,[11] about 3.3 million black, 225,000 American Indian, 1.3 million Asian American, 36,000 Native Hawaiian or Pacific Islander, and 261,000 were two or more races combined.

**Table 2.14**  Annual Estimates of the U.S. Population 65 Years and Over by Race, 2008

| Race | Frequency (f) | Percentage (%) |
|---|---|---|
| White alone | 33,737,000 | 86.8 |
| Black alone | 3,314,000 | 8.5 |
| American Indian alone | 225,000 | 0.6 |
| Asian alone | 1,294,000 | 3.3 |
| Native Hawaiian or Pacific Islander | 36,000 | 0.1 |
| Two or more races combined | 261,000 | 0.7 |
| Total | 38,867,000 | 100.0 |

Source: U.S. Census Bureau, *The 2010 Statistical Abstract*, Table 10.

A **pie chart** shows the differences in frequencies or percentages among the categories of a nominal or an ordinal variable. The categories are displayed as segments of a circle whose pieces add up to 100% of the total frequencies. The pie chart shown in Figure 2.4 displays the same information that Table 2.14 presents. Although you can inspect these data in Table 2.14, you can interpret the information more easily by seeing it presented in the pie chart in Figure 2.4. It shows that the elderly population is predominantly white (86.8%), followed by black (8.5%).

**Figure 2.4**    Annual Estimates of the U.S. Population 65 Years and Over by Race, 2008

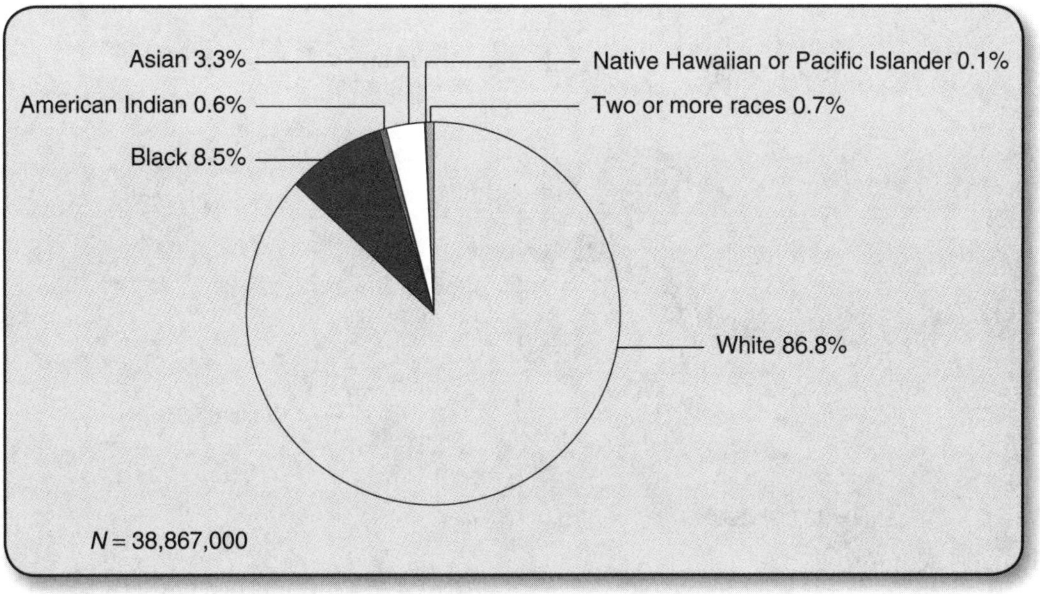

*Source:* U.S. Census Bureau, *The 2010 Statistical Abstract*, Table 10.

> Notice that the pie chart contains all the information presented in the frequency distribution. Like the frequency distribution, charts have an identifying number, a title that describes the content of the figure, and a reference to a source. The frequency or percentage is represented both visually and in numbers.

✓ *Learning Check*

---

***Pie chart***    A graph showing the differences in frequencies or percentages among the categories of a nominal or an ordinal variable. The categories are displayed as segments of a circle whose pieces add up to 100% of the total frequencies.

---

Note that the percentages for several of the racial groups are about 3% or less. It might be better to combine categories—American Indian, Asian, Native Hawaiian—into an "other races" category. This will leave us with three distinct categories: White, Black, and Other and Two or More Races. The revised pie chart is presented in Figure 2.5. We can highlight the diversity of the elderly population by "exploding" the pie chart, moving the segments representing these groups slightly outward to draw them to the viewer's attention.

## ▣ THE BAR GRAPH: LIVING ARRANGEMENTS AND LABOR FORCE PARTICIPATION OF THE ELDERLY

The **bar graph** provides an alternative way to present nominal or ordinal data graphically. It shows the differences in frequencies or percentages among categories of a nominal or an ordinal variable. The

**Figure 2.5**   Annual Estimates of U.S. Population 65 Years and Over by Race, 2008

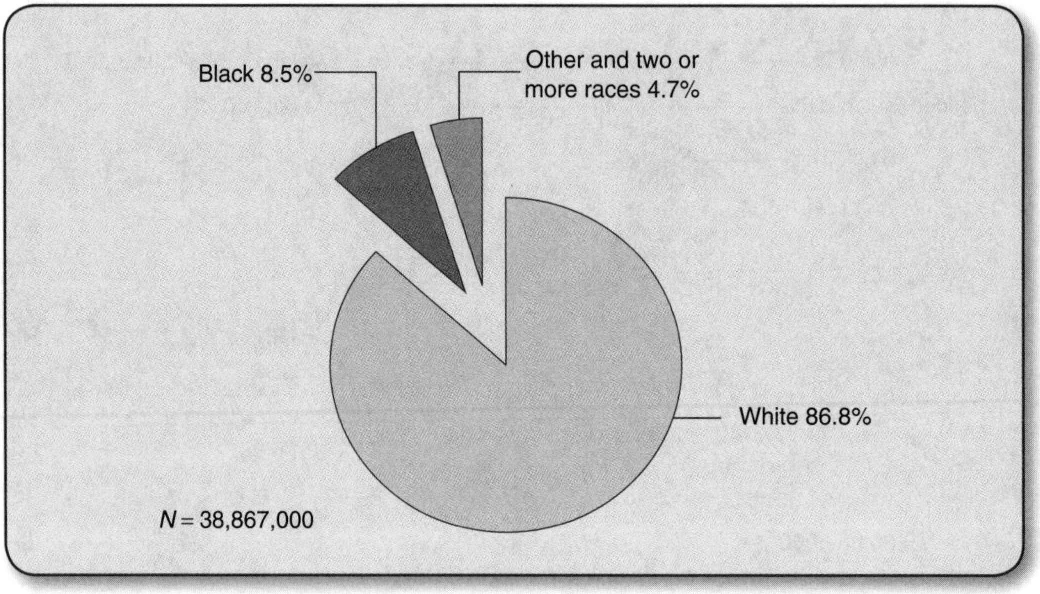

*Source:* U.S. Census Bureau, *The 2010 Statistical Abstract*, Table 10.

categories are displayed as rectangles of equal width with their height proportional to the frequency or percentage of the category.

---

**Bar graph**   A graph showing the differences in frequencies or percentages among the categories of a nominal or an ordinal variable. The categories are displayed as rectangles of equal width with their height proportional to the frequency or percentage of the category.

---

Let's illustrate the bar graph with an overview of the living arrangements of the elderly. Living arrangements change considerably with advancing age—an increasing number of the elderly live alone or with other relatives. Figure 2.6 is a bar graph displaying the percentage distribution of elderly males by living arrangements in 2009. This chart is interpreted similar to a pie chart except that the categories of the variable are arrayed along the horizontal axis (sometimes referred to as the *X*-axis) and the percentages along the vertical axis (sometimes referred to as the *Y*-axis). It shows that in 2009, 18.7% of elderly males lived alone, 72% were married and living with their spouses, 6.1% were living with other relatives, and the remaining 3.2% were living with nonrelatives.

**Figure 2.6**  Living Arrangement of Males (65 and older) in the United States, 2009

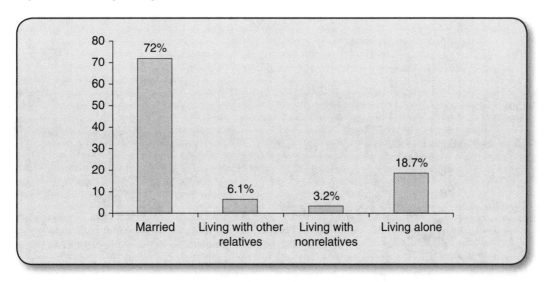

*Source:* U.S. Census Bureau, Current Population Survey, 2009, Annual Social and Economic Supplement, Tables A1 and A2.

Construct a bar graph by first labeling the categories of the variables along the horizontal axis. For these categories, construct rectangles of equal width, with the height of each proportional to the frequency or percentage of the category. Note that a space separates each of the categories to make clear that they are nominal categories.

Bar graphs are often used to compare one or more categories of a variable among different groups. For example, there is an increasing likelihood that women will live alone as they age. The longevity of women is the major factor in the gender differences in living arrangements.[12] In addition, elderly widowed men are more likely to remarry than elderly widowed women. Also, it has been noted that the current generation of elderly women has developed more protective social networks and interests.[13]

Suppose we want to show how the patterns in living arrangements differ between men and women. Figure 2.7 compares the percentage of women and men 65 years and older who lived with others or alone in 2009. It clearly shows that elderly women are more likely than elderly men to live alone.

We can also construct bar graphs horizontally, with the categories of the variable arrayed along the vertical axis and the percentages or frequencies displayed on the horizontal axis. This format is illustrated in Figure 2.8, which compares the percentage of men and women 55 years and over in the civilian labor force for 2008. We see that for all age categories, men were more likely to be employed than women.

## ▣ THE HISTOGRAM

The **histogram** is used to show the differences in frequencies or percentages among categories of an interval-ratio variable. The categories are displayed as contiguous bars, with width proportional to the width of the category and height proportional to the frequency or percentage of that category. A histogram looks very similar to a bar chart except that the bars are contiguous to each other (touching) and may not be of equal width. In a bar chart, the spaces between the bars visually indicate that the categories are

**Figure 2.7**   Living Arrangement of U.S. Elderly (65 and older) by Gender, 2009

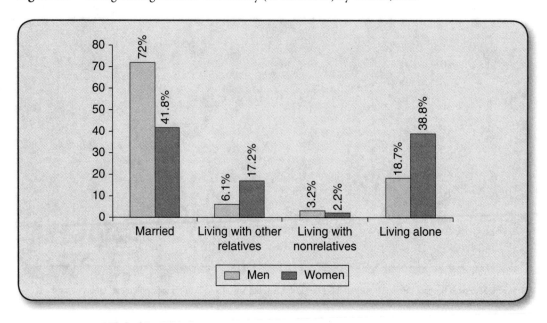

*Source:* U.S. Census Bureau, Current Population Survey, 2009, Annual Social and Economic Supplement, Tables A1 and A2.

**Figure 2.8**   Percentage of Men and Women 55 Years and Over Who Are Employed, 2008

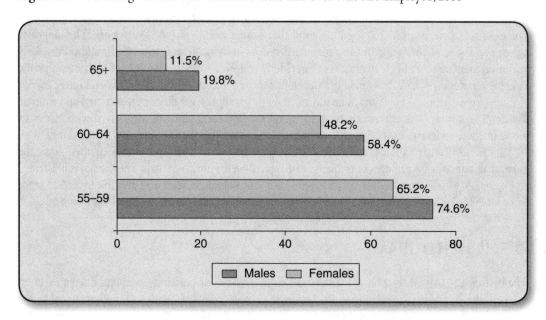

*Source:* U.S. Census Bureau, American Community Survey 2006–2008, Table B23001.

separate. Examples of variables with separate categories are *marital status* (married, single), *gender* (male, female), and *employment status* (employed, unemployed). In a histogram, the touching bars indicate that the categories or intervals are ordered from low to high in a meaningful way. For example, the categories of the variables *hours spent studying, age,* and *years of school completed* are contiguous, ordered intervals.

---

*Histogram*    A graph showing the differences in frequencies or percentages among the categories of an interval-ratio variable. The categories are displayed as contiguous bars, with width proportional to the width of the category and height proportional to the frequency or percentage of that category.

---

Figure 2.9 is a histogram displaying the percentage distribution of the population 65 years and over by age. The data on which the histogram is based are presented in Table 2.15. To construct the histogram of Figure 2.9, arrange the age intervals along the horizontal axis and the percentages (or frequencies) along the vertical axis. For each age category, construct a bar with the height corresponding to the percentage of the elderly in the population in that age category. The width of each bar corresponds to the number of years that the age interval represents. The area that each bar occupies tells us the proportion of the population that falls into a given age interval. The histogram is drawn with the bars touching each other to indicate that the categories are contiguous.

**Figure 2.9**    Age Distribution of U.S. Population 65 Years and Over, 2008

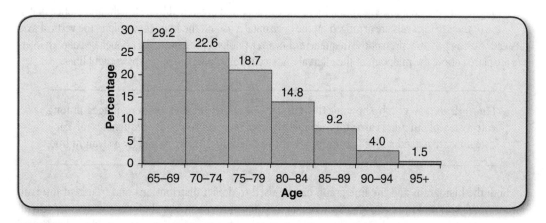

*Source:* U.S. Census Bureau, *The 2010 Statistical Abstract*, Table 10.

*When bar charts or histograms are used to display the frequencies of the categories of a single variable, the categories are shown on the X-axis and the frequencies on the Y-axis. In a horizontal bar chart or histogram, this is reversed.*

✓ *Learning Check*

**Table 2.15**   Percentage Distribution of U.S. Population 65 Years and Over by Age, 2008

| Age (years) | Percentage (%) |
|---|---|
| 65–69 | 29.2 |
| 70–74 | 22.6 |
| 75–79 | 18.7 |
| 80–84 | 14.8 |
| 85–89 | 9.2 |
| 90–94 | 4.0 |
| 95+ | 1.5 |
| Total (*N*) | 100.0 (38,870,000) |

## ▣ THE LINE GRAPH

Numerical growth of the elderly population is taking place worldwide, occurring in both developed and developing countries. In 1994, 30 nations had elderly populations of at least 2 million; demographic projections indicate that there will be 55 such nations by 2020. Japan is one of the nations experiencing dramatic growth of its elderly population. Figure 2.10 is a line graph displaying the elderly population of Japan by age.

The **line graph** is another way to display interval-ratio distributions; it shows the differences in frequencies or percentages among categories of an interval-ratio variable. Points representing the frequencies of each category are placed above the midpoint of the category and are joined by a straight line. Notice that in Figure 2.10 the age intervals are arranged on the horizontal axis and the frequencies along the vertical axis. Instead of using bars to represent the frequencies, however, points representing the frequencies of each interval are placed above the midpoint of the intervals. Adjacent points are then joined by straight lines.

---

*Line graph*   A graph showing the differences in frequencies or percentages among categories of an interval-ratio variable. Points representing the frequencies of each category are placed above the midpoint of the category and are joined by a straight line.

---

Both the histogram and the line graph can be used to depict distributions and trends of interval-ratio variables. How do you choose which one to use? To some extent, the choice is a matter of individual preference, but in general, line graphs are better suited for comparing how a variable is distributed across two or more groups or across two or more time periods. For example, Figure 2.11 compares the elderly population in Japan for 2000 with the projected elderly population for the years 2010 and 2020.

Let's examine this line graph. It shows that Japan's population of age 65 and over is expected to grow dramatically in the coming decades. According to projections, Japan's oldest-old population, those 80 years or older, is also projected to grow rapidly, from about 4.8 million (less than 4% of the total population) to 10.8 million (8.9%) by 2020. This projected rise has already led to a reduction in retirement benefits and other adjustments to prepare for the economic and social impact of a rapidly aging society.[14]

**Figure 2.10**   Population of Japan, Age 55 and Above, 2009

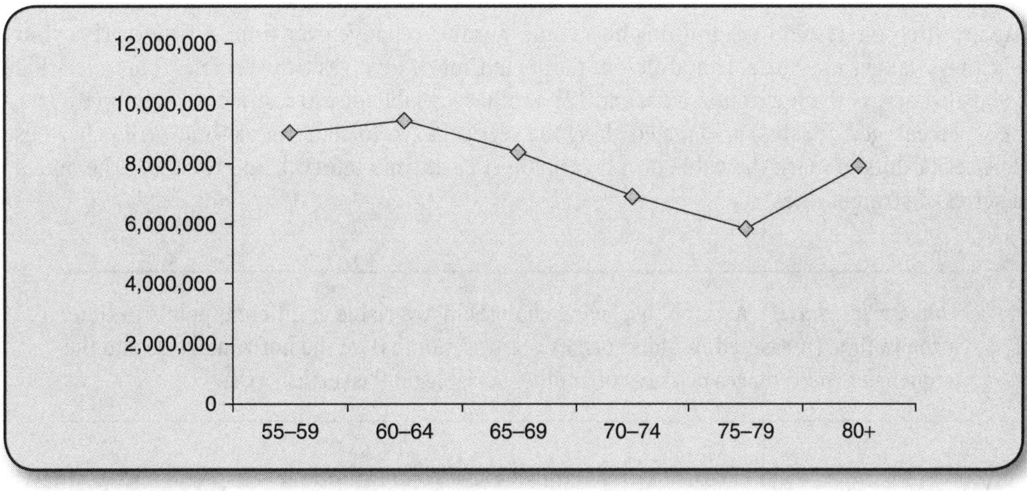

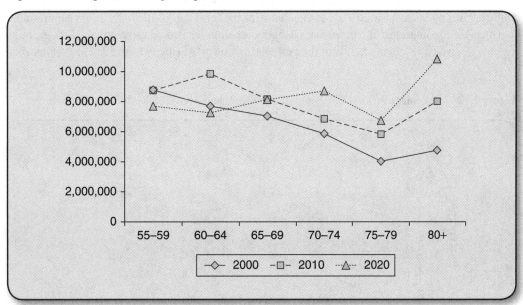

*Source:* Adapted from Ministry of Internal Affairs and Communications of Japan, Statistics Bureau, *Monthly Report April 2010*, Population Estimates.

**Figure 2.11**   Population of Japan, Age 55 and Over, 2000, 2010, and 2020

*Source:* Adapted from U.S. Census Bureau, Center for International Research, International Database, 2007.

*Look closely at the line graph shown in Figure 2.11, comparing 2010 and 2020 data. How would you characterize the population increase among the Japanese elderly?*

✓ *Learning Check*

## ▣ TIME-SERIES CHARTS

We are often interested in examining how some variables change over time. A **time-series chart** displays changes in a variable at different points in time. It involves two variables: (1) *time*, which is labeled across the horizontal axis, and (2) another variable of interest whose values (frequencies, percentages, or rates) are labeled along the vertical axis. To construct a time-series chart, use a series of dots to mark the value of the variable at each time interval, and then join the dots by a series of straight lines.

---

*Time-series chart*    A graph displaying changes in a variable at different points in time. It shows time (measured in units such as years or months) on the horizontal axis and the frequencies (percentages or rates) of another variable on the vertical axis.

---

Figure 2.12 shows a time series from 1900 to 2050 of the percentage of the total U.S. population that is 65 years or older (the figures for the years 2000 through 2050 are projections made by the Social Security Administration, as reported by the U.S. Census Bureau). The number of elderly increased from a little less than 5% in 1900 to about 12.4% in 2000. The rate is expected to increase to 20% of the total population. This dramatic increase in the elderly population, especially beginning in the year 2010, is associated with the "graying" of the baby boom generation.

Often, we are interested in comparing changes over time for two or more groups. Let's examine Figure 2.13, which charts the trends in the percentage of divorced elderly from 1960 to 2050 for men

**Figure 2.12**    Percentage of Total U.S. Population 65 Years and Above, 1900–2050

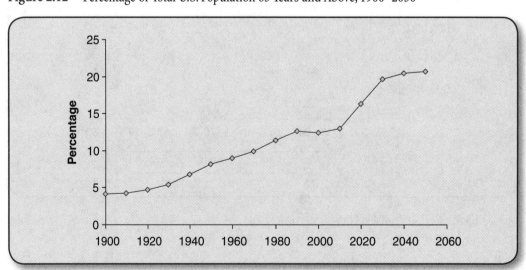

*Source:* Federal Interagency Forum on Aging Related Statistics, *Older Americans 2004: Key Indicators of Well Being*, 2004.

**Figure 2.13**   Percentage Currently Divorced Among U.S. Population 65 Years and Over, by Gender, 1960–2050

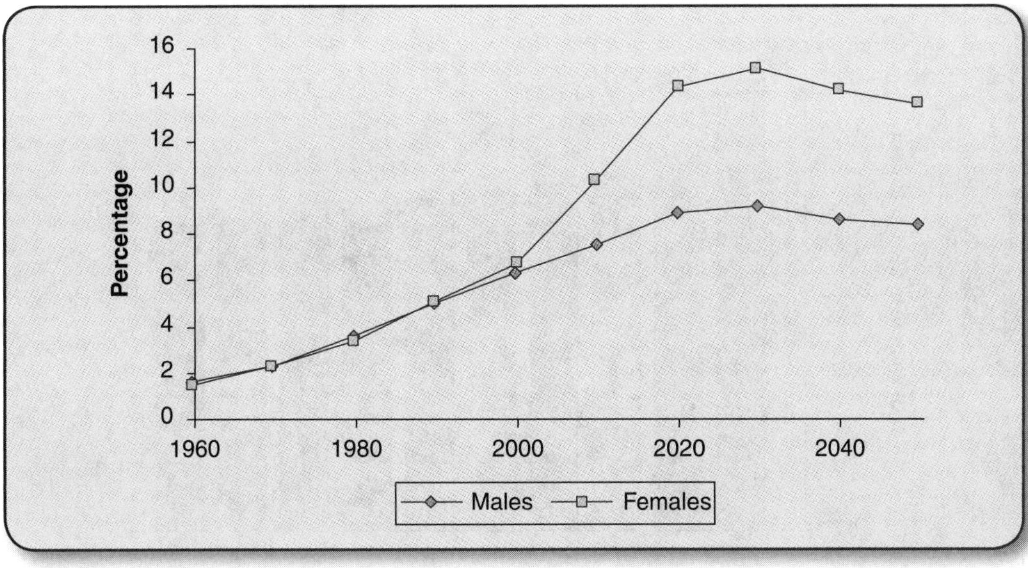

*Source:* U.S. Bureau of the Census, *65+ in America, Current Population Reports, Special Studies,* P23–190, 1996, Table 6-1.

and women. This time-series graph shows that the percentage of divorced elderly men and elderly women was about the same until 2000. For both groups, the percentage increased from less than 2% in 1960 to about 5% in 1990.[15] According to projections, however, there will be significant increases in the percentage of men and especially women who are divorced: from 5% of all the elderly in 1990 to 8.4% of all elderly men and 13.6% of all elderly women by the year 2050. This sharp upturn and the gender divergence are clearly emphasized in Figure 2.13.

> *How does the time-series chart differ from a line graph? The difference is that line graphs display frequency distributions of a single variable, whereas time-series charts display two variables. In addition, time is always one of the variables displayed in a time-series chart.*

✓ *Learning Check*

## ▣ A CAUTIONARY NOTE: DISTORTIONS IN GRAPHS

In this chapter, we have seen that statistical graphs can give us a quick sense of the main patterns in the data. However, graphs can not only quickly inform us, they can also quickly deceive us. Because we are often more interested in general impressions than in detailed analyses of the numbers, we are more vulnerable to being swayed by distorted graphs. Edward Tufte in his 1983 book *The Visual Display of Quantitative Information* not only demonstrates the advantages of working with graphs but also offers a detailed discussion of some of the pitfalls in the application and interpretation of graphics.

**Figure 2.14**    Female Representation in National Parliaments, 2009: (a) Using 0 as the Baseline and (b) Using 30 as the Baseline

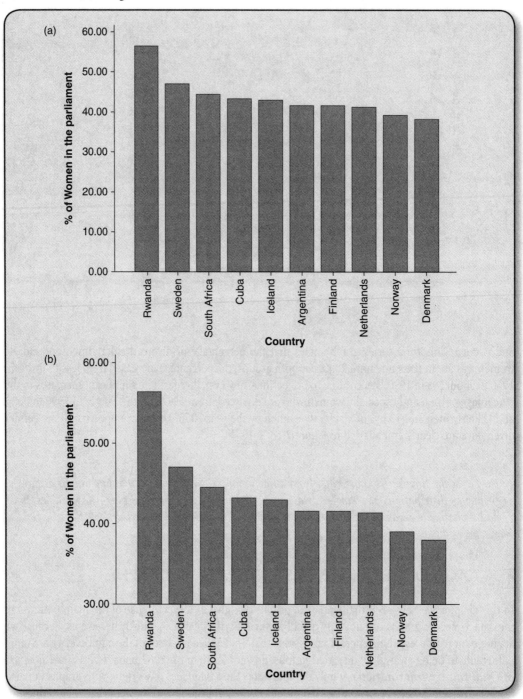

*Source:* Inter-Parliamentary Union. 2009. Women in National Parliaments. Retrieved December 22, 2009, from www.ipu.org/wmn-e/classif.htm.

Probably the most common distortions in graphical representations occur when the distance along the vertical or horizontal axis is altered either by not using 0 as the baseline (as demonstrated in Figures 2.14a and b) or in relation to the other axis. Axes may be stretched or shrunk to create any desired result to exaggerate or disguise a pattern in the data. In Figures 2.14a and b, 2009 international data on female representation in national parliaments are presented. Without altering the data in any way, notice how the difference between the countries is exaggerated by using 30 as a baseline (as in Figure 2.14b).

Remember: Always interpret the graph in the context of the numerical information the graph represents.

## ▣ STATISTICS IN PRACTICE: DIVERSITY AT A GLANCE

We now illustrate some additional ways in which graphics can be used to highlight diversity visually. In particular, we show how graphs can help us (a) explore the differences and similarities among the many social groups coexisting within American society and (b) emphasize the rapidly changing composition of the U.S. population. Indeed, because of the heterogeneity of American society, the most basic question to ask when you look at data is "compared with what?" This question is not only at the heart of quantitative thinking[16] but underlies inclusive thinking as well.

Three types of graphs—the bar chart, the line graph, and the time-series chart—are particularly suitable for making comparisons among groups. Let's begin with the bar chart displayed in Figure 2.15. It compares the college degree attainment of those 35 years of age and over by gender.

**Figure 2.15**    Percentage of College Graduates Among People 35 Years and Over by Age and Gender, 2008

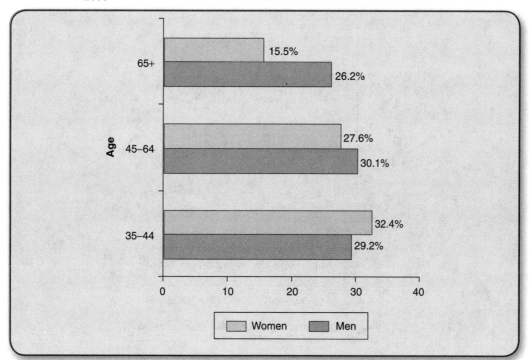

*Source:* U.S. Census Bureau, American Community Survey, 2008, Table B15001.

It shows that the percentage of men with a bachelor's degree or higher is greater than women in the older age groups (45 to 64 and 65+), whereas the percentage of women having at least a bachelor's degree exceeds that of men in the younger (35 to 44) age group. The smallest gap is among those who are 35 to 44 years, of whom 29.2% of men had a bachelor's degree or higher compared with 32.4% of women.

The line graph provides another way of looking at differences based on gender, race/ethnicity, or other attributes such as class, age, or sexual orientation. For example, Figure 2.16 compares years of school completed by black Americans aged 25 to 64 and 65 years and older with that of all Americans in the same age groups.

The data illustrate that in the United States the percentage of Americans who have completed only 8 years of education has declined dramatically from about 10.7% among Americans 65 years and older to 4.2% for those 25 to 64 years old. The decline for black Americans is even more dramatic, from 17% of the black elderly to about 2.2% for those 25 to 64 years old. The corresponding trend illustrated in Figure 2.16 is the increase in the percentage of Americans (all races as well as black Americans) who have completed 13 to 15 years of schooling, or 16 years or more. For example, about 17.3% of black Americans 65 years or older completed 13 to 15 years of schooling compared with 31.3% of those 25 to 64 years.

The trends shown in Figure 2.16 reflect the development of mass education in the United States during the past 50 years. The percentage of Americans who have completed 4 years of high school or

**Figure 2.16**   Educational Attainment in the United States by Race and Age, 2008

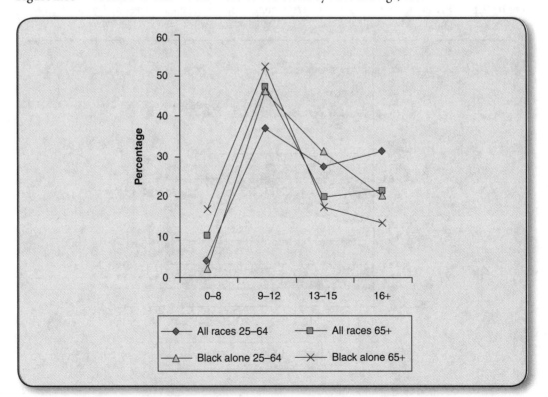

*Source:* The U.S. Census Bureau, *Educational Attainment in the United States: 2009*, Detailed Tables, Table 1.

more has risen from about 40% in 1940 to 86.6% in 2008 (record high according to the U.S. Census Bureau). Similarly, in 1940 only about 5% of Americans completed 4 years or more of college, compared with 29.4% in 2008.[17]

> *Figure 2.16 illustrates that overall, younger Americans (25–64 years old) are better educated than elderly Americans. However, despite these overall trends, there are differences between the number of years of schooling completed by "blacks" and "all races." Examine Figure 2.16, and find these differences. What do they tell you about schooling in America?*

✓ *Learning Check*

Finally, Figure 2.17 is a time-series chart showing changes over time in the teenage birthrate (number of births per 1,000 women) among white, black, and Hispanic women. It shows that between 1990 and 2006, birthrates for teenagers declined; however, rates remain highest for Hispanic and non-Hispanic black women and lowest for non-Hispanic white women.

To conclude, the three examples of graphs in this section as well as other examples throughout this chapter have illustrated how graphical techniques can portray the complexities of the social world by emphasizing the distinct characteristics of age, gender, and ethnic groups. By depicting similarities and differences, graphs help us better grasp the richness and complexities of the social world.

**Figure 2.17**   Birthrates for U.S. Teenagers by Racial/Ethnic Group, 1990, 1995, 2000, 2004, 2005, 2006

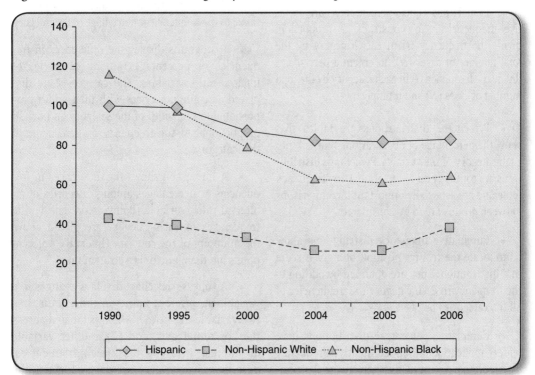

*Source:* U.S. Census Bureau, *The 2010 Statistical Abstract*, Table 84.

*Note:* Rates per 1,000 women in specified group.

## MAIN POINTS

- The most basic method for organizing data is to classify the observations into a frequency distribution—a table that reports the number of observations that fall into each category of the variable being analyzed.

- Constructing a frequency distribution is usually the first step in the statistical analysis of data.

- To obtain a frequency distribution for nominal and ordinal variables, count and report the number of cases that fall into each category of the variable along with the total number of cases ($N$).

- To construct a frequency distribution for interval-ratio variables that have a wide range of values, first combine the scores into a smaller number of groups—known as class intervals—each containing a number of scores.

- Proportions and percentages are relative frequencies. To construct a proportion, divide the frequency ($f$) in each category by the total number of cases ($N$). To obtain a percentage, divide the frequency ($f$) in each category by the total number of cases ($N$) and multiply by 100.

- Percentage distributions are tables that show the percentage of observations that fall into each category of the variable. Percentage distributions are routinely added to almost any frequency table and are especially important if comparisons between groups are to be considered.

- Cumulative frequency distributions allow us to locate the relative position of a given score in a distribution. They are obtained by adding to the frequency in each category the frequencies of all the categories below it.

- Cumulative percentage distributions have wider applications than cumulative frequency distributions. A cumulative percentage distribution is constructed by adding to the percentages in each category the percentages of all the categories below it.

- One other method of expressing raw frequencies in relative terms is known as a rate. Rates are defined as the number of actual occurrences in a given time period divided by the number of possible occurrences. Rates are often multiplied by some power of 10 to eliminate decimal points and make the number easier to interpret.

- A pie chart shows the differences in frequencies or percentages among categories of nominal or ordinal variable. The categories of the variable are segments of a circle whose pieces add up to 100% of the total frequencies.

- A bar graph shows the differences in frequencies or percentages among categories of a nominal or an ordinal variable. The categories are displayed as rectangles of equal width with their height proportional to the frequency or percentage of the category.

- Histograms display the differences in frequencies or percentages among categories of interval-ratio variables. The categories are displayed as contiguous bars with their width proportional to the width of the category and height proportional to the frequency or percentage of that category.

- A line graph shows the differences in frequencies or percentages among categories of an interval-ratio variable. Points representing the frequencies of each category are placed above the midpoint of the category (interval). Adjacent points are then joined by a straight line.

- A time-series chart displays changes in a variable at different points in time. It displays two variables: (1) time, which is labeled across the horizontal axis, and (2) another variable of interest whose values (e.g., frequencies, percentages, or rates) are labeled along the vertical axis.

## KEY TERMS

bar graph
cumulative frequency
   distribution
cumulative percentage
   distribution

frequency
   distribution
histogram
line graph
percentage

percentage distribution
pie chart
proportion
rate
time-series chart

## ON YOUR OWN

Log on to the web-based student study site at **www.sagepub.com/ssdsessentials** for additional study questions, web quizzes, web resources, flashcards, codebooks and datasets, web exercises, appendices, and links to social science journal articles reflecting the statistics used in this chapter.

## CHAPTER EXERCISES

1. Suppose you have surveyed 30 people and asked them whether they are white (W) or nonwhite (N), and how many traumas (serious accidents, rapes, or crimes) they have experienced in the past year. You also asked them to tell you whether they perceive themselves as being in the upper, middle, working, or lower class. Your survey resulted in the raw data presented in the table below:
   a. What level of measurement is the variable race? Class?
   b. Construct raw frequency tables for race and for class.
   c. What proportion of the 30 individuals is nonwhite? What percentage is white?
   d. What proportion of the 30 individuals identified themselves as middle class?

| Race | Class | Trauma | Race | Class | Trauma |
|------|-------|--------|------|-------|--------|
| W | L | 1 | W | W | 0 |
| W | M | 0 | W | M | 2 |
| W | M | 1 | W | W | 1 |
| N | M | 1 | W | W | 1 |
| N | L | 2 | N | W | 0 |
| W | W | 0 | N | M | 2 |
| N | W | 0 | W | M | 1 |
| W | M | 0 | W | M | 0 |
| W | M | 1 | N | W | 1 |
| N | W | 1 | W | W | 0 |
| N | W | 2 | W | W | 0 |
| N | M | 0 | N | M | 0 |
| N | L | 0 | N | W | 0 |
| W | U | 0 | N | W | 1 |
| W | W | 1 | W | W | 0 |

*Source:* Data based on General Social Survey files for 1987 to 1991.

*Notes:* Race: W, white; N, nonwhite; Class: L, lower class; M, middle class; U, upper class; W, working class.

2. We selected a sample of people from the International Social Survey Program (ISSP) 2000. Raw data are presented for their sex (SEX), social class (CLASS), and number of household members (HOMPOP). CLASS is a subjective measure, with respondents indicating L = *lower*, W = *working*, M = *middle*, and U = *upper*.

| Sex | Hompop | Class | Sex | Hompop | Class |
|-----|--------|-------|-----|--------|-------|
| F | 1 | L | F | 2 | U |
| F | 3 | W | M | 2 | M |
| F | 1 | M | F | 4 | W |
| M | 2 | M | M | 2 | U |
| F | 3 | M | M | 4 | M |
| M | 2 | U | M | 4 | W |
| F | 7 | L | M | 2 | M |
| M | 3 | M | F | 1 | W |
| M | 4 | M | M | 2 | U |
| F | 1 | M | M | 4 | W |
| F | 2 | M | F | 7 | M |
| M | 5 | L | M | 3 | M |
| M | 4 | M | M | 4 | M |
| M | 4 | L | F | 4 | M |
| M | 3 | W | F | 5 | L |

a. Construct a pie chart depicting the percentage distribution of sex. (*Hint:* Remember to include a title, percentages, and appropriate labels.)
b. Construct a pie chart showing the percentage distribution of social class.
c. Construct two pie charts comparing the percentage distribution of social class membership by sex.

3. Using the data from Exercise 1, construct a frequency distribution for trauma.
a. What level of measurement is used for the trauma variable?
b. Are people more likely to have experienced no traumas or only one trauma in the past year?
c. What proportion has experienced one or more traumas in the past year?

4. Using the data from Exercise 2, construct bar graphs showing percentage distributions for sex and class. Remember to include appropriate titles, percentages, and labels.

5. The Gallup Organization conducted a survey in June 2007, asking Americans whether they approve or disapprove of the U.S. government's recent efforts to deal with illegal immigration. Results are provided in the table below, noting the percentage by ethnic background (not all responses are reported here so totals

will not add up to 100%). Do these data support the statement that people's views on illegal immigration are related to their ethnic background? Why or why not?

|              | Non-Hispanic Whites | Blacks | Hispanics |
|--------------|:---:|:---:|:---:|
| Approve      | 32 | 29 | 37 |
| Disapprove   | 60 | 60 | 58 |

6. During the 2008 presidential election, health care and health insurance were identified by voters as important issues. Policy analysts have noted that the number of uninsured is increasing in the United States. Data from the National Center for Health Statistics are presented in Figure 2.18 and in the percentage table. What can be said about who did not have health insurance during 2005? How does the percentage of those without health insurance vary by ethnicity/race, age, marital, and poverty status?

**Figure 2.18**   The Uninsured Population Below 65 Years of Age by Selected Characteristics, 2005

| Characteristic | Percentage |
|---|---|
| **Age** | |
| Below 18 years | 16.2 |
| 18–44 Years | 61.6 |
| 45–64 Years | 22.2 |
| **Poverty** | |
| Below 100% | 25.1 |
| 100 to less than 200% | 31.6 |
| 200 to less than 400% | 28.7 |
| 400% or more | 14.6 |
| **Race and Hispanic origin** | |
| Hispanic | 31.3 |
| White only | 48.0 |
| Black | 13.9 |
| Asian | 4.3 |
| Other | 2.4 |
| **Marital status** | |
| Single | 40.2 |
| Married | 41.8 |
| Separated/divorced | 16.3 |
| Widowed | 1.7 |

*(Continued)*

Exercises

(Figure 2.18 Continued)

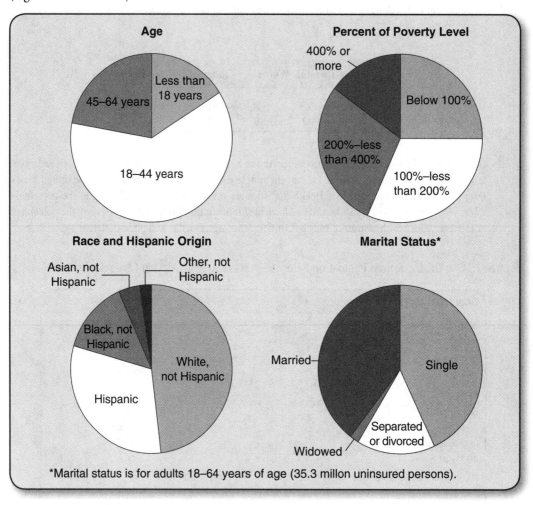

*Marital status is for adults 18–64 years of age (35.3 millon uninsured persons).

7. The following tables present the frequency distributions for education by gender and race based on the GSS 2008. Use them to answer the following questions.

| | Gender | |
| Education | Male (f) | Female (f) |
|---|---|---|
| Less than high school | 100 | 116 |
| High school graduate | 327 | 407 |
| Some college | 68 | 76 |
| College graduate | 190 | 216 |

*Source:* General Social Survey, 2008.

|  | Race/Ethnicity | | |
| Education | White (f) | Black (f) | Hispanic (f) |
| Less than high school | 131 | 50 | 21 |
| High school graduate | 567 | 109 | 25 |
| Some college | 113 | 19 | 5 |
| College graduate | 335 | 28 | 5 |

*Source:* General Social Survey, 2008.

a. Construct tables based on percentages and cumulative percentages of educational attainment for gender and race.

b. What percentage of males has continued their education beyond high school? What is the comparable percentage for females?

c. What percentage of whites has completed high school or less? What is the comparable percentage for blacks? For Hispanics?

d. Are the cumulative percentages more similar for men and women or for the racial and ethnic groups? (In other words, where is there more inequality?) Explain.

8. You are writing a research paper on grandparents who had one or more of their grandchildren living with them. In 2000, 2.4 million grandparents were defined as caregivers by the U.S. Census, meaning that they had primary responsibility for raising their grandchildren below the age of 18. You discover the following information from the U.S. Census Report, "Grandparents Living With Grandchildren: 2000" (C2KBR-31, October 2003): Among grandparent caregivers, 12% cared for a grandchild for less than 6 months, 11% for 6 to 11 months, 23% for 1 to 2 years, 15% for 3 to 4 years, and 39% for 5 or more years.

a. Construct a graph or chart that best displays this information on how long grandparents care for their grandchildren.

b. Explain why the graph you selected is appropriate.

9. From the MTF 2008, we present data on the number of moving violation tickets in the past 12 months for males and females.

|  | Males (%) | Females (%) |
| None | 71.3 | 80.4 |
| 1 | 17.1 | 13.0 |
| 2 | 7.4 | 5.0 |
| 3 | 2.1 | 0.7 |
| 4 | 2.1 | 0.9 |
| Total | 100.0 | 100.0 |

*Source:* Monitoring the Future, 2008.

a. What is the level of measurement for number of moving violation tickets?

b. What percentage of males and females had 3 tickets or more? Calculate the percentages for males and females separately.

c. Convert the percentages to frequencies. The total number of male respondents is 624 and the total number of female respondents is 699.

d. What can be said about the differences in the number of tickets between males and females?

Exercises

10. The 2008 General Social Survey (GSS) data on educational level can be further broken down by race as follows:
    a. Construct two histograms for education, one for whites ($N = 636$) and one for blacks ($N = 136$).
    b. Now use the two graphs to describe the differences in educational attainment by race.

| Years of Education | Whites | Blacks |
|---|---|---|
| 0 | 2 | 0 |
| 1 | 0 | 0 |
| 2 | 1 | 1 |
| 3 | 2 | 0 |
| 4 | 0 | 0 |
| 5 | 3 | 1 |
| 6 | 5 | 0 |
| 7 | 2 | 0 |
| 8 | 13 | 1 |
| 9 | 8 | 6 |
| 10 | 17 | 8 |
| 11 | 27 | 6 |
| 12 | 180 | 36 |
| 13 | 56 | 18 |
| 14 | 101 | 33 |
| 15 | 27 | 5 |
| 16 | 103 | 13 |
| 17 | 32 | 1 |
| 18 | 32 | 2 |
| 19 | 12 | 1 |
| 20 | 13 | 4 |

11. A Gallup Poll (June 9–30, 2004) compared 2,250 Americans' attitudes on minority rights and how much of a role the government should play in helping minorities improve their social and economic positions. Results from the survey are presented below (not all responses are reported). How would you characterize the differences between the attitudes of whites and minorities?

| What Should Be the Government's Role in Improving Economic and Social Position of Minorities? | Non-Hispanic Whites (%) | Blacks (%) | Hispanics (%) |
|---|---|---|---|
| Major role | 32 | 69 | 67 |
| Minor role | 51 | 22 | 21 |
| No role | 16 | 9 | 8 |

*Source:* Jeffrey Jones, "Blacks More Pessimistic Than Whites About Economic Opportunities," *Gallup Poll News Service*, The Gallup Organization. July 9, 2004. Used with permission.

12. Examine the bar chart representing the percentage of people speaking a language other than English at home as shown in Figure 2.19.
    a. Overall, which age group had the lowest percentage speaking a language other than English at home? Which age group had the highest?
    b. Describe the differences between the 2002 and 2008 percentages.

**Figure 2.19**  Percentage of People Speaking a Language Other Than English at Home Among the Population Aged 5 and Over, by Age, 2002 and 2008

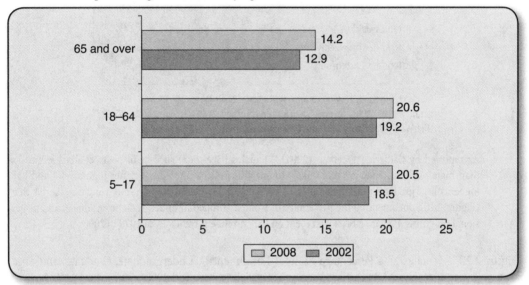

*Source:* U.S. Census Bureau, American Community Survey, 2002, Table P035, and American Community Survey, 2008, Table C16007.

13. The Gallup Organization reported that public tolerance for gay rights reached a "high watermark" in 2007 (in contrast with attitudes measured over the past three decades). In its report, the organization examined the difference between different demographic groups to the question of whether homosexuality is an acceptable alternative lifestyle. Which groups support homosexuality as an acceptable alternative lifestyle? Review each demographic variable and summarize the pattern or level of support.

|              | Yes (%) | No (%) |
|--------------|---------|--------|
| Sex          |         |        |
|   Men        | 53      | 44     |
|   Women      | 61      | 35     |
| Age          |         |        |
|   18–34 years | 75      | 23     |
|   35–54 years | 58      | 39     |
|   55+ years   | 45      | 51     |

*(Continued)*

(Continued)

|  | Yes (%) | No (%) |
|---|---|---|
| Political affiliation |  |  |
| Republican | 36 | 58 |
| Independent | 60 | 36 |
| Democrat | 72 | 27 |
| Religious service attendance |  |  |
| Attend weekly | 33 | 64 |
| Attend nearly weekly/monthly | 57 | 40 |
| Attend less often/never | 74 | 22 |

*Source:* Lydia Saad, "Tolerance for Gay Rights at High-Water Mark," 2007. Retrieved January 19, 2008, from www.gallup.com/poll/27694/Tolerance-Gay-Rights-HighWater-Mark.aspx. Used with permission.

14. As reported by Catherine Freeman (2004),[18] females have more success in postsecondary education than male students. They are more likely to enroll in college immediately after high school and have higher college graduate rates than males. In her report, Freeman provides the following time-series chart (Figure 2.20), documenting the percentage of women enrolled in undergraduate, graduate, and professional programs. Prepare a brief statement on the enrollment trends from 1970 to 2000.

**Figure 2.20** Females as a Percentage of Total Enrollment in Undergraduate, Graduate, and First-Professional Education: Various Years, Fall 1970 to Fall 2000

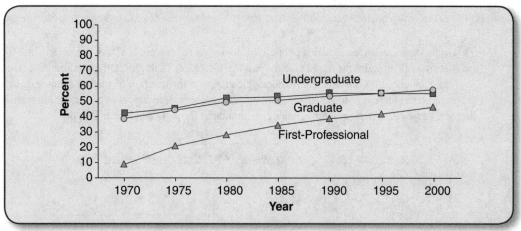

*Source:* Catherine Freeman, *Trends in Educational Equity for Girls and Women: 2004* (NCES 2005–016). U.S. Department of Education, National Center for Education Statistics (Washington, DC: Government Printing Office, 2004).

# Chapter 3

# Measures of Central Tendency

**Chapter Learning Objectives**

❖ Defining all measures of central tendency, explaining their differences, relative strengths and weaknesses
❖ Determining the mode in a given distribution
❖ Finding or calculating the median and percentiles
❖ Calculating the mean
❖ Determining the shape of a distribution

I n Chapter 2, we learned that frequency distributions and graphical techniques are useful tools for describing data. Another way of describing a distribution is by selecting a single number that describes or summarizes the distribution more concisely. Such numbers describe what is typical about the distribution, for example, the average income among Latinos who are college graduates or the most common party identification among the rural poor. These are called **measures of central tendency**.

---

*Measures of central tendency*   Categories or scores that describe what is average or typical of the distribution.

---

In this chapter, we will learn about three measures of central tendency: the *mode*, the *median*, and the *mean*. The choice of an appropriate measure of central tendency for representing a distribution depends on three factors: (1) the way the variables are measured (their level of measurement), (2) the shape of the distribution, and (3) the purpose of the research.

## ▣ THE MODE

The **mode** is the category or score with the largest frequency or percentage in the distribution.

We can use the mode to determine, for example, the most common foreign language spoken in the United States today. What is the most common foreign language spoken in the United States today? Look at Table 3.1, which lists the 10 most commonly spoken foreign languages in the United States and the number of people who speak each language. The table shows that Spanish is the most common; more than 30 million people speak Spanish. In this example, we refer to "Spanish" as the mode—the category with the largest frequency in the distribution.

The mode is used to describe nominal variables, those variables which allow classification based on only a qualitative and not a quantitative property. By describing the most commonly occurring category of a nominal variable (such as Spanish in our example), the mode thus reflects the most important element of the distribution of a variable measured at the nominal level. The mode is the only measure of central tendency that can be used with nominal-level variables. It can also be used to describe the most commonly occurring category in any distribution.

---

*Mode*    The category or score with the highest frequency (or percentage) in the distribution.

---

**Table 3.1**    Ten Most Common Foreign Languages Spoken in the United States, 2007

| Language | Number of Speakers |
|----------|-------------------:|
| Spanish | 34,547,000 |
| Chinese | 2,465,000 |
| French | 1,985,000 |
| Tagalog | 1,480,000 |
| Vietnamese | 1,207,000 |
| German | 1,104,000 |
| Korean | 1,062,000 |
| Russian | 851,000 |
| Italian | 799,000 |
| Arabic | 767,000 |

*Source:* U.S. Census Bureau, *The 2010 Statistical Abstract*, Table 53.

Listed below are the political party affiliations of 15 individuals. Find the mode.

| | | | | |
|---|---|---|---|---|
| *Democrat* | *Republican* | *Democrat* | *Republican* | *Republican* |
| *Independent* | *Democrat* | *Democrat* | *Democrat* | *Republican* |
| *Independent* | *Democrat* | *Independent* | *Republican* | *Democrat* |

## ▣ THE MEDIAN

The **median** is a measure of central tendency that can be calculated for variables that are at least at an ordinal level of measurement. The median represents the exact middle of a distribution; it is the score that divides the distribution into two equal parts so that half the cases are above it and half below it. For example, according to the U.S. Bureau of Labor Statistics, the median weekly earnings of full-time wage and salary workers in 2009 was $739.[1] This means that half the workers in the United States earned more than $739 a week and half earned less than $739. Since many variables used in social research are ordinal, the median is an important measure of central tendency.

---

*Median*   The score that divides the distribution into two equal parts so that half the cases are above it and half below it.

---

The median is a suitable measure for those variables whose categories or scores can be arranged in order of magnitude from the lowest to the highest. Therefore, the median can be used with ordinal or interval-ratio variables, for which scores can be at least rank ordered, but cannot be calculated for variables measured at the nominal level.

### Finding the Median in Sorted Data

It is very easy to find the median. In most cases, it can be done by a simple inspection of the sorted data. The location of the median score differs somewhat, depending on whether the number of observations is odd or even. Let's first consider an example with an odd number of cases.

#### An Odd Number of Cases

Suppose we are looking at the responses of five people to the question, "Thinking about the economy, how would you rate economic conditions in this country today?" Following are the responses of these five hypothetical persons:

Poor
Good
Only fair
Poor
Excellent

Total ($N$) = 5

First arrange the responses in order from the lowest to the highest (or the highest to the lowest):

| Poor |
| Poor |
| Only fair |
| Good |
| Excellent |
| Total ($N$) = 5 |

The median is the response associated with the middle case. Find the middle case when $N$ is odd by adding 1 to $N$ and dividing by 2: $(N + 1)/2$. Since $N$ is 5, you calculate $(5 + 1)/2 = 3$. The middle case is thus the third case, and the median is "only fair," the response associated with the third case. Notice that the median divides the distribution exactly in half so that there are two respondents who are more satisfied and two respondents who are less satisfied.

When working with an even number of cases, the middle case will not be a whole number. If the data are ordinal, simply report both of the category names that comprise the middle cases. If the data are interval ratio, average the two middle cases together and report that as the median.

✓ *Learning Check*  *Find the median of the following distribution of an interval-ratio variable: 22, 15, 18, 33, 17, 5, 11, 28, 40, 19, 8, 20.*

## Finding the Median in Frequency Distributions

Often our data are arranged in frequency distributions. Take, for instance, the frequency distribution displayed in Table 3.2. It shows the political views of GSS respondents in 2008.

**Table 3.2**  Political Views of GSS Respondents, 2008

| Political View | Frequency (f) | Cf | Percentage | C% |
|---|---|---|---|---|
| Extremely liberal | 52 | 52 | 3.6 | 3.6 |
| Liberal | 175 | 227 | 12.2 | 15.8 |
| Slightly liberal | 172 | 399 | 12.0 | 27.8 |
| Moderate | 545 | 944 | 38.0 | 65.8 |
| Slightly conservative | 203 | 1,147 | 14.1 | 79.9 |
| Conservative | 241 | 1,388 | 16.8 | 96.7 |
| Extremely conservative | 48 | 1,436 | 3.3 | 100.0 |
| Total ($N$) | 1,436 | | 100.0 | |

We begin by specifying $N$, the total number of respondents. In this particular example, $N = 1,436$. We then use the formula $(N + 1)/2$, or $(1,436 + 1)/2 = 718.5$. The median is the value of the category associated with the average of 718th and 719th cases. The cumulative frequencies ($Cf$) of both 718 and 719 fall in the category "moderate"; thus, the median is "moderate." This may seem strange; however, the median is always the value of the response category, not the frequency.

A second approach to locating the median in a frequency distribution is to use the cumulative percentages column, as shown in the last column of Table 3.2. The median is the value of the category associated with the cumulative percentage reaching at least 50.[2] Looking at Table 3.2, the percentage value equal to 50% falls within the category "moderate." The median for this distribution is therefore "moderate." If you are not sure why the middle of the distribution—the 50% point—is associated with the category "moderate," look again at the cumulative percentage column ($C\%$). Notice that 27.8% of the observations are accumulated below the category "moderate" and that 65.8% are accumulated up to and including the category "moderate." We know, then, that the percentage value equal to 50% is located somewhere within the "moderate" category.

*For a review of cumulative distributions, refer to Chapter 2.*

✓ *Learning Check*

## Statistics in Practice: Gendered Income Inequality

We can use the median to compare groups, such as the income levels of men and women in the United States over the past few decades. Income levels profoundly influence our lives both socially and economically. Higher income is associated with increased education and work experience for both men and women.

Figure 3.1 compares the median incomes for men and women in 1973 and in 2008. Because the median is a single number summarizing central tendency in the distribution, we can use it to note differences between subgroups of the population or changes over time. In this example, the increase in median income from 1973 to 2008 clearly shows a significant income gain for women. However, that said, in 2008 women still made, on average, about $10,000 less than men.

## Locating Percentiles in a Frequency Distribution

The median is a special case of a more general set of measures of location called *percentiles*. A **percentile** is a score at or below which a specific percentage of the distribution falls. The $n$th percentile is a score below which $n\%$ of the distribution falls. For example, the 75th percentile is a score that divides the distribution so that 75% of the cases are below it. The median is the 50th percentile. It is a score that divides the distribution so that 50% of the cases fall below it. Like the median, percentiles require that data be ordinal or higher in level of measurement. Percentiles are easy to identify when the data are arranged in frequency distributions.

---

*Percentile*    A score below which a specific percentage of the distribution falls.

---

**Figure 3.1**   Median Income for Men and Women, 1973 and 2008

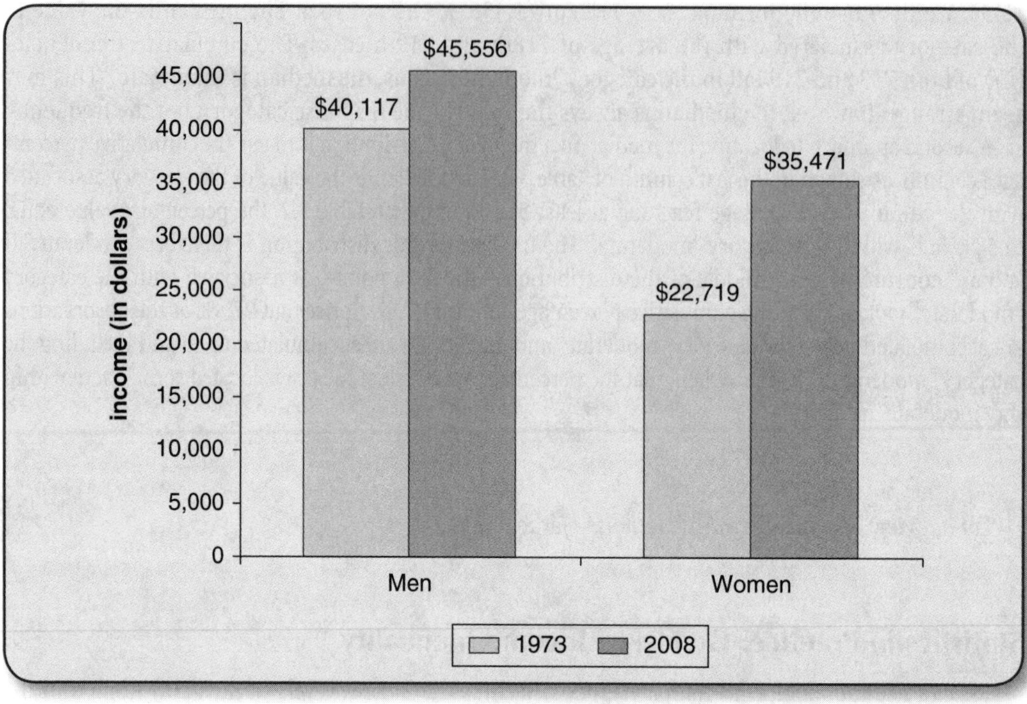

*Source:* Data from 1973 obtained from the U.S. Census Bureau, Current Population Reports P60–226, *Income, Poverty and Health Insurance Coverage in the United States: 2003*. Data from 2008 obtained from the U.S. Census Bureau, American Community Survey, 2008.

To help illustrate how to locate percentiles in a frequency distribution, we display in Table 3.3 the frequency distribution, the percentage distribution, and the cumulative percentage distribution of opinion about police job performance of respondents for the 2008 Monitoring the Future (MTF) survey.

**Table 3.3**   Frequency Distribution for Police Job Performance, 2008 MTF Respondents

| Police Job Performance | Frequency (f) | Percentage | C% |
|---|---|---|---|
| Very poor | 125 | 9.9 | 9.9 |
| Poor | 201 | 15.9 | 25.8 |
| Fair | 452 | 35.7 | 61.5 |
| Good | 380 | 30.0 | 91.5 |
| Very good | 108 | 8.5 | 100.0 |
| Total (N) | 1,266 | 100.0 | |

The 50th percentile (the median) is "Fair," meaning that 50% of the respondents view police job performance above "Fair" and 50% of the respondents view police job performance as "Fair" or below "Fair" (as you can see from the cumulative percentage column, 50% falls somewhere in the third category, associated with the category "Fair"). Similarly, the 20th percentile is "Poor" because 20% of the respondents view police job performance as "Poor" or below "Poor."

Percentiles are widely used to evaluate relative performance on standardized achievement tests, such as the SAT or ACT. Let's suppose that your ACT score was 29. To evaluate your performance for the college admissions officer, the testing service translated your score into a percentile rank. Your percentile rank was determined by comparing your score with the scores of all other seniors who took the test at the same time. Suppose for a moment that 90% of all students received a lower ACT score than you (and 10% scored above you). Your percentile rank would have been 90. If, however, there were more students who scored better than you—let's say that 15% scored above you and 85% scored lower than you—your percentile rank would have been only 85.

Another widely used measure of location is the *quartile*. The lower quartile is equal to the 25th percentile and the upper quartile is equal to the 75th percentile. (Can you locate the upper quartile in Table 3.3?) A college admissions office interested in accepting the top 25% of its applicants based on their SAT scores could calculate the upper quartile (the 75th percentile) and accept everyone whose score is equivalent to the 75th percentile or higher. (Note that they would be calculating percentiles based on the scores of their applicants, not of all students in the nation who took the SAT.)

## ▣ THE MEAN

The arithmetic **mean** is by far the best known and most widely used measure of central tendency. The mean is what most people call the "average." The mean is typically used to describe central tendency in interval-ratio variables such as income, age, and education. You are probably already familiar with how to calculate the mean. Simply add up all the scores and divide by the total number of scores.

---

*Mean*   A measure of central tendency that is obtained by adding up all the scores and dividing by the total number of scores. It is the arithmetic average.

---

Crime statistics, for example, are often analyzed using the mean. Each year about 25% of U.S. households are victims of some form of crime. Although violent crimes are the least common types of crimes in the United States, the U.S. rate of violent crime is nonetheless the highest of any industrialized nation. For instance, murder rates in the United States are approximately five times as high as those in Europe.

Table 3.4 shows the 2007 murder rates (per 100,000 population) for the 15 largest cities in the United States. Because the variable "murder rate" is an interval-ratio variable, we will select the arithmetic mean as our measure of central tendency to summarize the information.

**Table 3.4**  2007 Murder Rates (per 100,000 population) for the 15 Largest Cities in the United States

| City | Murder Rate per 100,000 |
| --- | --- |
| New York | 6.0 |
| Los Angeles | 10.2 |
| Chicago | 15.7 |
| Houston | 16.2 |
| Philadelphia | 27.3 |
| Phoenix | 13.8 |
| San Diego | 4.7 |
| Las Vegas | 8.9 |
| San Antonio | 9.3 |
| Dallas | 16.1 |
| Detroit | 45.8 |
| San Jose | 3.5 |
| Honolulu | 2.1 |
| Indianapolis | 14.3 |
| Jacksonville | 15.4 |

Source: U.S. Census Bureau, *The 2010 Statistical Abstract*, Table 298.

To find the mean murder rate for the data presented in Table 3.4, add up the murder rates for all the cities and divide the sum by the number of cities:

$$\text{Mean} = \frac{\begin{array}{l}(6.0 + 10.2 + 15.7 + 16.2 + 27.3 + 13.8 + 4.7+ \\ 8.9 + 9.3 + 16.1 + 45.8 + 3.5 + 2.1 + 14.3 + 15.4)\end{array}}{15} = \frac{209.3}{15} = 13.95$$

The mean murder rate for the 15 largest cities in the United States is 13.95.[3]

## Calculating the Mean

Another way to calculate the arithmetic mean is to use a formula. Beginning with this section, we introduce a number of formulas that will help you calculate some of the statistical concepts that we are going to present. A formula is a shorthand way to explain what operations we need to follow to obtain a certain result. So instead of saying "add all the scores together and then divide by the number of scores," we can define the mean by the following formula:

$$\bar{Y} = \frac{\Sigma Y}{N} \tag{3.1}$$

Let's take a moment to consider these new symbols because we continue to use them in later chapters. We use $Y$ to represent the raw scores in the distribution of the variable of interest; $\bar{Y}$ is pronounced "Y-bar" and is the mean of the variable of interest. The symbol represented by the Greek letter $\Sigma$ is pronounced "sigma," and it is used often from now on. It is a summation sign (just like the + sign) and directs us to sum whatever comes after it. Therefore, $\Sigma Y$ means "add up all the raw $Y$ scores." Finally, the letter $N$, as you know by now, represents the number of cases (or observations) in the distribution.

Let's summarize the symbols as follows:

$Y =$ the raw scores of the variable $Y$

$\bar{Y} =$ the mean of $Y$

$\Sigma Y =$ the sum of all the $Y$ scores

$N =$ the number of observations or cases

Now that we know what the symbols mean, let's work through another example. The following are the ages of the 10 students in a graduate research methods class:

$$21, 32, 23, 41, 20, 30, 36, 22, 25, 27$$

What is the mean age of the students?

For these data, the ages included in this group are represented by $Y$; $N = 10$, the number of students in the class; and $\Sigma Y$ is the sum of all the ages:

$$\Sigma Y = 21 + 32 + 23 + 41 + 20 + 30 + 36 + 22 + 25 + 27 = 277$$

Thus, the mean age is

$$\bar{Y} = \frac{\Sigma Y}{N} = \frac{277}{10} = 27.7$$

## ▣ FINDING THE MEAN IN A FREQUENCY DISTRIBUTION

When data are arranged in a frequency distribution, we must give each score its proper weight by multiplying it by its frequency. We can use the following modified formula to calculate the mean:

$$\bar{Y} = \frac{\Sigma fY}{N}$$

where

$Y$ = the raw scores of the variable $Y$

$\bar{Y}$ = the mean of $Y$

$\Sigma fY$ = the sum of all the $fY$s

$N$ = the number of observations or cases

✓ *Learning Check*

*The following distribution is the same as the one you used to calculate the median in an earlier Learning Check: 22, 15, 18, 33, 17, 5, 11, 28, 40, 19, 8, 20. Calculate the mean. Is it the same as the median, or is it different?*

### Understanding Some Important Properties of the Arithmetic Mean

The following three mathematical properties make the mean the most important measure of central tendency. It is, in fact, a concept that is basic to numerous and more complex statistical operations.

#### Interval-Ratio Level of Measurement

Because it requires the mathematical operations of addition and division, the mean can be calculated only for variables measured at the interval-ratio level.

#### Center of Gravity

Because the mean (unlike the mode and the median) incorporates all the scores in the distribution, we can think of it as the center of gravity of the distribution. That is, the mean is the point that perfectly balances all the scores in the distribution. If we subtract the mean from each score and add up all the differences, the sum will always be zero!

#### Sensitivity to Extremes

The examples we have used to show how to compute the mean demonstrate that, unlike with the mode or the median, every score enters into the calculation of the mean. This property makes the mean sensitive to extreme scores in the distribution. The mean is pulled in the direction of either very high or very low values. A glance at Figure 3.2 should convince you of that. Figures 3.2a and b show the incomes of 10 individuals. In Figure 3.2b, the income of one individual has shifted from $5,000 to $35,000. Notice the effect it has on the mean; it shifts from $3,000 to $6,000! The

**Figure 3.2** The Value of the Mean Is Affected by Extreme Scores: (a) No Extreme Scores and (b) One Extreme Score

(a) No extreme scores: The mean is $3,000

| Income (*Y*) | Frequency(*f*) | *fY* |
|---|---|---|
| 1,000 | 1 | 1,000 |
| 2,000 | 2 | 4,000 |
| 3,000 | 4 | 12,000 |
| 4,000 | 2 | 8,000 |
| 5,000 | 1 | 5,000 |
| | *N* = 10 | Σ*fY* = 30,000 |

$$\text{Mean} = \frac{\Sigma fY}{N} = \frac{30,000}{10} = \$3,000$$

Median = $3,000

(b) One extreme score: The mean is $6,000

| Income (*Y*) | Frequency (*f*) | *fY* |
|---|---|---|
| 1,000 | 1 | 1,000 |
| 2,000 | 2 | 4,000 |
| 3,000 | 4 | 12,000 |
| 4,000 | 2 | 8,000 |
| 35,000 | 1 | 35,000 |
| | *N* = 10 | Σ*fY* = 60,000 |

$$\text{Mean} = \frac{\Sigma fY}{N} = \frac{60,000}{10} = \$6,000$$

Median = $3,000

mean is disproportionately affected by the relatively high income of $35,000 and is misleading as a measure of central tendency for this distribution. Notice that the median's value is not affected by this extreme score; it remains at $3,000. Thus, the median gives us better information on the typical income for this group. In the next section, we will see that because of the sensitivity of the mean, it is not suitable as a measure of central tendency in distributions that have a few very extreme values on one side of the distribution. (A few extreme values are no problem if they are not mostly on one side of the distribution.)

## ▣ THE SHAPE OF THE DISTRIBUTION: TELEVISION, EDUCATION, AND SIBLINGS

In this chapter, we have looked at the way in which the mode, median, and mean reflect central tendencies in the distribution. Distributions (this discussion is limited to distributions of interval-ratio variables) can also be described by their general shape, which can be easily represented visually. A distribution can be either symmetrical or skewed, depending on whether there are a few extreme values at one end of the distribution.

A distribution is **symmetrical** (Figure 3.3a) if the frequencies at the right and left tails of the distribution are identical, so that if it is divided into two halves, each will be the mirror image of the other. In a unimodal, symmetrical distribution, the mean, median, and mode are identical.

In **skewed** distributions, there are a few extreme values on one side of the distribution. Distributions that have a few extremely low values are referred to as **negatively** skewed (Figure 3.3b). A defining feature of a negatively skewed distribution is that the value of the mean is always less than the value of the median; in other words, the mean is pulled in the direction of the lower scores. Alternatively, distributions with a few extremely high values are said to be **positively** skewed (Figure 3.3c). A defining feature of a positively skewed distribution is that the value of the mean is always greater than the value of the median; the mean is pulled toward the higher scores.

**Figure 3.3**   Types of Frequency Distributions

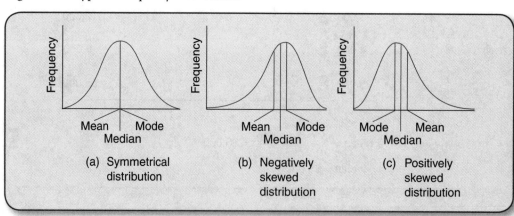

*Symmetrical distribution*    The frequencies at the right and left tails of the distribution are identical; each half of the distribution is the mirror image of the other.

*Skewed distribution*    A distribution with a few extreme values on one side of the distribution.

*Negatively skewed distribution*    A distribution with a few extremely low values.

*Positively skewed distribution*    A distribution with a few extremely high values.

We can illustrate the differences among the three types of distributions by examining three variables in the 2008 GSS.[4] The frequency distributions for these three variables are presented in Tables 3.5 through 3.7, and the corresponding graphs are depicted in Figures 3.4 through 3.6.

Table 3.5    Hours Spent per Day Watching Television

| Hours Spent per Day Watching TV | Frequency (f) | fY | Percentage | C% |
|---|---|---|---|---|
| 1 | 201 | 201 | 33.0 | 33.0 |
| 2 | 261 | 522 | 42.9 | 75.9 |
| 3 | 147 | 441 | 24.1 | 100.0 |
| Total | 609 | 1,164 | 100.0 | |

$$\bar{Y} = \frac{\Sigma fY}{N} = \frac{1,164}{609} = 1.91$$

Median = 2.0
Mode = 2.0

## The Symmetrical Distribution

First, let's examine Table 3.5 and Figure 3.4, displaying the distribution of the number of hours per day spent watching television. Notice that the largest number (261) watch television 2 hours/day (mode = 2.0), and about a fairly similar number (201 and 147, respectively) reported either 1 or 3 hours of watching television per day. As shown in Figure 3.4, the mode, the median, and the mean are almost identical, and they coincide at about the middle of the distribution: the distribution is roughly symmetrical.

## The Positively Skewed Distribution

Now let's examine Table 3.6 and Figure 3.5, displaying the distribution of the number of siblings of a respondent.

**Figure 3.4** Hours Spent per Day Watching Television

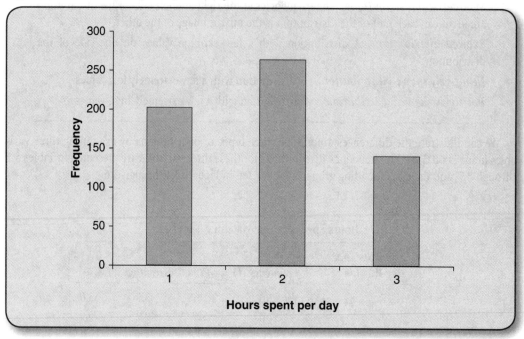

**Table 3.6** Number of Brothers and Sisters

| Number of Siblings | Frequency (f) | fY | Percentage | C% |
|---|---|---|---|---|
| 0 | 68 | 0 | 4.8 | 4.8 |
| 1 | 286 | 286 | 20.2 | 25.0 |
| 2 | 301 | 602 | 21.3 | 46.3 |
| 3 | 239 | 717 | 16.9 | 63.2 |
| 4 | 189 | 756 | 13.4 | 76.6 |
| 5 | 117 | 585 | 8.3 | 84.9 |
| 6 | 87 | 522 | 6.2 | 91.1 |
| 7 | 70 | 490 | 4.9 | 96.0 |
| 8 | 57 | 456 | 4.0 | 100.0 |
| Total | 1,414 | 4,414 | 100.0 | |

$$\bar{Y} = \frac{\Sigma fY}{N} = \frac{4,414}{1,414} = 3.12$$

Median = 3.0
Mode = 2.0

**Figure 3.5**   Number of Brothers and Sisters

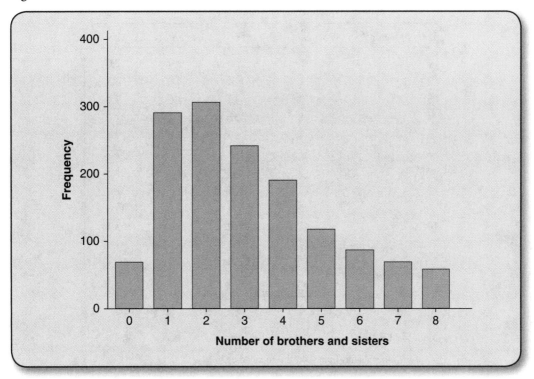

Notice that in this distribution, the mean, the median, and the mode have different values, with the mode having the lowest value (mode = 2.00), the median having the second lowest value (median = 3.0), and the mean having the highest value (mean = 3.12).

The distribution as depicted in Table 3.6 and Figure 3.5 is positively skewed. As a general rule, for skewed distributions the mean, the median, and the mode do not coincide. The mean, which is always pulled in the direction of extreme scores, falls closest to the tail of the distribution where a small number of extreme scores are located.

## The Negatively Skewed Distribution

Now examine Table 3.7 and Figure 3.6 for the number of years spent in school among those respondents who did not finish high school. Here you can see the opposite pattern. The distribution is negatively skewed. First, note that the largest number of years spent in school are concentrated at the high end of the scale (11 years) and that there are fewer respondents at the low end. The mean, the median, and the mode also differ in values as they did in the previous example. However, here the mode has the highest value (mode = 11.0), the median has the second highest value (median = 10.0), and the mean has the lowest value (mean = 8.89).

**Table 3.7** Years of School Among Respondents Without a High School Degree

| Years of School | Frequency (f) | fY | Percentage | C% |
|---|---|---|---|---|
| 0 | 5 | 0 | 2.0 | 2.0 |
| 2 | 2 | 4 | 0.8 | 2.8 |
| 3 | 4 | 12 | 1.6 | 4.4 |
| 4 | 8 | 32 | 3.2 | 7.6 |
| 5 | 4 | 20 | 1.6 | 9.2 |
| 6 | 21 | 126 | 8.5 | 17.7 |
| 7 | 9 | 63 | 3.6 | 21.3 |
| 8 | 28 | 224 | 11.3 | 32.6 |
| 9 | 35 | 315 | 14.2 | 46.8 |
| 10 | 41 | 410 | 16.6 | 63.4 |
| 11 | 90 | 990 | 36.4 | 99.8[a] |
| Total | 247 | 2,196 | 99.8 | |

$$\bar{Y} = \frac{\Sigma fY}{N} = \frac{2,196}{247} = 8.89$$

Median = 10.0
Mode = 11.0

a. Answer varies slightly from 100.0 due to rounding.

**Figure 3.6** Years of School Among Respondents Without a High School Degree

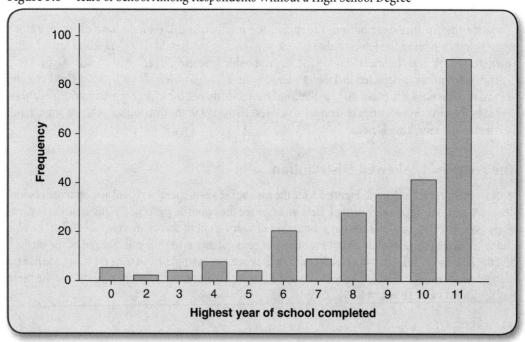

## Guidelines for Identifying the Shape of a Distribution

Following are some useful guidelines for identifying the shape of a distribution.

1. In unimodal distributions, when the mode, the median, and the mean coincide or are almost identical, the distribution is symmetrical.

2. When the mean is higher than the median (or is positioned to the right of the median), the distribution is positively skewed.

3. When the mean is lower than the median (or is positioned to the left of the median), the distribution is negatively skewed.

## ▣ CONSIDERATIONS FOR CHOOSING A MEASURE OF CENTRAL TENDENCY

So far, we have considered three basic kinds of averages: the mode, the median, and the mean. Each can represent the central tendency of a distribution. But which one should we use? There is no simple answer to this question. However, in general, we tend to use only one of the three measures of central tendency, and the choice of the appropriate one involves a number of considerations. These considerations and how they affect our choice of the appropriate measure are presented in the form of a decision tree in Figure 3.7.

**Figure 3.7**   How to Choose a Measure of Central Tendency

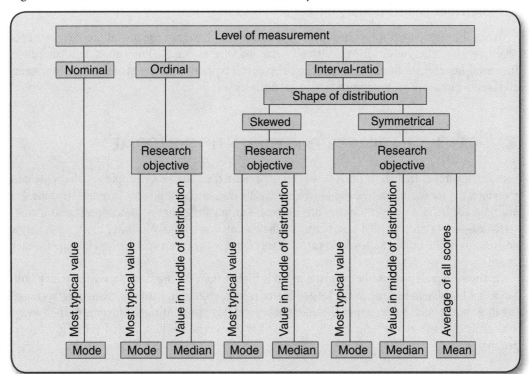

## Level of Measurement

One of the most basic considerations in choosing a measure of central tendency is the variable's level of measurement. The valid use of any of the three measures requires that the data be measured at the level appropriate for that measure or higher. Thus, as shown in Figure 3.7, with nominal variables our choice is restricted to the mode as a measure of central tendency.

However, with ordinal data, we have two choices: the mode or the median (or sometimes both). Our choice depends on what we want to know about the distribution. If we are interested in showing what is the most common or typical value in the distribution, then our choice is the mode. If, however, we want to show which value is located exactly in the middle of the distribution, then the median is our measure of choice.

When the data are measured on an interval-ratio level, the choice between the appropriate measures is a bit more complex and is restricted by the shape of the distribution.

## Skewed Distribution

When the distribution is skewed, the mean may give misleading information on the central tendency because its value is affected by extreme scores in the distribution. The median (see, e.g., the following statistics in practice section) or the mode can be chosen as the preferred measure of central tendency because neither is influenced by extreme scores. Thus, either one could be used as an "average," depending on the research objective.

## Symmetrical Distribution

When the distribution we want to analyze is symmetrical, we can use any of the three averages. Again, our choice depends on the research objective and what we want to know about the distribution. In general, however, the mean is our best choice because it contains the greatest amount of information and is easier to use in more advanced statistical analyses.

## ▣ STATISTICS IN PRACTICE: REPRESENTING INCOME

Personal income is frequently positively skewed because there are fewer people with high income; therefore, the mean may not be the most appropriate measure to represent "average" income. For example, the 2008 American Community Survey—an ongoing survey of economic and income statistics—reported the 2008 mean and median annual earnings of white, black, and Latino households in the United States. In Figure 3.8, we compare the mean and median income for each group.

As shown, for all groups, the reported mean is higher than the median. This discrepancy indicates that household income in the United States is highly skewed, with the mean overrepresenting those households in the upper-income bracket and misrepresenting the income of the average household. A preferable alternative (also shown) is to use the median annual earnings of these groups.

**Figure 3.8** Mean and Median Income Compared

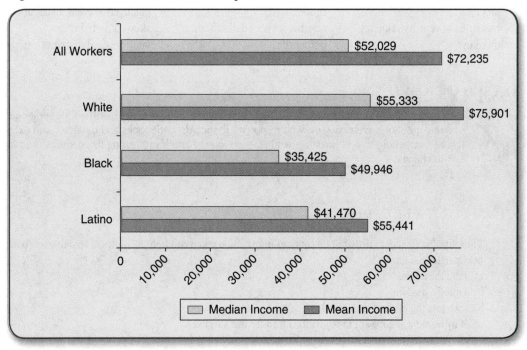

All Workers — $52,029 / $72,235
White — $55,333 / $75,901
Black — $35,425 / $49,946
Latino — $41,470 / $55,441

Median Income ☐   Mean Income ■

*Source:* U.S. Census Bureau, *American Community Survey,* 2008.

Since the earnings of whites are the highest in comparison with all other groups, it is useful to look at each group's median earnings relative to the earnings of whites. For example, blacks were paid just 64 cents ($35,425/$55,333) and Latinos ($41,470/$55,333) were paid 75 cents for every $1 paid to whites.

## MAIN POINTS

• The mode, the median, and the mean are measures of central tendency—numbers that describe what is average or typical about the distribution.

• The mode is the category or score with the largest frequency (or percentage) in the distribution. It is often used to describe the most commonly occurring category of a nominal-level variable.

• The median is a measure of central tendency that represents the exact middle of the distribution. It is calculated for variables measured on at least an ordinal level of measurement.

• The mean is typically used to describe central tendency in interval-ratio variables, such as income, age, or education. We obtain the mean by summing all the scores and dividing by the total ($N$) number of scores.

• In a symmetrical distribution, the frequencies at the right and left tails of the distribution are identical. In skewed distributions, there are either a few extremely high (positive skew) or a few extremely low (negative skew) values.

## KEY TERMS

| | | |
|---|---|---|
| mean | mode | positively skewed distribution |
| measures of central tendency | negatively skewed distribution | skewed distribution |
| median | percentile | symmetrical distribution |

## ON YOUR OWN

 Log on to the web-based student study site at **www.sagepub.com/ssdsessentials** for additional study questions, web quizzes, web resources, flashcards, codebooks and datasets, web exercises, appendices, and links to social science journal articles reflecting the statistics used in this chapter.

## CHAPTER EXERCISES

1. The following frequency distribution contains information about children's attitudes of smoking one pack of cigarettes per day.
   a. Find the mode.
   b. Find the median.
   c. Interpret the mode and the median.
   d. Why would you not want to report the mean for this variable?

| Attitude Toward Smoking One Pack of Cigarettes per Day | Frequency | Percentage | Cumulative Percentage |
|---|---|---|---|
| 1. Don't disapprove | 288 | 19.8 | 19.8 |
| 2. Disapprove | 418 | 28.7 | 48.5 |
| 3. Strongly disapprove | 749 | 51.5 | 100.0 |
| Total | 1,455 | 100.0 | |

*Source:* Monitoring the Future, 2008.

2. Same-sex unions have increasingly become a heated political issue. The 2008 GSS asked respondents' opinions on homosexual relations. Four response categories ranged from "Always Wrong" to "Not Wrong at All." See the following frequency distribution:

| Homosexual Relations | Frequency | Percentage | Cumulative Percentage |
|---|---|---|---|
| 1. Always wrong | 468 | 49.3 | 49.3 |
| 2. Almost always wrong | 35 | 3.7 | 53.0 |
| 3. Sometimes wrong | 73 | 7.7 | 60.7 |
| 4. Not wrong at all | 374 | 39.3 | 100.0 |
| Total | 950 | 100.0 | |

Exercises

a. At what level is this variable measured? What is the mode for this variable?

b. Calculate the median for this variable. In general, how would you characterize the public's attitude about homosexual relations?

3. The following frequency distribution contains information on the number of weeks worked last year for a sample of 65 Latino adults.

| Weeks Worked Last Year | Frequency | Percentage | Cumulative Percentage |
|:---:|:---:|:---:|:---:|
| 0 | 10 | 15.4 | 15.4 |
| 8 | 1 | 1.5 | 16.9 |
| 12 | 2 | 3.1 | 20.0 |
| 15 | 1 | 1.5 | 21.5 |
| 24 | 1 | 1.5 | 23.0 |
| 25 | 1 | 1.5 | 24.5 |
| 26 | 1 | 1.5 | 26.0 |
| 30 | 1 | 1.5 | 27.5 |
| 32 | 1 | 1.5 | 29.0 |
| 36 | 2 | 3.1 | 32.1 |
| 38 | 1 | 1.5 | 33.6 |
| 39 | 1 | 1.5 | 35.1 |
| 40 | 5 | 7.7 | 42.8 |
| 44 | 2 | 3.1 | 45.9 |
| 45 | 1 | 1.5 | 47.4 |
| 47 | 2 | 3.1 | 50.5 |
| 48 | 1 | 1.5 | 52.0 |
| 49 | 1 | 1.5 | 53.5 |
| 50 | 2 | 3.1 | 56.6 |
| 51 | 1 | 1.5 | 58.1 |
| 52 | 27 | 41.5 | 99.6 |
| Total | 65 | 99.6 | |

*Source:* General Social Survey, 2008.

a. What is the level of measurement for "weeks worked last year"? What is the mode for "weeks worked last year"? What is the median for "weeks worked last year"?

b. Construct quartiles for weeks worked last year. What is the 25th percentile? The 50th percentile? The 75th percentile? Why don't you need to calculate the 50th percentile to answer this question?

4. Using data from the 2008 MTF survey, the following is the frequency distribution for number of friends who drink alcohol:

| Number of Friends Who Drink Alcohol | Frequency |
|---|---|
| None | 125 |
| A few | 177 |
| Some | 262 |
| Most | 521 |
| All | 216 |

a. Calculate the median category for the number of friends who drink alcohol.
b. Also, report which category contains the 20th and 80th percentiles.

5. Does health status vary with age? The following table, taken from the 2008 GSS, depicts the health status across various age groups (not all ages are displayed).

| Health Status | Age Group | | | |
|---|---|---|---|---|
| | 18–29 | 30–39 | 40–49 | 50–59 |
| Excellent | 64 | 57 | 46 | 37 |
| Good | 86 | 87 | 94 | 83 |
| Fair | 36 | 27 | 37 | 47 |
| Poor | 2 | 5 | 8 | 15 |

a. Calculate the median and mode for each age group.
b. Use this information to characterize whether health status varies by age. Does the median or mode provide a better description of the data? Do the statistics support the idea that some age groups are healthier than others?

6. The number of Americans on Medicare is increasing as expected with the aging baby boomer population. The following table shows the number of Americans on Medicare in 2000 and 2005 for eight U.S. states. *Note:* The numbers listed are in thousands.

| State | 2000 | 2005 |
|---|---|---|
| Alabama | 685 | 740 |
| Delaware | 112 | 125 |
| Florida | 2,804 | 3,008 |
| Illinois | 1,635 | 1,674 |
| Minnesota | 654 | 691 |
| New Hampshire | 170 | 185 |
| New York | 2,715 | 2,758 |
| Washington | 736 | 807 |

*Source:* U.S. Census Bureau, *The 2008 Statistical Abstract*, Table 135.

a. Calculate the mean number of Americans on Medicare in these eight states for both 2000 and 2005. How would you characterize the difference in the number of Americans on Medicare between 2000 and 2005? Does the mean adequately represent the central tendency of the distribution of Americans on Medicare in each year for these eights states? Why or why not?

b. Recalculate the mean for each year after removing Florida, Illinois, and New York from the table. Is the mean now a better representation of central tendency for the remaining five states? Explain.

7. U.S. households have become smaller over the years. The following table from the 2008 GSS contains information on the number of people currently aged 18 years or older living in a respondent's household. Calculate the mean number of people living in a U.S. household in 2008.

| Household Size | Frequency |
| --- | --- |
| 1 | 384 |
| 2 | 541 |
| 3 | 228 |
| 4 | 195 |
| 5 | 98 |
| 6 | 36 |
| 7 | 12 |
| 8 | 4 |
| 9 | 1 |
| 11 | 1 |
| Total | 1,500 |

8. In Exercise 6, you calculated the mean number of Americans on Medicare. We now want to test whether the distribution of Americans on Medicare is symmetrical or skewed.
   a. Calculate the median and mode for each year, using all eight states. Based on these results and the means, how would you characterize the distribution of Americans on Medicare for each year?
   b. Does the mean or median best represent the central tendency of each distribution? Why?
   c. If you found the distributions to be skewed, what might be the statistical cause?

9. In Exercise 7, you examined U.S. household size in 2008. Using these data, construct a histogram to represent the distribution of household size.
   a. From the appearance of the histogram, would you say the distribution is positively or negatively skewed? Why?
   b. Now calculate the median for the distribution and compare this value with the value of the mean from Exercise 7. Do these numbers provide further evidence to support your decision about how the distribution is skewed? Why do you think the distribution of household size is asymmetrical?

10. Exercise 3 used GSS data on the number of weeks worked per year for a sample of 65 Latino adults.
    a. Calculate the mean number of weeks worked per year.
    b. Compare the value of the mean with those you have already calculated for the median. Without constructing a histogram, describe whether and how the distribution of weeks worked per year is skewed.

**Exercises**

11. You listen to a debate between two politicians discussing the economic health of the United States. One politician says that the average income of all workers in the United States is $72,235; the other says that American workers make, on average, only $52,029, so Americans are not as well-off as the first politician claims. Is it possible for both these politicians to be correct? If so, explain how.

12. Do murder rates in cities vary with city size? Investigate this question using the following data for selected U.S. cities grouped by population size, the top 10 cities and the bottom 10 (all among the largest U.S. cities). Calculate the mean and median for each group of cities. Where is murder rate the highest? Do the mean and median have the same pattern for the two groups?

| Murder Rate per 100,000 in 2005 | | | |
| --- | --- | --- | --- |
| Top 10 by Population | Murder Rate | Bottom 10 by Population | Murder Rate |
| New York, NY | 6.6 | Corpus Christi, TX | 2.8 |
| Los Angeles, CA | 12.6 | Bakersfield, CA | 11.2 |
| Chicago, IL | 15.6 | Buffalo, NY | 19.8 |
| Houston, TX | 16.3 | Stockton, CA | 14.6 |
| Philadelphia, PA | 25.6 | Newark, NJ | 34.5 |
| Phoenix, AZ | 15.0 | St. Paul, MN | 8.6 |
| Las Vegas, NV | 11.3 | Anchorage, AK | 5.8 |
| San Diego, CA | 4.0 | Lexington, KY | 5.6 |
| San Antonio, TX | 6.8 | St. Petersburg, FL | 11.8 |
| Dallas, TX | 16.4 | Mobile, AL | 14.0 |

Source: U.S. Census Bureau, The 2008 Statistical Abstract, Table 302.

13. Many policymakers are interested in understanding adolescent drug use. Use the following information from the MTF survey to explore whether or not attitudes toward trying marijuana vary by sex.

| Attitude Toward Trying Marijuana | Males | Females |
| --- | --- | --- |
| Don't disapprove | 325 | 302 |
| Disapprove | 160 | 157 |
| Strongly disapprove | 185 | 260 |
| Total | 670 | 719 |

Source: Monitoring the Future, 2008.

Calculate the appropriate measures of central tendency for males and females. How would you describe their attitudes toward trying marijuana? Are they similar or different and why? Use the appropriate measures of central tendency that you calculated to support your answer.

14. Identity theft is a growing concern. The following table shows the number of identity theft cases reported to the U.S. Federal Trade Commission in 2006 for 11 states selected at random.

| State | Number of Reported Identity Theft Cases |
|---|---|
| Arkansas | 384 |
| Florida | 17,780 |
| Illinois | 10,080 |
| Michigan | 6,784 |
| New Jersey | 6,394 |
| New York | 16,452 |
| North Dakota | 189 |
| Oregon | 2,815 |
| Rhode Island | 615 |
| Texas | 26,006 |
| Wisconsin | 2,536 |

*Source:* U.S. Census Bureau, *The 2008 Statistical Abstract*, Table 314.

a. Calculate the mean and median number of identity theft cases reported for the 11 states presented above.
b. Is this distribution skewed? If yes, how do you know?
c. Provide two possible explanations that might explain why some states report more identity theft cases than others.

# Measures of Variability

I n the previous chapter, we looked at measures of central tendency: the mean, the median, and the mode. With these measures, we can use a single number to describe what is average for or typical of a distribution. Although measures of central tendency can be very helpful, they tell only part of the story. In fact, when used alone, they may mislead rather than inform. Another way of summarizing a distribution of data is by selecting a single number that describes how much variation and diversity there is in the distribution. Numbers that describe diversity or variation are called measures of variability. Researchers often use measures of central tendency along with **measures of variability** to describe their data.

---

*Measures of variability*    Numbers that describe diversity or variability in the distribution of a variable.

---

In this chapter, we discuss four measures of variability: the range, the interquartile range, the standard deviation, and the variance. Before we discuss these measures, let's explore why they are important.

## ▣ THE IMPORTANCE OF MEASURING VARIABILITY

The importance of looking at variation and diversity can be illustrated by thinking about the differences in the experiences of U.S. women. Are women united by their similarities or divided by their differences?

The answer is *both*. To address the similarities without dealing with differences is "to misunderstand and distort that which separates as well as that which binds women together."[1] Even when we focus on one particular group of women, it is important to look at the differences as well as the commonalities. Take, for example, Asian American women. As a group, they share a number of characteristics.

> Their participation in the workforce is higher than that of women in any other ethnic group. Many . . . live life supporting others, often allowing their lives to be subsumed by the needs of the extended family. . . . However, there are many circumstances when these shared experiences are not sufficient to accurately describe the condition of a particular Asian-American woman. Among Asian-American women there are those who were born in the United States . . . and . . . those who recently arrived in the United States. Asian-American women are diverse in their heritage or country of origin: China, Japan, the Philippines, Korea . . . and . . . India. . . . Although the majority of Asian-American women are working class—contrary to the stereotype of the "ever successful" Asians—there are poor, "middle-class," and even affluent Asian-American women.[2]

As this example illustrates, one basis of stereotyping is treating a group as if it were totally represented by its central value, ignoring the diversity within the group. Sociologists often contribute to this type of stereotyping when their empirical generalizations, based on a statistical difference between averages, are interpreted in an overly simplistic way. All this argues for the importance of using measures of variability as well as central tendency whenever we want to characterize or compare groups. Whereas the similarities and commonalities in the experiences of Asian American women are depicted by a measure of central tendency, the diversity of their experiences can be described only by using measures of variation.

The concept of variability has implications not only for describing the diversity of social groups such as Asian American women but also for issues that are important in your everyday life. Let's consider the issue of statistics instruction on the college level.

Let's suppose that a university committee is examining the issue of how to better respond to the needs of students. In its attempt to evaluate statistics courses offered by different instructors, the committee compares the grading policy in two courses. The first is taught by Professor Brown; the second is taught by Professor Yamato. The committee finds that over the years, the average grade for Professor Brown's class has been C+. The average grade in Professor Yamato's class is also C+. We could easily be misled by these statistics into thinking that the grading policy of both instructors is about the same. However, we need to look more closely into how the grades are distributed in each of the classes. The differences in the distribution of grades are illustrated in Figure 4.1, which displays the line graph for the two classes.

Compare the shapes of these two distributions. Notice that while both distributions have the same mean, they are shaped very differently. The grades in Professor Yamato's class are more spread out ranging from A to F, whereas the grades for Professor Brown's class are clustered around the mean and range only from B to C. Although the means for both distributions are identical, the grades in Professor Yamato's class vary considerably more than the grades given by Professor Brown. The comparison between the two classes is more complex than we first thought it would be.

As this example demonstrates, information on how scores are spread from the center of a distribution is as important as information about the central tendency in a distribution. This type of information is obtained by measures of variability.

**Figure 4.1**   Distribution of Grades for Professors Brown's and Yamato's Statistics Classes

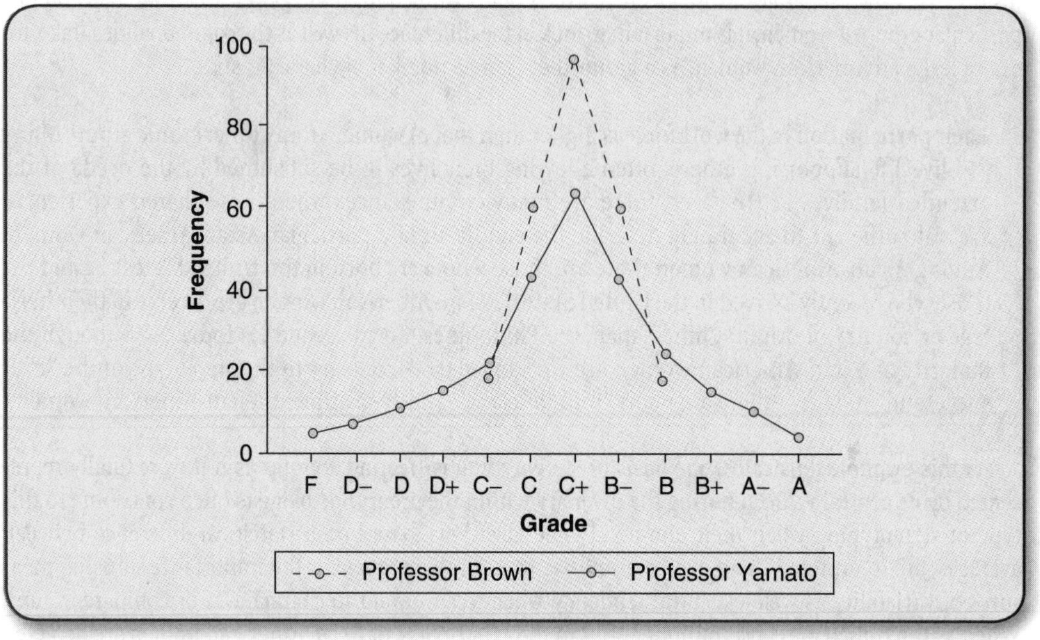

✓ *Learning*
*Check*

*Look closely at Figure 4.1. Whose class would you choose to take? If you were worried that you might fail statistics, your best bet would be Professor Brown's class where no one fails. However, if you want to keep up your GPA and are willing to work, Professor Yamato's class is the better choice. If you had to choose one of these classes based solely on the average grades, your choice would not be well informed.*

## ▣ THE RANGE

The simplest and most straightforward measure of variation is the **range**, which measures variation in interval-ratio variables. It is the difference between the highest (maximum) and the lowest (minimum) scores in the distribution:

$$\text{Range} = \text{Highest score} - \text{Lowest score}$$

In the 2008 General Social Survey (GSS), the oldest person included in the study was 89 years old and the youngest was 18. Thus, the range was $89 - 18 = 71$ years.

---

*Range*   A measure of variation in interval-ratio variables. It is the difference between the highest (maximum) and the lowest (minimum) scores in the distribution.

---

The range can also be calculated on percentages. For example, since the 1980s, relatively large communities of the elderly have become noticeable not just in the traditional retirement meccas of the Sun Belt[3] but also in the Ozarks of Arkansas and the mountains of Colorado and Montana. The number of elderly persons increased in every state during the 1990s and the 2000s (Washington, D.C., is the exception), but by different amounts. As the baby boomers age into retirement, we would expect this trend to continue. Table 4.1 displays the percentage change in the elderly population from 2008 to 2015 by region and by state as predicted by the U.S. Census Bureau.[4]

**Table 4.1** Projected Percentage Change in the Population 65 Years and Over by Region and State, 2008–2015

| *Region and State* | *Percentage Change* | *Region and State* | *Percentage Change* |
|---|---|---|---|
| Northeast | 16.0 | South (cont.) | |
| Connecticut | 20.9 | Georgia | 21.1 |
| Delaware | 22.3 | Kentucky | 12.5 |
| Maine | 25.6 | Louisiana | 23.0 |
| Massachusetts | 17.7 | Maryland | 23.1 |
| New Hampshire | 28.4 | Mississippi | 16.4 |
| New Jersey | 20.4 | North Carolina | 20.7 |
| New York | 12.8 | Oklahoma | 12.8 |
| Pennsylvania | 12.5 | South Carolina | 22.1 |
| Rhode Island | 18.2 | Tennessee | 18.2 |
| Vermont | 31.4 | Texas | 25.9 |
| | | Virginia | 26.8 |
| Midwest | 14.0 | Washington, D.C. | −14.1 |
| Indiana | 11.3 | West Virginia | 15.4 |
| Illinois | 12.9 | | |
| Iowa | 11.2 | West | 27.0 |
| Kansas | 14.2 | Alaska | 50.0 |
| Michigan | 15.5 | Arizona | 36.8 |
| Minnesota | 19.2 | California | 27.1 |
| Missouri | 14.5 | Colorado | 22.7 |
| Nebraska | 12.4 | Hawaii | 18.8 |
| North Dakota | 12.6 | Idaho | 20.2 |
| Ohio | 12.4 | Montana | 27.0 |
| South Dakota | 9.4 | Nevada | 42.1 |
| Wisconsin | 17.6 | New Mexico | 31.9 |
| | | Oregon | 17.1 |
| South | 22.8 | Utah | 13.8 |
| Alabama | 15.1 | Washington | 23.2 |
| Arkansas | 14.7 | Wyoming | 36.9 |
| Florida | 29.7 | | |

*Source:* U.S. Census Bureau, *The 2010 Statistical Abstract*, Tables 16 and 18.

To find the ranges in a distribution, simply pick out the highest and the lowest scores in the distribution and subtract. Alaska has the highest percentage increase, with 50%, and Washington, D.C., has the lowest increase, with −14.1%. The range is 64.1 percentage points, or 50% to −14.1%.

Although the range is simple and quick to calculate, it is a rather crude measure because it is based on only the lowest and the highest scores. These two scores might be extreme and rather atypical, which might make the range a misleading indicator of the variation in the distribution. For instance, note that among the 50 states and Washington, D.C., listed in Table 4.1, no state has a percentage decrease as that of Washington, D.C., and only Nevada has a percentage increase nearly as high as Alaska's. The range of 64.1 percentage points does not give us information about the variation in states between Washington, D.C., and Alaska.

✓ *Learning Check*

> *Why can't we use the range to describe diversity in nominal variables? The range can be used to describe diversity in ordinal variables (e.g., we can say that responses to a question ranged from "somewhat satisfied" to "very dissatisfied"), but it has no quantitative meaning. Why not?*

## ▣ THE INTERQUARTILE RANGE: INCREASES IN ELDERLY POPULATIONS

To remedy the limitation of the range, we can employ an alternative—the *interquartile range*. The **interquartile range (IQR)**, a measure of variation for interval-ratio variables, is the width of the middle 50% of the distribution. It is defined as the difference between the lower and upper quartiles ($Q_1$ and $Q_3$).

$$IQR = Q_3 - Q_1$$

Recall that the first quartile ($Q_1$) is the 25th percentile, the point at which 25% of the cases fall below it. The third quartile ($Q_3$) is the 75th percentile, the point at which 75% of the cases fall below it. The IQR, therefore, defines variation for the middle 50% of the cases.

---

*Interquartile range (IQR)*   The width of the middle 50% of the distribution. It is defined as the difference between the lower and upper quartiles (*Q1* and *Q3*).

---

Like the range, the IQR is based on only two scores. However, because it is based on intermediate scores, rather than on the extreme scores in the distribution, it avoids some of the instability associated with the range.

These are the steps for calculating the IQR:

1. To find $Q_1$ and $Q_3$, order the scores in the distribution from the highest to the lowest score, or vice versa. Table 4.2 presents the data of Table 4.1 arranged in order from Alaska (50.0%) to Washington, D.C. (−14.1%).

**Table 4.2**   Projected Percentage Change in the Population 65 Years and Over, 2008–2015, by State, Ordered From the Highest to the Lowest

| State | Percentage Change | State | Percentage Change | State | Percentage Change |
|---|---|---|---|---|---|
| Alaska | 50.0 | Delaware | 22.3 | Alabama | 15.1 |
| Nevada | 42.1 | South Carolina | 22.1 | Arkansas | 14.7 |
| Wyoming | 36.9 | Georgia | 21.1 | Missouri | 14.5 |
| Arizona | 36.8 | Connecticut | 20.9 | Kansas | 14.2 |
| New Mexico | 31.9 | North Carolina | 20.7 | Utah | 13.8 |
| Vermont | 31.4 | New Jersey | 20.4 | Illinois | 12.9 |
| Florida | 29.7 | Idaho | 20.2 | New York | 12.8 |
| New Hampshire | 28.4 | Minnesota | 19.2 | Oklahoma | 12.8 |
| California | 27.1 | Hawaii | 18.8 | North Dakota | 12.6 |
| Montana | 27.0 | Rhode Island | 18.2 | Kentucky | 12.5 |
| Virginia | 26.8 | Tennessee | 18.2 | Pennsylvania | 12.5 |
| Texas | 25.9 | Massachusetts | 17.7 | Nebraska | 12.4 |
| Maine | 25.6 | Wisconsin | 17.6 | Ohio | 12.4 |
| Washington | 23.2 | Oregon | 17.1 | Indiana | 11.3 |
| Maryland | 23.1 | Mississippi | 16.4 | Iowa | 11.2 |
| Louisiana | 23.0 | Michigan | 15.5 | South Dakota | 9.4 |
| Colorado | 22.7 | West Virginia | 15.4 | Washington, D.C. | −14.1 |

*Source:* U.S. Census Bureau, *The 2010 Statistical Abstract*, Tables 16 and 18.

2. Next, we need to identify the first quartile, $Q_1$ or the 25th percentile. We have to identify the percentage change in the elderly population associated with the state that divides the distribution so that 25% of the states are below it. To find $Q_1$, we multiply $N$ by 0.25:

$$(N)(0.25) = (51)(0.25) = 12.75$$

The first quartile falls between the 12th and the 13th states. Counting from the bottom, the 12th state is Illinois, and the percentage increase associated with it is 12.9. The 13th state is Utah, with a percentage increase of 13.8. To find the first quartile, we take the average of 12.9 and 13.8. Therefore, $(12.9 + 13.8)/2 = 13.35$ is the first quartile ($Q_1$).

3. To find $Q_3$, we have to identify the state that divides the distribution in such a way that 75% of the states are below it. We multiply $N$ this time by 0.75:

$$(N)(0.75) = (51)(0.75) = 38.25$$

The third quartile falls between the 38th and the 39th states. Counting from the bottom, the 38th state is Washington, and the percentage increase associated with it is 23.2. The 39th state is Maine, with a percentage increase of 25.6. To find the third quartile, we take the average of 23.2 and 25.6. Therefore, $(23.2 + 25.6)/2 = 24.4$ is the third quartile $(Q_3)$.

4. We are now ready to find the IQR:

$$IQR = Q_3 - Q_1 = 24.4 - 13.35 = 11.05$$

The IQR of percentage change in the elderly population is 11.05 percentage points.

Notice that the IQR gives us better information than the range. The range gave us a 64.1-point spread, from 50% to −14.1%, but the IQR tells us that half the states are clustered between 24.4 and 13.35—a much narrower spread. The extreme scores represented by Alaska and Washington, D.C., have no effect on the IQR because they fall at the extreme ends of the distribution. This difference between the range and the interquartile range is also illustrated in Figure 4.2 where the extreme values of 10 children affects the value of the range but not of the IQR.

✓ *Learning Check*

*Why is the IQR better than the range as a measure of variability, especially when there are extreme scores in the distribution? To answer this question, you may want to examine Figure 4.2.*

## ▣ THE VARIANCE AND THE STANDARD DEVIATION: CHANGES IN THE ELDERLY POPULATION

The elderly population in the United States today is 10 times as large as in 1900, and it is projected to continue to increase. The pace and direction of these demographic changes will create compelling social, economic, and ethical choices for individuals, families, and governments.

Table 4.3 presents the projected percentage change in the elderly population for all regions of the United States.

Table 4.3 shows that between 2008 and 2015, the size of the elderly population in the United States is projected to increase by an average of 19.95%. But this average increase does not inform us about the regional variation in the elderly population. For example, will the northeastern states show a smaller-than-average increase because of the out-migration of the elderly population to the warmer climate of the Sun Belt states? Is the projected increase higher in the South because of the immigration of the elderly?

Although it is important to know the average projected percentage change for the nation as a whole, you may also want to know whether regional increases might differ from the national average. If the regional

**Figure 4.2**   The Range Versus the Interquartile Range: Number of Children Among Two Groups of Women

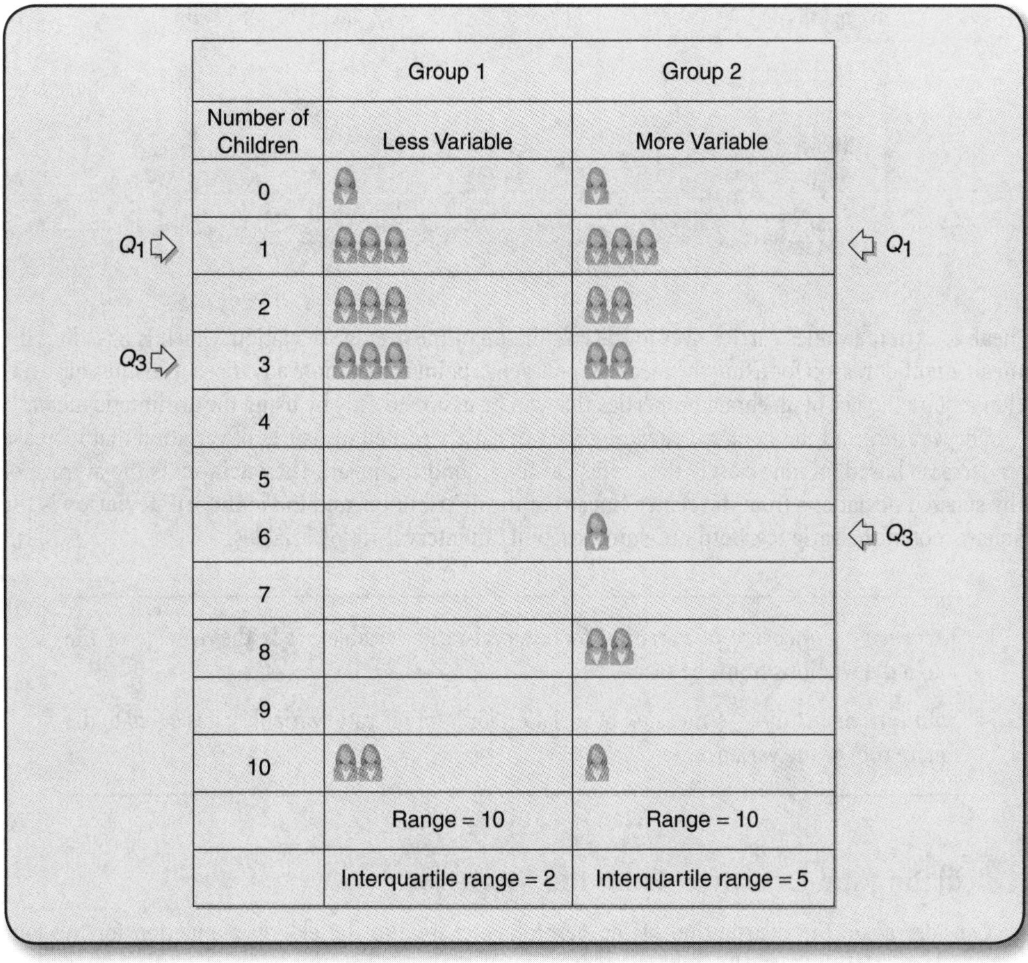

| Number of Children | Group 1 — Less Variable | Group 2 — More Variable |
|---|---|---|
| 0 | | |
| 1 | | |
| 2 | | |
| 3 | | |
| 4 | | |
| 5 | | |
| 6 | | |
| 7 | | |
| 8 | | |
| 9 | | |
| 10 | | |
| | Range = 10 | Range = 10 |
| | Interquartile range = 2 | Interquartile range = 5 |

projected increases are close to the national average, the figures will cluster around the mean, but if the regional increases deviate much from the national average, they will be widely dispersed around the mean.

Table 4.3 suggests that there is considerable regional variation. The percentage change ranges from 27.0% in the West to 14.0% in the Midwest, so the range is 13.0% (27.0% − 14.0% = 13.0%). Moreover, most of the regions are projected to deviate considerably from the national average of 19.95%. We want a measure that will give us information about the overall variations among all regions in the United States and, unlike the range or the IQR, will not be based on only two scores.

Such a measure will reflect how much, on the average, each score in the distribution deviates from some central point, such as the mean. We use the mean as the reference point rather than other kinds of averages (the mode or the median) because the mean is based on all the scores in the distribution. Therefore, it is more useful as a basis from which to calculate average deviation. The sensitivity of the

Table 4.3   Projected Percentage Change in the Elderly Population
by Region, 2008–2015

| Region | Percentage |
|---|---|
| Northeast | 16.0 |
| South | 22.8 |
| Midwest | 14.0 |
| West | 27.0 |
| Mean ($\bar{Y}$) | 19.95 |

Source: U.S. Census Bureau, *The 2010 Statistical Abstract*, Tables 16 and 18.

mean to extreme values carries over to the calculation of the average deviation, which is based on the mean. Another reason for using the mean as a reference point is that more advanced measures of variation require the use of algebraic properties that can be assumed only by using the arithmetic mean.

The *variance* and the *standard deviation* are two closely related measures of variation that increase or decrease based on how closely the scores cluster around the mean. The **variance** is the average of the squared deviations from the center (mean) of the distribution, and the **standard deviation** is the square root of the variance. Both measure variability in interval-ratio variables.

---

*Variance*   A measure of variation for interval-ratio variables; it is the average of the squared deviations from the mean.

*Standard deviation*   A measure of variation for interval-ratio variables; it is equal to the square root of the variance.

---

## Calculating the Deviation From the Mean

Consider again the distribution of the percentage change in the elderly population for the four regions of the United States. Because we want to calculate the average difference of all the regions from the national average (the mean), it makes sense to first look at the difference between each region and the mean. This difference, called a deviation from the mean, is symbolized as $(Y - \bar{Y})$ The sum of these deviations can be symbolized as $\Sigma(Y - \bar{Y})$.

The calculations of these deviations for each region are displayed in Table 4.4 and Figure 4.3. We have also summed these deviations. Note that each region has either a positive or a negative deviation score. The deviation is positive when the percentage change in the elderly home population is above the mean. It is negative when the percentage change is below the mean. Thus, for example, the Northeast's deviation score of −3.95 means that its percentage change in the elderly population was 3.95 percentage points below the mean.

You may wonder if we could calculate the average of these deviations by simply adding up the deviations and dividing them. Unfortunately we cannot, because the sum of the deviations of scores from the mean is always zero, or algebraically $\Sigma(Y - \bar{Y}) = 0$. In other words, if we were to subtract the mean

**Figure 4.3**  Illustrating Deviations From the Mean

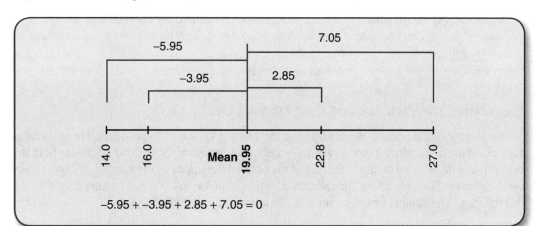

$$-5.95 + -3.95 + 2.85 + 7.05 = 0$$

**Table 4.4**  Projected Percentage Change in the Elderly Population, 2008–2015, by Region, Deviation From the Mean, and Deviation From the Mean Squared

| *Region* | *Percentage* | $(Y - \bar{Y})$ | $(Y - \bar{Y})^2$ |
|---|---|---|---|
| Northeast | 16.0 | 16.0–19.95 = –3.95 | 15.60 |
| South | 22.8 | 22.8–19.95 = 2.85 | 8.12 |
| Midwest | 14.0 | 14.0–19.95 = –5.95 | 35.40 |
| West | 27.0 | 27.0–19.95 = 7.05 | 49.70 |
| | $\Sigma Y = 79.8$ | $\Sigma(Y - \bar{Y}) = 0$ | $\Sigma(Y - \bar{Y})^2 = 108.82$ |

$$\text{Mean} = \bar{Y} = \frac{\Sigma Y}{N} = \frac{79.8}{4} = 19.95$$

from each score and then add up all the deviations as we did in the second to last column in Table 4.4, the sum would be zero, which in turn would cause the average deviation (i.e., average difference) to compute to zero. This is always true because the mean is the center of gravity of the distribution.

Mathematically, we can overcome this problem by squaring the deviations—that is, multiplying each deviation by itself to get rid of the negative sign.

The final column in Table 4.4 squares the actual deviations from the mean and adds together the squares. The sum of the squared deviations is symbolized as $\Sigma(Y - \bar{Y})^2$. Note that by squaring the deviations, we end up with a sum representing the deviation from the mean, which is positive. (Note that this sum will equal zero if all the cases have the same value as the mean.) In our example, this sum is $\Sigma(Y - \bar{Y})^2 = 108.82$.

✓ Learning
Check

*Examine Table 4.4 again and note the disproportionate contribution of the western region to the sum of the squared deviations from the mean (it actually accounts for about 45% of the sum of squares). Can you explain why? (Hint: It has something to do with the sensitivity of the mean to extreme values.)*

## Calculating the Variance and the Standard Deviation

The average of the squared deviations from the mean is known as the *variance*. The variance is symbolized as $S_Y^2$. Remember that we are interested in the *average* of the squared deviations from the mean. Therefore, we need to divide the sum of the squared deviations by the number of scores ($N$) in the distribution. However, unlike the calculation of the mean, we will use $N - 1$ rather than $N$ in the denominator.[5] The formula for the variance can be stated as

$$S_Y^2 = \frac{\Sigma(Y - \bar{Y})^2}{N - 1} \tag{4.1}$$

where

$S_Y^2$ = the variance

$(Y - \bar{Y})$ = the deviation from the mean

$\Sigma(Y - \bar{Y})^2$ = the sum of the squared deviations from the mean

$N$ = the number of scores

Note that the formula incorporates all the symbols we defined earlier. This formula means that the variance is equal to the average of the squared deviations from the mean.

Follow these steps to calculate the variance:

1. Calculate the mean, $\bar{Y} = \Sigma(Y)/N$.

2. Subtract the mean from each score to find the deviation, $Y - \bar{Y}$.

3. Square each deviation, $(Y - \bar{Y})^2$.

4. Sum the squared deviations, $\Sigma(Y - \bar{Y})^2$.

5. Divide the sum by $N - 1$, $\Sigma(Y - \bar{Y})^2/(N - 1)$.

6. The answer is the variance.

To assure yourself that you understand how to calculate the variance, go back to Table 4.4 and follow this step-by-step procedure for calculating the variance. Now plug the required quantities into Formula 4.1. Your result should look like this:

$$S_Y^2 = \frac{\Sigma(Y - \bar{Y})^2}{N - 1} = \frac{108.82}{3} = 36.27$$

One problem with the variance is that it is based on squared deviations and therefore is no longer expressed in the original units of measurement. For instance, it is difficult to interpret the variance of 36.27, which represents the distribution of the percentage change in the elderly population, because this figure is expressed in squared percentages. Thus, we often take the square root of the variance and interpret it instead. This gives us the *standard deviation*, $S_Y$.

The standard deviation, symbolized as $S_Y$, is the square root of the variance, or

$$S_Y = \sqrt{S_Y^2}$$

The standard deviation for our example is

$$S_Y = \sqrt{S_Y^2} = \sqrt{36.27} = 6.02$$

The formula for the standard deviation uses the same symbols as the formula for the variance:

$$S_Y = \sqrt{\frac{\Sigma(Y - \bar{Y})^2}{N - 1}} \tag{4.2}$$

As we interpret the formula, we can say that the standard deviation is equal to the square root of the average of the squared deviations from the mean.

The advantage of the standard deviation is that unlike the variance, it is measured in the same units as the original data. For instance, the standard deviation for our example is 6.02. Because the original data were expressed in percentages, this number is expressed as a percentage as well. In other words, you could say, "The standard deviation is 6.02%." But what does this mean? The actual number tells us very little by itself, but it allows us to evaluate the dispersion of the scores around the mean.

In a distribution where all the scores are identical, the standard deviation is zero (0). Zero is the lowest possible value for the standard deviation; in an identical distribution, all the points would be the same, with the same mean, mode, and median. There is no variation or dispersion in the scores.

The more the standard deviation departs from zero, the more variation there is in the distribution. There is no upper limit to the value of the standard deviation. In our example, we can conclude that a standard deviation of 6.02% means that the projected percentage change in the elderly population for the four regions of the United States is widely dispersed around the mean of 19.95%.

The standard deviation can be considered a standard against which we can evaluate the positioning of scores relative to the mean and to other scores in the distribution. As we will see in more detail in Chapter 5, in most distributions, unless they are highly skewed, about 34% of all scores fall between the mean and 1 standard deviation above the mean. Another 34% of scores fall between the mean and 1 standard deviation below it. Thus, we would expect the majority of scores (68%) to fall within 1 standard deviation of the mean. For example, let's consider the distribution of grade point average (GPA) for Monitoring the Future (MTF) survey respondents in 2008. The mean GPA on a 4.0 scale is 3.13, with a standard deviation of 0.68. We can expect about 68% of MTF survey respondents to have a GPA within the range of 2.45 (3.13 − 0.68) to 3.81 (3.13 + 0.68). Hence, based on the mean and the standard deviation, we have a pretty good indication of what would be considered a typical GPA for the majority of

**Figure 4.4**    Illustrating the Means and Standard Deviations for Age Characteristics

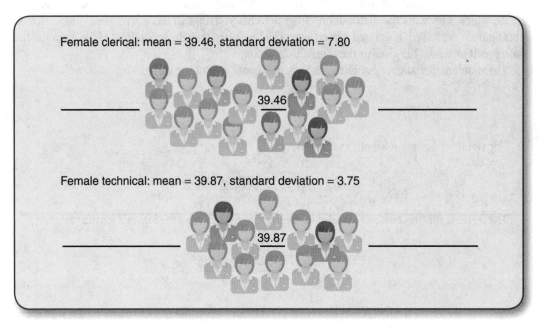

Female clerical: mean = 39.46, standard deviation = 7.80

39.46

Female technical: mean = 39.87, standard deviation = 3.75

39.87

*Source:* Adapted from Marjorie Armstrong-Stassen, "The Effect of Gender and Organizational Level on How Survivors Appraise and Cope with Organizational Downsizing," *Journal of Applied Behavioral Science* 34, no. 2 (June 1998): 125–142. Reprinted by permission of SAGE.

**Table 4.5**    Age Characteristics of Female Clerical and Technical
Employees

| Characteristics | Female Clerical (N = 22) | Female Technical (N = 39) |
|---|---|---|
| Mean age | 39.46 | 39.87 |
| Standard deviation | 7.80 | 3.75 |

*Source:* Adapted from Marjorie Armstrong-Stassen, "The Effect of Gender and Organizational Level on How Survivors Appraise and Cope with Organizational Downsizing," *Journal of Applied Behavioral Science* 34, no. 2 (June 1998): 125–142. Reprinted with permission.

respondents. For example, we would consider a respondent with a 3.9 GPA to be an exceptional student in comparison with the other students. More than two thirds of all respondents fall closer to the mean than a student with such a GPA.

Another way to interpret the standard deviation is to compare it with another distribution. For instance, Table 4.5 displays the means and standard deviations of employee age for two samples drawn from a *Fortune* 100 corporation. Samples are divided into female clerical and female technical. Note that the mean ages for both samples are about the same—approximately 39 years of age. However, the standard deviations suggest that the distribution of age is dissimilar between the two groups. Figure 4.4 loosely illustrates this dissimilarity in the two distributions.

The relatively low standard deviation for female technical indicates that this group is relatively homogeneous in age. That is to say, most of the women's ages, while not identical, are fairly similar. The average deviation from the mean age of 39.87 is 3.75 years. In contrast, the standard deviation for female clerical employees is about twice the standard deviation for female technical. This suggests a wider dispersion or greater heterogeneity in the ages of clerical workers. We can say that the average deviation from the mean age of 39.46 is 7.80 years for clerical workers. The larger standard deviation indicates a wider dispersion of points below or above the mean. On average, clerical employees are farther in age from their mean of 39.46.

# ▣ CONSIDERATIONS FOR CHOOSING A MEASURE OF VARIATION

So far, we have considered four measures of variation: (1) the range, (2) the IQR, (3) the variance, and (4) the standard deviation. Each measure can represent the degree of variability in a distribution. But which one should we use? There is no simple answer to this question. However, in general, we tend to use only one measure of variation, and the choice of the appropriate one involves a number of considerations. These considerations and how they affect our choice of the appropriate measure are presented in the form of a decision tree in Figure 4.5.

As in choosing a measure of central tendency, one of the most basic considerations in choosing a measure of variability is the variable's level of measurement. Valid use of any of the measures requires that the data are measured at the level appropriate for that measure or higher, as shown in Figure 4.5.

*Ordinal level:* You can use the IQR. However, the IQR relies on distance between two scores to express variation, information that cannot be obtained from ordinal-measured scores. The compromise is to use the IQR (reporting $Q_1$ and $Q_3$) alongside the median, interpreting the IQR as the range of rank-ordered values that includes the middle 50% of the observations.[6]

*Interval-ratio level:* For interval-ratio variables, you can choose the variance (or standard deviation), the range, or the IQR. Because the range, and to a lesser extent the IQR, is based on only two scores in the distribution (and therefore tends to be sensitive if either of the two points is extreme), the variance and/or standard deviation is usually preferred. However, if a distribution is extremely skewed so that the mean is no longer representative of the central tendency in the distribution, the range and the IQR can be used. The range and the IQR will also be useful when you are reading tables or quickly scanning data to get a rough idea of the extent of dispersion in the distribution.

# ▣ READING THE RESEARCH LITERATURE: DIFFERENCES IN COLLEGE ASPIRATIONS AND EXPECTATIONS AMONG LATINO ADOLESCENTS

In Chapter 2, we discussed how frequency distributions are presented in the professional literature. We noted that most statistical tables presented in the social science literature are considerably more

**Figure 4.5**  How to Choose a Measure of Variation

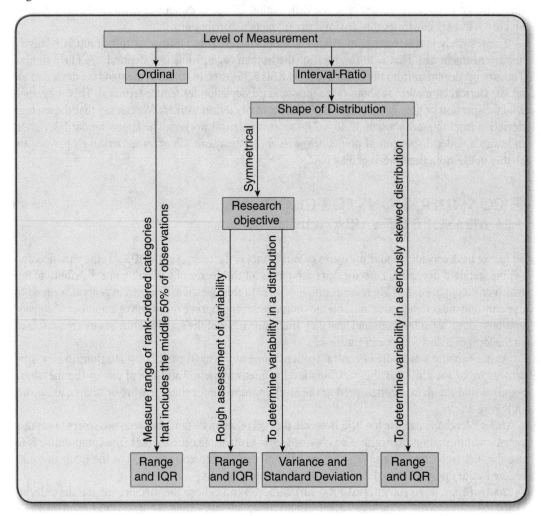

complex than those we describe in this book. The same can be said about measures of central tendency and variation. Most research articles use measures of central tendency and variation in ways that go beyond describing the central tendency and variation of a single variable. In this section, we refer to both the mean and the standard deviation because in most research reports the standard deviation is given along with the mean.

Table 4.6 displays data taken from a research article published in *Social Problems*.[7] This table illustrates a common research application of the mean and standard deviation. The authors of this article examine how ethnicity plays into one's college aspirations and expectations. Their major focus is to explore "potential differences in college aspirations and expectations across the three largest Latino groups and the potential sources of such differences."[8] We focus on only their data for Cubans and Mexicans to present a simplified example of the mean and standard deviation. Understanding the relationship between ethnicity and college aspirations and expectations is, nonetheless, critical in that

**Table 4.6**    Ethnicity and College Aspirations and Expectations

|  | *Cubans* | | *Mexicans* | |
| --- | --- | --- | --- | --- |
|  | *Mean* | *Standard Deviation* | *Mean* | *Standard Deviation* |
| I. How much respondent wants to go to college | 4.50 | 1.80 | 4.20 | 1.30 |
| II. How likely respondent will go to college | 4.30 | 2.00 | 3.70 | 1.40 |

*Source:* Adapted from Stephanie A. Bohon, Monica Kirkpatrick Johnson, and Bridget K. Gorman, "College Aspirations and Expectations Among Latino Adolescents in the United States," *Social Problems* 53, no. 2 (2006): 207–225. Published by the University of California Press.

*Note:* The authors examine variation among other ethnicities in this paper. However, to simplify this example, we focus on the descriptive for Cubans and Mexicans.

the U.S. Latino population is growing faster than any other minority group; yet Latinos remain the least educated of all other people of color.

Data for this study come from the National Longitudinal Study of Adolescent Health survey. This survey is a representative sample of American adolescents in Grades 7 to 12. Complex sampling strategies are employed to ensure a representative sample. Thus, factors such as variation in geographic location, type of school, racial makeup, and so on are accounted for during data collection.

Respondents were asked a variety of questions, but the authors focused specifically on questions about college aspirations and expectations. Their measure of college aspirations is based on a scale of 1 to 5 derived from the question, "How much do you want to go to college?" An answer of 1 indicated a low desire to go to college, while an answer of 5 indicated a high desire to go to college. Their measure of college expectations was also based on the same scale ranging from 1 (*low*) to 5 (*high*). However, it was derived from the question, "How likely is it that you will go to college?"

We should first examine whether the means are similar or different. For both the expectations and aspirations measure, we can see that Mexicans have slightly lower aspirations (4.20) and expectations (3.70) than Cubans (4.50 and 4.30, respectively). Furthermore, the standard deviations indicate that there is more variability in each of these measures for Cubans than for Mexicans.

The researchers of this study described the data displayed in Table 4.6 as follows:

They show strong aspirations for and expectations of college attendance across each of the five groups. Important differences across ethnic groups exist, however. As anticipated, Mexicans have weaker than average . . . and Cubans have stronger than average aspirations and expectations.[9]

Why might this be? The authors conclude their discussion of the data presented in Table 4.6 by arguing as follows:

Differential aspirations and expectations may be explained by the considerable differences in family and household characteristics, parental hopes for their child's educational success, and academic skills and disengagement.[10]

## MAIN POINTS

- Measures of variability are numbers that describe how much variation or diversity there is in a distribution.

- The range measures variation in interval-ratio variables and is the difference between the highest (maximum) and the lowest (minimum) scores in the distribution. To find the range, subtract the lowest from the highest score in a distribution. For an ordinal variable, just report the lowest and the highest values without subtracting.

- The interquartile range (IQR) measures the width of the middle 50% of the distribution. It is defined as the difference between the lower and upper quartiles ($Q_1$ and $Q_3$). For an ordinal variable, just report $Q_1$ and $Q_3$ without subtracting.

- The variance and the standard deviation are two closely related measures of variation for interval-ratio variables that increase or decrease based on how closely the scores cluster around the mean. The variance is the average of the squared deviations from the center (mean) of the distribution; the standard deviation is the square root of the variance.

## KEY TERMS

interquartile range (IQR)                    standard deviation
measures of variability                      variance
range

## ON YOUR OWN

Log on to the web-based student study site at **www.sagepub.com/ssdsessentials** for additional study questions, web quizzes, web resources, flashcards, codebooks and datasets, web exercises, appendices, and links to social science journal articles reflecting the statistics used in this chapter.

## CHAPTER EXERCISES

1. Public corruption continues to be a concern. Let's examine data from the U.S. Department of Justice to explore the variability in public corruption in the years 1990 and 2007. All the numbers below are of those convicted of public corruption.
   a. What is the range of convictions in 1990? In 2007? Which is greater?
   b. What is the mean number of convictions in 1990 and 2007?
   c. Calculate the standard deviation for 1990 and 2007.
   d. Which year appears to have more variability in number of convictions as measured by the standard deviation? Are the results consistent with what you found using the range?

**Number of Public Corruption Convictions by Year**

| 1990 | | 2007 | |
|---|---|---|---|
| *Govt. Level* | *No. of Convictions* | *Govt. Level* | *No. of Convictions* |
| Federal | 583 | Federal | 405 |
| State | 79 | State | 85 |
| Local | 225 | Local | 275 |

*Source:* U.S. Census Bureau, *The 2010 Statistical Abstract*, Table 327.

2. Use Table 4.1 from the chapter for this exercise to continue comparisons by region. Use only the information for states in the West and Midwest.
   a. Compare the western states with those in the Midwest on the projected percentage increase in the elderly population by calculating the range. Which region had a greater range?
   b. Calculate the IQR for each region. Which is greater?
   c. Use the statistics to characterize the variability in population increase of the elderly in the two regions. Does one region have more variability than another? If yes, why do you think that is?

3. The U.S. Census Bureau collects information about divorce rates. The following table summarizes the divorce rate for 10 U.S. states in 2007. Use the table to answer the questions that follow.

| *State* | *Divorce Rate per 1,000 Population* |
|---|---|
| Alaska | 4.3 |
| Florida | 4.7 |
| Idaho | 4.9 |
| Maine | 4.5 |
| Maryland | 3.1 |
| Nevada | 6.5 |
| New Jersey | 3.0 |
| Texas | 3.3 |
| Vermont | 3.8 |
| Wisconsin | 2.9 |

*Source:* U.S. Census Bureau, *The 2010 Statistical Abstract*, Table 126.

   a. Calculate and interpret the range and the IQR. Which is a better measure of variability? Why?
   b. Calculate and interpret the mean and standard deviation.
   c. Identify two possible explanations for the variation in divorce rates across the 10 states.

4. The respondents of the 2007 HINTS reported their psychological distress on a scale between 0 and 24. In the following table, you will see separate data on two groups of respondents' distress scores: those who have ever been diagnosed as having cancer, and those who have not.

|  | Psychological Distress Score | |
| --- | --- | --- |
|  | Diagnosed | Not Diagnosed |
| $\bar{Y}$ | 3.9 | 4.87 |
| $\Sigma Y$ | 729 | 5,849 |
| $\Sigma(Y - \bar{Y})^2$ | 3,059.14 | 25,180.20 |
| $N$ | 187 | 1,200 |

a. Calculate the variance and standard deviation from these statistics for both groups.

b. What can you say about the variability in the distress scores for those respondents who have been diagnosed as having cancer and those who have not? Why might there be a difference? Why might there be more variability for one group than for the other?

c. Was it necessary in this problem to provide you with the mean value to calculate the variance and standard deviation?

5. You are interested in studying the variability of crimes committed (including violent and property crimes) and police expenditures in the eastern and midwestern United States. The U.S. Census Bureau collected the following statistics on these two variables for 21 states in the East and Midwest in 2006.

| State | Number of Crimes per 100,000 Population | Police Protection Expenditures (in millions of dollars) |
| --- | --- | --- |
| Maine | 2,635 | 221 |
| New Hampshire | 2,013 | 274 |
| Vermont | 2,442 | 136 |
| Massachusetts | 2,838 | 1,673 |
| Rhode Island | 2,815 | 286 |
| Connecticut | 2,785 | 905 |
| New York | 2,488 | 7,585 |
| New Jersey | 2,644 | 3,010 |
| Pennsylvania | 2,883 | 2,535 |
| Ohio | 4,029 | 2,689 |
| Indiana | 3,817 | 1,039 |
| Illinois | 3,562 | 3,761 |
| Michigan | 3,775 | 2,333 |
| Wisconsin | 3,102 | 1,434 |
| Minnesota | 3,398 | 1,294 |
| Iowa | 3,087 | 570 |

| State | Number of Crimes per 100,000 Population | Police Protection Expenditures (in millions of dollars) |
|---|---|---|
| Missouri | 4,373 | 1,179 |
| North Dakota | 2,128 | 106 |
| South Dakota | 1,791 | 132 |
| Nebraska | 3,623 | 334 |
| Kansas | 4,175 | 635 |

*Source:* U.S. Census Bureau, *The 2010 Statistical Abstract*, Tables 297 and 431.

    a. Calculate the mean for each variable.

    b. Calculate the standard deviation for each variable.

    c. Compare the mean with the standard deviation for each variable. Does there appear to be more variability in the number of crimes or in police expenditures per capita in these states? Which states contribute more to this greater variability?

    d. Suggest why one variable has more variability than the other. In other words, what social forces would cause one variable to have a relatively large standard deviation?

6. Use the data in Table 4.2 from the chapter for this exercise. Calculate the standard deviation for the projected percentage change in the elderly population from 2008 to 2015.

7. The following table summarizes the racial differences in education and the ideal number of children for Chinese Americans and Filipino Americans. Based on the means and standard deviations (in parentheses), what conclusions can be drawn about differences in the ideal number of children?

| | Chinese Americans | Filipino Americans |
|---|---|---|
| Education (years) | 15.83 (2.691) | 13.87 (3.335) |
| Ideal number of children | 2.20 (0.422) | 4.00 (2.769) |

*Source:* General Social Survey, 2008.

8. Average life expectancy for females in 2006 is reported for 10 countries. Calculate the appropriate measures of central tendency and variability for both European countries and non-European countries. Is there more variability in life expectancy for European countries or non-European countries? If so, what might explain these differences?

| Country | Life Expectancy at Birth |
|---|---|
| European countries | |
| France | 84.4 |
| Germany | 82.4 |
| Netherlands | 81.9 |
| Spain | 84.4 |
| Turkey | 74.0 |
| Non-European countries | |
| Japan | 85.8 |
| Australia | 82.7 |
| Mexico | 78.1 |
| Iceland | 83.0 |
| New Zealand | 82.2 |

*Source:* U.S. Census Bureau, *The 2010 Statistical Abstract*, Table 1304.

9. You have been asked to prepare a brief statement about young drivers for a public policymaker. Using statistics from the following table, write a paragraph or two about young drivers in America. In your answer, be sure to identify at least two explanations for your findings.

| Number of Driving Offenses in the Past 12 Months | N | Mean | Standard Deviation |
|---|---|---|---|
| Moving violations | 1,357 | 0.38 | 0.804 |
| Automobile collisions/accidents | 1,348 | 0.28 | 0.610 |

*Source:* Monitoring the Future, 2008.

# Chapter 5

# The Normal Distribution

## Chapter Learning Objectives

- ❖ Recognizing the importance and the use of the normal distribution in statistics
- ❖ Describing the properties of the normal distribution
- ❖ Transforming a raw score into standard (Z) score and vice versa
- ❖ Using the standard normal table
- ❖ Transforming a Z score into proportion (or percentage) and vice versa

W hile in the preceding chapters we have described empirical distributions that are based on real data, in this chapter we will describe a theoretical distribution known as the *normal curve* or the **normal distribution.** A *theoretical distribution* is similar to an empirical distribution in that it can be organized into frequency distributions, displayed using graphs, and described by its central tendency and variation using measures such as the mean and the standard deviation. However, unlike an empirical distribution, a theoretical distribution is based on theory rather than on real data. The value of the theoretical normal distribution lies in the fact that many empirical distributions that we study seem to approximate it. We can often learn a lot about the characteristics of these empirical distributions based on our knowledge of the theoretical normal distribution.

---

*Normal distribution*   A bell-shaped and symmetrical theoretical distribution with the mean, the median, and the mode all coinciding at its peak and with the frequencies gradually decreasing at both ends of the curve.

---

## ▣ PROPERTIES OF THE NORMAL DISTRIBUTION

The normal curve (Figure 5.1) looks like a bell-shaped line graph. Because of this property, it is sometimes called the *bell-shaped curve*. One of the most striking characteristics of the normal distribution is its perfect symmetry. Each half is the mirror image of the other. This means that precisely half the observations fall on each side of the middle of the distribution. In addition, the midpoint of the normal curve is the point at which three measures coincide: the mode (the point of the highest frequency), the median (the point that divides the distribution into two equal halves), and the mean (the average of all the scores). Notice also that most of the observations are clustered around the middle, with the frequencies gradually decreasing at both ends of the distribution.

**Figure 5.1**   The Normal Curve

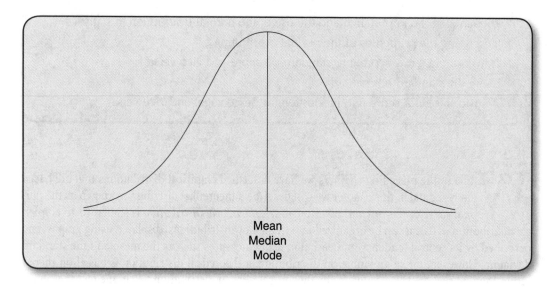

Mean
Median
Mode

## Empirical Distributions Approximating the Normal Distribution

The normal curve is a theoretical ideal, and real-life distributions never match this model perfectly. However, researchers study many variables (e.g., standardized tests such as the GRE) that closely resemble this theoretical model. When we say that a variable is "normally distributed," we mean that the graphic display will reveal an approximately bell-shaped and symmetrical distribution closely resembling the idealized model shown in Figure 5.1. This property makes it possible for us to describe many empirical distributions based on our knowledge of the normal curve.

For example, let's examine the frequencies and the bar chart presented in Table 5.1. These data are the final scores of 1,200 students of social statistics. We overlaid a normal curve on the distribution shown in Table 5.1. Notice how closely our empirical distribution of statistics scores approximates the normal curve.

**Table 5.1**   Final Grades in Social Statistics of 1,200 Students (1983–1993): A Near Normal Distribution

| Midpoint Score | Frequency Bar Chart | Freq. | Cum. Freq. | % | Cum. % |
|---|---|---|---|---|---|
| 40 | * | 4 | 4 | 0.33 | 0.33 |
| 50 | ******** | 78 | 82 | 6.50 | 6.83 |
| 60 | ****************** | 275 | 357 | 22.92 | 29.75 |
| 70 | ***************************** | 483 | 840 | 40.25 | 70.00 |
| 80 | ****************** | 274 | 1,114 | 22.83 | 92.83 |
| 90 | ******** | 81 | 1,195 | 6.75 | 99.58 |
| 100 | * | 5 | 1,200 | 0.42 | 100.00 |

| 10 | 50 | 100 | 200 | 300 | 400 | 500 |

Mean $(\bar{Y}) = 70.07$   Median $= 70.00$   Mode $= 70.00$
Standard deviation $(S_Y) = 10.27$

Note that 70 is the most frequent score obtained by the students, and therefore, it is the mode of the distribution. Because about half the students are either above (49.99%) or below (50.01%) this score (based on raw frequencies), both the mean (70.07) and the median (70) are approximately 70. Also shown in Table 5.1 is the gradual decrease in the number of students who scored either above or below 70. Very few students scored higher than 90 or lower than 50.

When we use the term *normal curve*, we are not referring to identical distributions. The shape of a normal distribution varies, depending on the mean and standard deviation of the particular distribution. (The symbol $\mu_y$ is the population notation for the mean while $\sigma_y$ stands for the population standard deviation. We will discuss these symbols in more detail in the next chapter.) For example, in Figure 5.2, we present two normally shaped distributions with identical means ($\mu_Y = 12$) but with different standard deviations ($\sigma_{Y_1} = 3$, and $\sigma_{Y_2} = 5$). Note that the distribution with the larger standard deviation appears relatively wider and flatter.

## Areas Under the Normal Curve

Regardless of the precise shape of the distribution, in all normal or nearly normal curves we find a constant proportion of the area under the curve lying between the mean and any given distance from the mean when measured in standard deviation units. The area under the normal curve may be

**Figure 5.2**   Two Normal Distributions With Equal Means but Different Standard Deviations

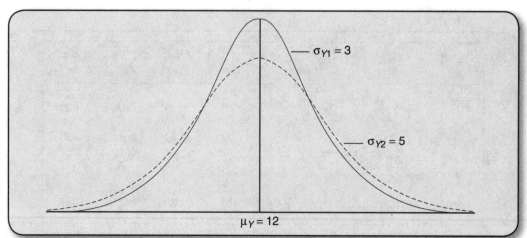

conceptualized as a proportion or percentage of the number of observations in the sample. Thus, the entire area under the curve is equal to 1.00 or 100% (1.00 × 100) of the observations. Because the normal curve is perfectly symmetrical, exactly 0.50 or 50% of the observations lie above or to the right of the center, which is the mean of the distribution, and 50% lie below or to the left of the mean.

In Figure 5.3, note the percentage of cases that will be included between the mean and 1, 2, and 3 standard deviations above and below the mean. The mean of the distribution divides it exactly into halves; 34.13% is included between the mean and 1 standard deviation to the right of the mean; and the same percentage is included between the mean and 1 standard deviation to the left of the mean. The plus signs indicate standard deviations above the mean; the minus signs denote standard deviations below the

**Figure 5.3**   Percentages Under the Normal Curve

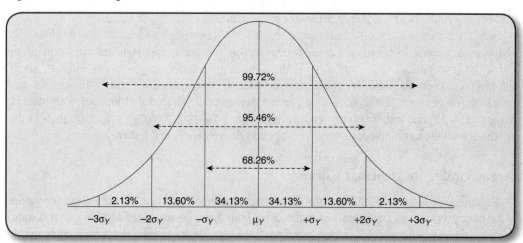

mean. Thus, between the mean and ±1 standard deviation, 68.26% of all the observations in the distribution occur; between the mean and ±2 standard deviations, 95.46% of all observations in the distribution occur; and between the mean and ±3 standard deviations, 99.72% of all the observations occur.

✓ *Learning Check*

> *Review and confirm the properties of the normal curve. What is the area underneath the curve equal to? What percentage of the distribution is within 1 standard deviation? Within 2 and 3 standard deviations? Verify the percentage of cases by summing the percentages in Figure 5.3.*

## Interpreting the Standard Deviation

The fixed relationship between the distance from the mean and the areas under the curve represents a property of the normal curve that has highly practical applications. As long as a distribution is normal and we know the mean and the standard deviation, we can determine the relative frequency (proportion or percentage) of cases that fall between any score and the mean.

This property provides an important interpretation for the standard deviation of empirical distributions that are approximately normal. For such distributions, when we know the mean and the standard deviation, we can determine the percentage of scores that are within any distance, measured in standard deviation units, from that distribution's mean. For example, we know that college entrance tests such as the SAT and ACT are normally distributed. The SAT, for instance, has a mean of 500 and a standard deviation of 100. This means that approximately 68% of the students who take the test obtain a score between 400 (1 standard deviation below the mean) and 600 (1 standard deviation above the mean). We can also anticipate that approximately 95% of the students who take the test will score between 300 (2 standard deviations below the mean) and 700 (2 standard deviations above the mean).

Not every empirical distribution is normal. We've learned that the distributions of some common variables, such as income, are skewed and therefore not normal. The fixed relationship between the distance from the mean and the areas under the curve applies only to distributions that are normal or approximately normal.

## ▣ STANDARD (*Z*) SCORES

We can express the difference between any score in a distribution and the mean in terms of *standard scores*, also known as *Z scores*. A **standard (*Z*) score** is the number of standard deviations that a given raw score (or the observed score) is above or below the mean. A raw score can be transformed into a *Z* score to find how many standard deviations it is above or below the mean.

---

*Standard (Z) score*   The number of standard deviations that a given raw score is above or below the mean.

---

## Transforming a Raw Score Into a *Z* Score

To transform a raw score into a *Z* score, we divide the difference between the score and the mean by the standard deviation. For instance, to transform a final score in the statistics class into a *Z* score, we subtract the mean of 70.07 from that score and divide the difference by the standard deviation of 10.27. Thus, the $Z$ score of 80 is

$$\frac{80 - 70.07}{10.27} = 0.97$$

or 0.97 standard deviations above the mean. Similarly, the $Z$ score of 60 is

$$\frac{60 - 70.07}{10.27} = -0.98$$

or 0.98 standard deviations below the mean; the negative sign indicates that this score is below the mean.

This calculation, in which the difference between a raw score and the mean is divided by the standard deviation, gives us a method of standardization known as *transforming a raw score into a Z score* (also known as a standard score). The $Z$-score formula is

$$Z = \frac{Y - \bar{Y}}{S_Y} \tag{5.1}$$

A $Z$ score allows us to represent a raw score in terms of its relationship to the mean and to the standard deviation of the distribution. It represents how far a given raw score is from the mean in standard deviation units. A positive $Z$ indicates that a score is larger than the mean, and a negative $Z$ indicates that it is smaller than the mean. The larger the $Z$ score, the larger the difference between the score and the mean (Table 5.2).

## Transforming a *Z* Score Into a Raw Score

For some normal curve applications, we need to reverse the process, transforming a $Z$ score into a raw score instead of transforming a raw score into a $Z$ score. A $Z$ score can be converted to a raw score

Table 5.2  Examples of Final Social Science Statistics Scores Converted to Z Score

| Final Score | Z Score | | | |
|---|---|---|---|---|
| 40 | $Z = \dfrac{40 - 70.07}{10.27}$ | $= \dfrac{-30.07}{10.27}$ | $= -2.93$ | |
| 60 | $Z = \dfrac{60 - 70.07}{10.27}$ | $= \dfrac{-10.07}{10.27}$ | $= -0.98$ | |
| 80 | $Z = \dfrac{80 - 70.07}{10.27}$ | $= \dfrac{9.93}{10.27}$ | $= 0.97$ | |
| 100 | $Z = \dfrac{100 - 70.07}{10.27}$ | $= \dfrac{29.93}{10.27}$ | $= 2.91$ | |
| $\bar{Y} = 70.07$ | $S_Y = 10.27$ | | | |

to find the score associated with a particular distance from the mean when this distance is expressed in standard deviation units. For example, suppose we are interested in finding out the final score in the statistics class that lies 1 standard deviation above the mean. To solve this problem, we begin with the $Z$ score formula:

$$Z = \frac{Y - \bar{Y}}{S_Y}$$

Note that for this problem, we have the following values for $Z$ ($Z = 1$), the mean ($\bar{Y} = 70.07$), and the standard deviation ($S_y = 10.27$), but we need to determine the value of $Y$:

$$1 = \frac{Y - 70.07}{10.27}$$

Through simple algebra, we solve for $Y$:

$$Y = 70.07 + 1(10.27) = 70.07 + 10.27 = 80.34$$

The score of 80.34 lies 1 standard deviation (or 1 $Z$ score) above the mean of 70.07.

The general formula for transforming a $Z$ score into a raw score is

$$Y = \bar{Y} + Z(S_Y) \tag{5.2}$$

Thus, to transform a $Z$ score into a raw score, multiply the $Z$ score by the standard deviation and add this product to the mean.

Now, what statistics score lies 1.5 standard deviations below the mean? Because the score lies below the mean, the $Z$ score is negative. Thus,

$$Y = 70.07 + (-1.5)(10.27) = 70.07 - 15.41 = 54.66$$

The score of 54.66 lies 1.5 standard deviations below the mean of 70.07.

> *Transform the Z scores in Table 5.2 back into raw scores. Your answers should agree with the raw scores listed in the table.*

✓ *Learning Check*

## ▣ THE STANDARD NORMAL DISTRIBUTION

When a normal distribution is represented in standard scores ($Z$ scores), we call it the **standard normal distribution**. Standard scores, or $Z$ scores, are the numbers that tell us the distance between an actual score and the mean in terms of standard deviation units. The standard normal distribution has a mean of 0.0 and a standard deviation of 1.0.

---

*Standard normal distribution*   A normal distribution represented in standard *(Z)* scores, with mean = 0 and standard deviation = 1.

---

Figure 5.4 shows a standard normal distribution with areas under the curve associated with 1, 2, and 3 standard scores above and below the mean. To help you understand the relationship between raw scores of a distribution and standard $Z$ scores, we also show the raw scores in the statistics class that correspond to these standard scores. For example, notice that the mean for the statistics score distribution is 70.07 and the corresponding $Z$ score—the mean of the standard normal distribution—is 0. The score of 80.34 is 1 standard deviation above the mean (70.07 + 10.27 = 80.34); therefore, its corresponding $Z$ score is +1. Similarly, the score of 59.80 is 1 standard deviation below the mean (70.07 − 10.27 = 59.80), and its $Z$-score equivalent is −1.

**Figure 5.4**   The Standard Normal Distribution

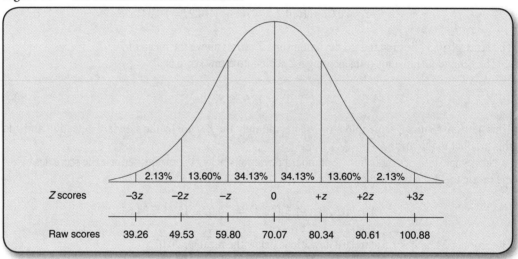

## ▣ THE STANDARD NORMAL TABLE

We can use $Z$ scores to determine the proportion of cases that are included between the mean and any $Z$ score in a normal distribution. The areas or proportions under the standard normal curve, corresponding to any $Z$ score or its fraction, are organized into a special table called the **standard normal table**. The table is presented in Appendix A. In this section, we discuss how to use this table.

---

*Standard normal table*   A table showing the area (as a proportion, which can be translated into a percentage) under the standard normal curve corresponding to any $Z$ score or its fraction.

---

**Figure 5.5** Areas Between Mean and *Z* (B) and Beyond *Z* (C)

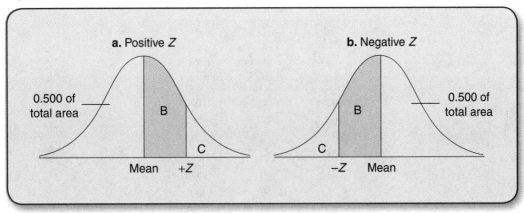

## The Structure of the Standard Normal Table

Table 5.3 reproduces a small part of the standard normal table. Note that the table consists of three columns. Column A lists positive *Z* scores. Because the normal curve is symmetrical, the proportions that correspond to positive *Z* scores are identical to the proportions corresponding to negative *Z* scores.

Column B shows the area included between the mean and the *Z* score listed in Column A. Note that when *Z* is positive, the area is located on the right side of the mean (see Figure 5.5a), whereas for a negative *Z* score, the same area is located left of the mean (Figure 5.5b).

Column C shows the proportion of the area that is beyond the *Z* score listed in Column A. Areas corresponding to positive *Z* scores are on the right side of the curve (see Figure 5.5a). Areas corresponding to negative *Z* scores are identical except that they are on the left side of the curve (Figure 5.5b).

## Transforming *Z* Scores Into Proportions (or Percentages)

We illustrate how to use Appendix A with some simple examples using our data on students' final statistics scores (see Table 5.1). The examples in this section are applications that require the transformation of *Z* scores into proportions (or percentages).

### Finding the Area Between the Mean and a Specified Positive Z Score

Use the standard normal table to find the area between the mean and a specified positive *Z* score. To find the percentage of students whose scores range between the mean (70.07) and 85, follow these steps.

1. Convert 85 to a *Z* score:

$$Z = \frac{85 - 70.07}{10.27} = 1.45$$

2. Look up 1.45 in Column A (in Appendix A) and find the corresponding area in Column B, 0.4265. We can translate this proportion into a percentage (0.4265 × 100 = 42.65%) of the area under the curve included between the mean and a *Z* score of 1.45 (Figure 5.6).

3. Thus, 42.65% of the students scored between 70.07 and 85.

**Table 5.3**   Excerpt of the Standard Normal Table

| A | B | C |
|---|---|---|
| Z | Area Between Mean and Z | Area Beyond Z |
| 0.00 | 0.0000 | 0.5000 |
| 0.01 | 0.0040 | 0.4960 |
| 0.02 | 0.0080 | 0.4920 |
| 0.03 | 0.0120 | 0.4880 |
| 0.04 | 0.0160 | 0.4840 |
| 0.05 | 0.0199 | 0.4801 |
| 0.06 | 0.0239 | 0.4761 |
| 0.07 | 0.0279 | 0.4721 |
| 0.08 | 0.0319 | 0.4681 |
| 0.09 | 0.0359 | 0.4641 |
| 0.10 | 0.0398 | 0.4602 |

**Figure 5.6**   Finding the Area Between the Mean and a Specified Positive $Z$ Score

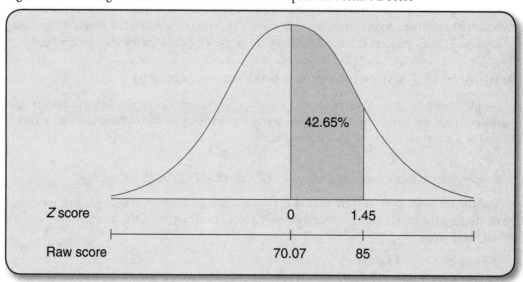

To find the actual number of students who scored between 70.07 and 85, multiply the proportion 0.4265 by the total number of students. Thus, approximately 512 students (0.4265 × 1,200 = 512) obtained a score between 70.07 and 85.

### Finding the Area Between the Mean and a Specified Negative Z Score

What is the percentage of students whose scores ranged between 65 and 70.07? We can use the standard normal table and the following steps to find out.

1. Convert 65 to a $Z$ score:

$$Z = \frac{65 - 70.07}{10.27} = -0.49$$

2. Because the normal distribution is symmetrical, we ignore the negative sign of $Z$ and look up 0.49 in Column A. The area corresponding to a $Z$ score of 0.49 is 0.1879. This indicates that 0.1879 of the area under the curve is included between the mean and a $Z$ of $-0.49$ (Figure 5.7). We convert this proportion to 18.79% ($0.1879 \times 100 = 18.79\%$).

3. Thus, approximately 225 ($0.1879 \times 1{,}200 = 225$) students obtained a score between 65 and 70.07.

**Figure 5.7**  Finding the Area Between the Mean and a Specified Negative $Z$ Score

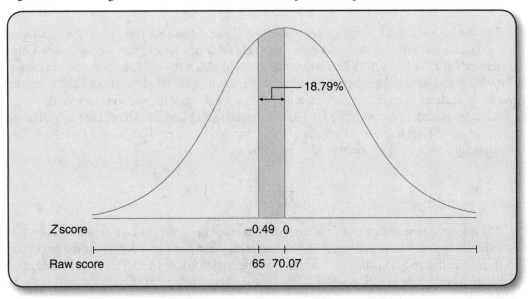

### Finding the Area Above a Positive Z Score or Below a Negative Z Score

We can compare students who have done very well or very poorly to get a better idea of how they compare with other students in the class.

To identify students who did very well, we selected all students who scored above 85. To find how many students scored above 85, first convert 85 to a $Z$ score:

$$Z = \frac{85 - 70.07}{10.27} = 1.45$$

**Figure 5.8**    Finding the Area Above a Positive $Z$ Score or Below a Negative $Z$ Score

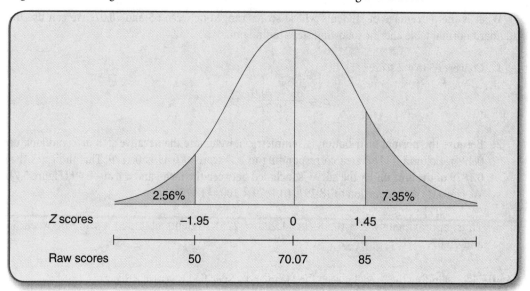

The area beyond a $Z$ of 1.45 includes all students who scored above 85. This area is shown in Figure 5.8. To find the proportion of students whose scores fall in this area, refer to the entry in Column C that corresponds to a $Z$ of 1.45, 0.0735. This means that 7.35% ($0.0735 \times 100 = 7.35\%$) of the students scored above 85. To find the actual number of students in this group, multiply the proportion 0.0735 by the total number of students. Thus, there were $1,200 \times 0.0735$, or about 88 students, who scored above 85.

A similar procedure can be applied to identify the number of students who did not do well in the class. The cutoff point for poor performance in this class was the score of 50. To determine how many students did poorly, we first converted 50 to a $Z$ score:

$$Z = \frac{50 - 70.07}{10.27} = -1.95$$

The $Z$ score corresponding to a final score of 50 is equal to $-1.95$. The area beyond a $Z$ of $-1.95$ includes all students who scored below 50. This area is also shown in Figure 5.8. Locate the proportion of students in this area in Column C in the entry corresponding to a $Z$ of 1.95. This proportion is equal to 0.0256. Thus, 2.56% ($0.0256 \times 100 = 2.56\%$) of the group, or about 31 ($0.0256 \times 1,200$) students, performed poorly in statistics.

## Transforming Proportions (or Percentages) Into $Z$ Scores

The examples in this section are applications that require transforming proportions (or percentages) into $Z$ scores.

Assuming that a grade of A is assigned to the top 10% of the students, what would it take to get an A in the class? To answer this question, we need to identify the cutoff point for the top 10% of the class. This problem involves two steps:

1. Find the $Z$ score that bounds the top 10% or 0.1000 ($0.1000 \times 100 = 10\%$) of all the students who took statistics (Figure 5.9).

**Figure 5.9** Finding a $Z$ Score Bounding an Area Above It

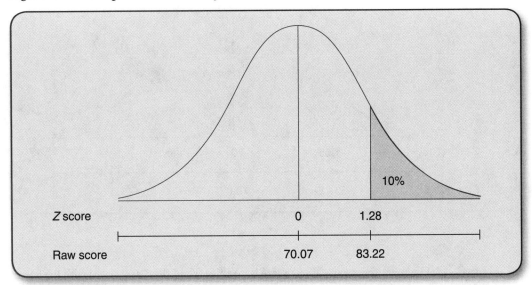

Refer to the areas under the normal curve shown in Appendix A. First, look for an entry of 0.1000 (or the value closest to it) in Column C. The entry closest to 0.1000 is 0.1003. Then, locate the $Z$ in Column A that corresponds to this proportion. The $Z$ score associated with the proportion 0.1003 is 1.28.

2. Find the final score associated with a $Z$ of 1.28.
   This step involves transforming the $Z$ score into a raw score. We learned earlier in this chapter (Formula 5.2) that to transform a $Z$ score into a raw score we multiply the score by the standard deviation and add that product to the mean. Thus,

$$Y = 70.07 + 1.28(10.27) = 70.07 + 13.15 = 83.22$$

The cutoff point for the top 10% of the class is a score of 83.22.

Now, let's assume that a grade of F was assigned to the bottom 5% of the class. What would be the cutoff point for a failing score in statistics? Again, this problem involves two steps:

1. Find the $Z$ score that bounds the lowest 5% or 0.0500 of all the students who took the class (Figure 5.10).
   Refer to the areas under the normal curve, and look for an entry of 0.0500 (or the value closest to it) in Column C. The entry closest to 0.0500 is 0.0495. Then, locate the $Z$ in Column A that corresponds to this proportion, 1.65. Because the area we are looking for is on the left side of the

**Figure 5.10** Finding a Z Score Bounding an Area Below It

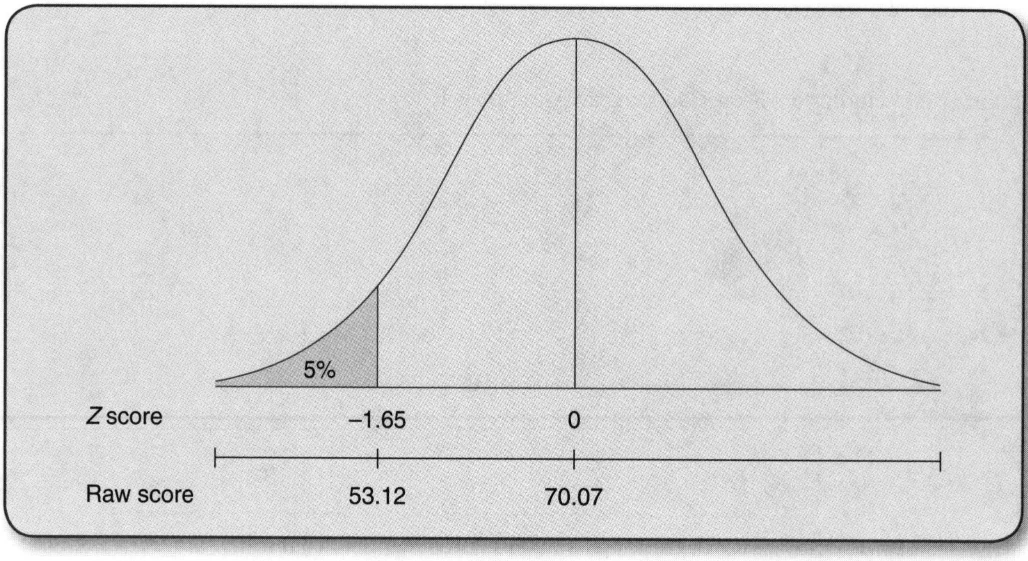

curve—that is, below the mean—the Z score is negative. Thus, the Z associated with the lowest 0.0500 (or 0.0495) is −1.65.

2. To find the final score associated with a Z of −1.65, convert the Z score to a raw score:

$$Y = 70.07 + (-1.65)(10.27) = 70.07 - 16.95 = 53.12$$

The cutoff for a failing score in statistics is 53.12.

✓ Learning Check

*Can you find the number of students who got a score of at least 90 in the statistics course? How many students got a score below 60?*

## MAIN POINTS

- The normal distribution is central to the theory of inferential statistics. It also provides a model for many empirical distributions that approximate normality.

- In all normal or nearly normal curves, we find a constant proportion of the area under the curve lying between the mean and any given distance from the mean when measured in standard deviation units.

- The standard normal distribution is a normal distribution represented in standard scores, or Z scores, with mean = 0 and standard deviation = 1. Z scores express the number of standard deviations that a given score is above or below the mean. The proportions corresponding to any Z score or its fraction are organized into a special table called the standard normal table.

## KEY TERMS

normal distribution
standard normal distribution

standard normal table
standard *(Z)* score

## ON YOUR OWN

Log on to the web-based student study site at **www.sagepub.com/ssdsessentials** for additional study questions, web quizzes, web resources, flashcards, codebooks and datasets, web exercises, appendices, and links to social science journal articles reflecting the statistics used in this chapter.

## CHAPTER EXERCISES

1.  We discovered that 978 GSS respondents in 2008 watched television for an average of 2.95 hr/day, with a standard deviation of 2.58 hr. Answer the following questions assuming the distribution of the number of television hours is normal.
    a.  What is the *Z* score for a person who watches more than 8 hr/day?
    b.  What proportion of people watch television less than 5 hr/day? How many does this correspond to in the sample?
    c.  What number of television hours per day corresponds to a *Z* score of +1?
    d.  What is the percentage of people who watch between 1 and 6 hr of television per day?

2.  If a particular distribution that you are studying is not normal, it may be difficult to determine the area under the curve of the distribution or to translate a raw score into a *Z* value. Is this statement true? Why or why not?

3.  Let's assume that education is normally distributed. Using GSS data, we find the mean number of years of education is 13.44 with a standard deviation of 3.1. A total of 1,497 respondents were included in the survey. Use these numbers to answer the following questions.
    a.  If you have 13.44 years of education, that is, the mean number of years of education, what is your *Z* score?
    b.  If your friend has 14.22 years of education, how many respondents have between your years of education and your friend's years of education?

4.  The 2008 GSS provides the following statistics for the average years of education for lower-, working-, middle-, and upper-class respondents and their associated standard deviations.
    a.  Assuming that years of education is normally distributed in the population, what proportion of working-class respondents have 12 to 16 years of education? What proportion of upper-class respondents have 12 to 16 years of education?
    b.  What is the probability that a working-class respondent, drawn at random from the population, will have more than 16 years of education? What is the equivalent probability for a middle-class respondent drawn at random?
    c.  What is the probability that a lower- or upper-class respondent will have less than 12 years of education?
    d.  Find the upper and lower limits, centered on the mean, that will include 50% of all working-class respondents.
    e.  If years of education is actually positively skewed in the population, how would that change your other answers?

|              | *Mean* | *Standard Deviation* | **N** |
| ------------ | ------ | -------------------- | ----- |
| Lower class  | 11.36  | 2.96                 | 121   |
| Working class| 12.73  | 2.79                 | 676   |
| Middle class | 14.40  | 3.04                 | 636   |
| Upper class  | 15.49  | 2.95                 | 53    |

**Exercises**

5. The following table displays unemployment information for each U.S. state and District of Columbia in 2009. The unemployment numbers listed in the table are in thousands. For example, Alabama was home to 212,000 unemployed residents when data were collected.

### Unemployment (in thousands) in the United States: 2009

| State | Number | State | Number | State | Number |
|---|---|---|---|---|---|
| Alabama | 212 | Kentucky | 218 | North Dakota | 16 |
| Alaska | 29 | Louisiana | 141 | Ohio | 611 |
| Arizona | 284 | Maine | 57 | Oklahoma | 114 |
| Arkansas | 100 | Maryland | 209 | Oregon | 217 |
| California | 2,086 | Massachusetts | 293 | Pennsylvania | 519 |
| Colorado | 208 | Michigan | 665 | Rhode Island | 64 |
| Connecticut | 156 | Minnesota | 236 | South Carolina | 255 |
| Delaware | 35 | Mississippi | 123 | South Dakota | 21 |
| District of Columbia | 34 | Missouri | 283 | Tennessee | 317 |
| Florida | 966 | Montana | 31 | Texas | 911 |
| Georgia | 457 | Nebraska | 45 | Utah | 90 |
| Hawaii | 43 | Nevada | 161 | Vermont | 25 |
| Idaho | 60 | New Hampshire | 47 | Virginia | 278 |
| Illinois | 665 | New Jersey | 418 | Washington | 314 |
| Indiana | 320 | New Mexico | 69 | West Virginia | 63 |
| Iowa | 100 | New York | 813 | Wisconsin | 262 |
| Kansas | 102 | North Carolina | 484 | Wyoming | 19 |

*Source:* U.S. Bureau of Labor Statistics, *News Release,* March 3, 2010, USDL-10-0231, Table 1.

a. What are the mean and standard deviation unemployment numbers for all states?

b. Using information from (a), how many states fall more than 1 standard above the mean? How does this number compare with the number expected from the theoretical normal curve distribution? Can you suggest anything that these states have in common that might cause them to have higher levels of unemployment?

c. Create a histogram representing the unemployment figures for all 50 states and the District of Columbia. Does the distribution appear to be normal? Explain your answer.

6. Information on the occupational prestige scores for blacks and whites are presented in the following table.

| | Mean | Standard Deviation | N |
|---|---|---|---|
| Blacks | 40.75 | 13.63 | 190 |
| Whites | 44.70 | 14.03 | 1,094 |

*Source:* General Social Survey, 2008.

a. What percentage of whites should have occupational prestige scores above 60? How many whites in the sample should have occupational prestige scores above 60?

b. What percentage of blacks should have occupational prestige scores above 60? How many blacks should have occupational prestige scores above 60?

c. What proportion of whites has prestige scores between 30 and 70? How many whites have prestige scores between 30 and 70?

d. How many blacks in the sample should have an occupational prestige score between 30 and 60?

7. SAT scores are normed so that, in any year, the mean of the verbal or math test should be 500 and the standard deviation 100. Assuming this is true (it is only approximately true, both because of variation from year to year and because scores have decreased since the SAT tests were first developed), answer the following questions.

a. What percentage of students score above 625 on the math SAT in any given year?

b. What percentage of students score between 400 and 600 on the verbal SAT?

8. The Hate Crime Statistics Act of 1990 requires the Attorney General to collect national data about crimes that manifest evidence of prejudice based on race, religion, sexual orientation, or ethnicity, including the crimes of murder and non-negligent manslaughter, forcible rape, aggravated assault, simple assault, intimidation, arson, and destruction, damage, or vandalism of property. The Hate Crime Data collected in 2007 reveals, based on a randomly selected sample of 300 incidents, that the mean number of victims in a particular type of hate crime was 1.28, with a standard deviation of 0.82. Assuming that the number of victims was normally distributed, answer the following questions.

a. What proportion of crime incidents had more than 2 victims?

b. What is the probability that there was more than 1 victim in an incident?

c. What proportion of crime incidents had fewer than 4 victims?

9. The number of hours people work each week varies widely for many reasons. Using the 2008 GSS, you find that the mean number of hours worked last week was 42.26, with a standard deviation of 14.34 hr, based on a sample size of 894.

a. Assume that hours worked is approximately normally distributed in the sample. What is the probability that someone in the sample will work 60 hr or more in a week? How many people in the sample of 894 should have worked 60 hr or more?

b. What is the probability that someone will work 30 hr or fewer in a week (i.e., work part-time)? How many people does this represent in the sample?

10. What is the value of the mean score for any standard normal distribution? What is the value of the standard deviation for any standard normal distribution? Explain why this is true for any standard normal distribution.

11. You are asked to do a study of shelters for abused and battered women to determine the necessary capacity in your city to provide housing for most of these women. After recording data for a whole year, you find that the mean number of women in shelters each night is 250, with a standard deviation of 75. Fortunately, the distribution of the number of women in the shelters each night is normal, so you can answer the following questions posed by the city council.

a. If the city's shelters have a capacity of 350, will that be enough places for abused women on 95% of all nights? If not, what number of shelter openings will be needed?

b. The current capacity is only 220 openings, because some shelters have been closed. What is the percentage of nights that the number of abused women seeking shelter will exceed current capacity?

12. Based on the chapter discussion,

a. What are the properties of the normal distribution? Why is it called "normal"?

b. What is the meaning of a positive (+) $Z$ score? What is the meaning of a negative (−) $Z$ score?

# Chapter 6

# Sampling, Sampling Distributions, and Estimation

### Chapter Learning Objectives

❖ Understanding the aims of sampling and basic principles of probability
❖ Understanding and applying the concept of the sampling distribution
❖ Understanding the nature of the central limit theorem
❖ Learning procedures for estimating confidence intervals

Until now, we have ignored the question of who or what should be observed when we collect data or whether the conclusions based on our observations can be generalized to a larger group of observations. The truth is that we are rarely able to study or observe everyone or everything we are interested in. Although we have learned about various methods to analyze observations, remember that these observations represent only a tiny fraction of all the possible observations we might have chosen. Consider the following examples.

*Example 1:* The Muslim Student Association on your campus is interested in conducting a study of experiences with campus diversity. You have enough funds to survey only 300 students on your campus. Given that your campus is home to more than 20,000 students, what should you do?

*Example 2:* Local environmental activists want to assess recycling practices on your campus to develop a proposal to reduce unnecessary waste. Since the university serves more than 30,000 students, faculty, and staff, how should the activists proceed?

*Example 3:* The student union on your campus is trying to find out how it can better address the needs of commuter students and has commissioned you to conduct a needs assessment survey. You have been given enough money to survey about 500 students. Given that there are nearly 15,000 commuters on your campus, is this an impossible task?

What do these problems have in common? In all situations, the major problem is that there is too much information and not enough resources to collect and analyze all of it.

# ▣ AIMS OF SAMPLING[1]

Researchers in the social sciences almost never have enough time or money to collect information about the entire group that interests them. Known as the **population**, this group includes all the cases (individuals, objects, or groups) in which the researcher is interested. For example, in our first illustration, there are more than 20,000 students; the population in the second illustration consists of all 30,000 faculty, staff, and students; and in the third illustration, the population is all 15,000 commuter students.

---

*Population*   A group that includes all the cases (individuals, objects, or groups) in which the researcher is interested.

---

Fortunately, we can learn a lot about a population if we carefully select a subset of it. This subset is called a **sample**. Through the process of *sampling*—selecting a subset of observations from the population of interest—we attempt to generalize the characteristics of the larger group (population) based on what we learn from the smaller group (the sample). This is the basis of *inferential statistics*—making predictions or inferences about a population from observations based on a sample.

---

*Sample*   A subset of cases selected from a population.

---

The term **parameter**, associated with the population, refers to measures used to describe the distribution of the population we are interested in. For instance, the average commuting time for *all* the 15,000 students on your campus is a population parameter because it refers to a population characteristic. In previous chapters, we have learned the many ways of describing a distribution, such as a proportion, a mean, or a standard deviation. When used to describe the population distribution, these measures are referred to as parameters.

---

*Parameter*   A measure (e.g., mean or standard deviation) used to describe the population distribution.

---

We use the term **statistic** when referring to a corresponding characteristic calculated for the sample. For example, the average commuting time for a *sample* of commuter students is a sample statistic.

---

*Statistic*   A measure (e.g., mean or standard deviation) used to describe the sample distribution.

---

In this chapter as well as in Chapter 7, we discuss some of the principles involved in generalizing results from samples to the population. In our discussion, we will use different notations when referring to sample statistics and population parameters. Table 6.1 presents the sample notation and the corresponding population notation.

The distinctions between a sample and a population and between a parameter and a statistic are illustrated in Figure 6.1. We've included for illustration the population parameter of 0.60—the proportion of white respondents in the population. However, since we almost never have enough resources to

**Table 6.1**   Sample and Population Notations

| Measure Notation | Sample Notation | Population |
|---|---|---|
| Mean | $\bar{Y}$ | $\mu_Y$ |
| Proportion | $p$ | $\pi$ |
| Standard deviation | $S_Y$ | $\sigma_Y$ |
| Variance | $S_Y^2$ | $\sigma_Y^2$ |

**Figure 6.1**   The Proportion of White Respondents in a Population and in a Sample

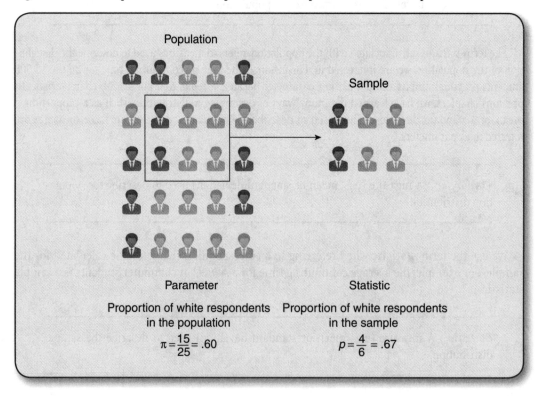

collect information about the population, it is rare that we know the value of a parameter. The goal of most research is to find the population parameter. Researchers usually select a sample from the population to obtain an estimate of the population parameter. Thus, the major objective of sampling theory and statistical inference is to provide estimates of unknown parameters from sample statistics that can be easily obtained and calculated.

> *It is important that you understand what the terms* population, sample, parameter, *and* statistic *mean. Use your own words so that the meaning makes sense to you. If you cannot clearly define these terms, review the preceding material. You will see these sample and population notations over and over again. If you memorize them, you will find it much easier to understand the formulas used in inferential statistics.*

✓ *Learning Check*

## ▣ PROBABILITY SAMPLING

Social researchers are usually more systematic in their effort to obtain samples that are representative of the population than we are when we gather information in our everyday life. One general approach, *probability sampling*, allows the researcher to use the principles of statistical inference to generalize from the sample to the population.

**Probability sampling** is a method that enables the researcher to specify for each case in the population the probability of its inclusion in the sample. The purpose of probability sampling is to select a sample that is as representative as possible of the population. The sample is selected in such a way as to allow the use of the principles of probability to evaluate the generalizations made from the sample to the population. A probability sample design enables the researcher to estimate the extent to which the findings based on one sample are likely to differ from what would be found by studying the entire population.

---

*Probability sampling*   A method of sampling that enables the researcher to specify for each case in the population the probability of its inclusion in the sample.

---

Although accurate estimates of sampling error can be made only from probability samples, social scientists often use nonprobability samples because they are more convenient and cheaper to collect. Nonprobability samples are useful under many circumstances for a variety of research purposes. Their main limitation is that they do not allow the use of the method of inferential statistics to generalize from the sample to the population. Because in this chapter we deal with only inferential statistics, we do not discuss nonprobability sampling. In the following section, we will learn about a sampling design that follows the principles of probability sampling: the simple random sample.[2]

## The Simple Random Sample

The *simple random sample* is the most basic probability sampling design, and it is incorporated into even more elaborate probability sampling designs. A **simple random sample** is a sample design chosen in such a way as to ensure that (1) every member of the population has an equal chance of being chosen and (2) every combination of $N$ members has an equal chance of being chosen.

---

*Simple random sample*   A sample designed in such a way as to ensure that (1) every member of the population has an equal chance of being chosen and (2) every combination of $N$ members has an equal chance of being chosen.

---

## ▣ THE CONCEPT OF THE SAMPLING DISTRIBUTION

Researchers usually select a sample from the population and use the principles of statistical inference to estimate the characteristics, or parameters, of that population based on the characteristics, or statistics, of the sample. In this section, we describe one of the most important concepts in statistical inference—*sampling distribution*. The sampling distribution helps estimate the likelihood of our sample statistics and, therefore, enables us to generalize from the sample to the population.

### The Population

To illustrate the concept of the sampling distribution, let's consider as our population the 20 individuals listed in Table 6.2.[3] Our variable, $Y$, is the income (in dollars) of these 20 individuals, and the parameter we are trying to estimate is the mean income.

We use the symbol $\mu_Y$ to represent the population mean; the Greek letter mu ($\mu$) stands for the mean, and the subscript $Y$ identifies the specific variable, income. Using Formula 3.1, we can calculate the population mean:

$$\mu_Y = \frac{\Sigma Y}{N} = \frac{Y_1 + Y_2 + Y_3 + Y_4 + Y_5 + \cdots + Y_{20}}{20}$$

$$= \frac{11,350 + 7,859 + 41,654 + 13,445 + 17,458 + \cdots + 25,671}{20}$$

$$= 22,766$$

Using Formula 4.2, we can also calculate the standard deviation for this population distribution. We use the Greek symbol sigma ($\sigma$) to represent the population's standard deviation and the subscript $Y$ to stand for our variable, income:

$$\sigma_Y = 14,687$$

Table 6.2   The Population: Personal Income (in dollars) for
            20 Individuals (hypothetical data)

| Individual | Income (Y) |
|---|---|
| Case 1 | 11,350 ($Y_1$) |
| Case 2 | 7,859 ($Y_2$) |
| Case 3 | 41,654 ($Y_3$) |
| Case 4 | 13,445 ($Y_4$) |
| Case 5 | 17,458 ($Y_5$) |
| Case 6 | 8,451 ($Y_6$) |
| Case 7 | 15,436 ($Y_7$) |
| Case 8 | 18,342 ($Y_8$) |
| Case 9 | 19,354 ($Y_9$) |
| Case 10 | 22,545 ($Y_{10}$) |
| Case 11 | 25,345 ($Y_{11}$) |
| Case 12 | 68,100 ($Y_{12}$) |
| Case 13 | 9,368 ($Y_{13}$) |
| Case 14 | 47,567 ($Y_{14}$) |
| Case 15 | 18,923 ($Y_{15}$) |
| Case 16 | 16,456 ($Y_{16}$) |
| Case 17 | 27,654 ($Y_{17}$) |
| Case 18 | 16,452 ($Y_{18}$) |
| Case 19 | 23,890 ($Y_{19}$) |
| Case 20 | 25,671 ($Y_{20}$) |
| Mean ($\mu_Y$) = 22,766 | Standard deviation ($\sigma_Y$) = 14,687 |

Of course, most of the time, we do not have access to the population. So instead, we draw one sample, compute the mean—the statistic—for that sample, and use it to estimate the population mean—the parameter.

## The Sample

Let's pretend that $\mu_Y$ is unknown and that we estimate its value by drawing a random sample of three individuals ($N = 3$) from the population of 20 individuals and calculate the mean income for that sample. The incomes included in that sample are as follows:

| | |
|---|---|
| Case  8 | 18,342 |
| Case 16 | 16,456 |
| Case 17 | 27,654 |

Now let's calculate the mean for that sample:

$$\bar{Y} = \frac{18{,}342 + 16{,}456 + 27{,}654}{3} = 20{,}817$$

Note that our sample mean, $(\bar{Y})$ = $20,817, differs from the actual population parameter, $22,766. This discrepancy is due to sampling error. **Sampling error** is the discrepancy between a sample estimate of a population parameter and the real population parameter. By comparing the sample statistic with the population parameter, we can determine the sampling error. The sampling error for our example is 1,949 (22,766 − 20,817 = 1,949).

---

*Sampling error*   The discrepancy between a sample estimate of a population parameter and the real population parameter.

---

## The Dilemma

Unfortunately, we rarely have information about the actual population parameter and few, if any, sample estimates correspond exactly to the actual population parameter. This presents a dilemma: If sample estimates vary and if most estimates result in some sort of sampling error, how much confidence can we place in the estimate? On what basis can we infer from the sample to the population?

## The Sampling Distribution

The answer to this dilemma is to use a device known as the sampling distribution. The **sampling distribution** is a theoretical probability distribution of all possible sample values for the statistic in which we are interested. If we were to draw all possible random samples of the same size from our population of interest, compute the statistic for each sample, and plot the frequency distribution for that statistic, we would obtain an approximation of the sampling distribution.[4]

---

*Sampling distribution*   The sampling distribution is a theoretical probability distribution of all possible sample values for the statistics in which we are interested.

---

## ▣ THE SAMPLING DISTRIBUTION OF THE MEAN

Sampling distributions are theoretical distributions, which means that they are never really observed. However, to help grasp the concept of the sampling distribution, let's illustrate how one could be generated from a limited number of samples.

## An Illustration

The **sampling distribution of the mean** is a theoretical distribution of sample means that would be obtained by drawing from the population all possible samples of the same size.

---

*Sampling distribution of the mean*   A theoretical probability distribution of sample means that would be obtained by drawing from the population all possible samples of the same size.

---

Let's go back to our example in which our population is made up of 20 individuals and their incomes. From that population (Table 6.2), we now randomly draw 50 possible samples of size 3, computing the mean income for each sample and replacing it before drawing another.

In our first sample of size 3, we draw three incomes: $8,451, $41,654, and $18,923. The mean income for this sample is

$$\bar{Y} = \frac{8,451 + 41,654 + 18,923}{3} = 23,009$$

We repeat this process 49 more times, each time computing the sample mean and restoring the sample to the original list. Table 6.3 lists the means of the first five and the 50th samples of $N = 3$ that were drawn from the population of 20 individuals. (Note that $\Sigma\bar{Y}$ refers to the sum of all the means computed for each of the samples and $M$ refers to the total number of samples that were drawn.)

The histogram for all 50 sample means ($M = 50$) is displayed in Figure 6.2. This distribution is an example of a sampling distribution of the mean. Note that in its structure, the sampling distribution

Table 6.3    Mean Income of 50 Samples of Size 3

| Sample | Mean ($\bar{Y}$) |
|---|---|
| First | 23,009 |
| Second | 19,079 |
| Third | 18,873 |
| Fourth | 26,885 |
| Fifth | 21,847 |
| ⋮ | ⋮ |
| Fiftieth | 26,645 |
| Total ($M$) = 50 | $\Sigma\bar{Y} = 1,237,482$ |

**Figure 6.2**   Sampling Distribution of Sample Means for Sample Size $N = 3$ Drawn From the Population of 20 Individuals' Incomes

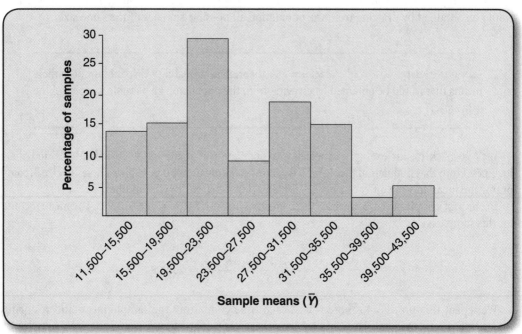

resembles a frequency distribution of raw scores, except that here each score is a sample mean, and the corresponding frequencies are the number of samples with that particular mean value.

Remember that the distribution depicted in Figure 6.2 is an empirical distribution, whereas the sampling distribution is a theoretical distribution.

## The Mean of the Sampling Distribution

Like the sample and population distributions, the sampling distribution can be described in terms of its mean and standard deviation. We use the symbol $\mu_{\bar{Y}}$ to represent the mean of the sampling distribution. The subscript $\bar{Y}$ indicates that the variable of this distribution is the mean. To obtain the mean of the sampling distribution, add all the individual sample means ($\Sigma \bar{Y} = 1,237,482$) and divide by the number of samples ($M = 50$). Thus, the mean of the sampling distribution of the mean is actually the mean of means:

$$\mu_{\bar{Y}} = \frac{\Sigma \bar{Y}}{M} = \frac{1,237,482}{50} = 24,750$$

## The Standard Error of the Mean

The standard deviation of the sampling distribution is also called the **standard error of the mean**. The standard error of the mean, $\sigma_{\bar{Y}}$, describes how much dispersion there is in the sampling distribution, or how much variability there is in the value of the mean from sample to sample:

$$\sigma_{\bar{Y}} = \frac{\sigma_Y}{\sqrt{N}}$$

This formula tells us that the standard error of the mean is equal to the standard deviation of the population $\sigma_Y$ divided by the square root of the sample size ($N$). For our example, because the population standard deviation is 14,687 and our sample size is 3, the standard error of the mean is

$$\sigma_{\bar{Y}} = \frac{14,687}{\sqrt{3}} = 8,480$$

---

*Standard error of the mean*　The standard deviation of the sampling distribution of the mean. It describes how much dispersion there is in the sampling distribution of the mean.

---

## ▣ THE CENTRAL LIMIT THEOREM

In Figures 6.3a and b, we compare the histograms for the population and sampling distributions of Tables 6.2 and 6.3. Figure 6.3a shows the population distribution of 20 incomes, with a mean $\mu_Y = 22{,}766$ and a standard deviation $\sigma_Y = 14{,}687$. Figure 6.3b shows the sampling distribution of the means from 50 samples of $N = 3$ with a mean $\mu_{\bar{Y}} = 24{,}749$ and a standard deviation (the standard error of the mean) $\sigma_{\bar{Y}} = 8{,}480$. These two figures illustrate some of the basic properties of sampling distributions in general and the sampling distribution of the mean in particular.

First, as can be seen from Figures 6.3a and b, the shapes of the two distributions differ considerably. Whereas the population distribution is skewed to the right, the sampling distribution of the mean is less skewed—that is, closer to symmetry and a normal distribution.

Second, whereas only a few of the sample means coincide exactly with the population mean, \$22,766, the sampling distribution centers on this value. The mean of the sampling distribution is a pretty good approximation of the population mean.

In the discussions that follow, we make frequent references to the mean and standard deviation of the two distributions. To distinguish among the different distributions, we use certain conventional symbols to refer to the means and standard deviations of the sample, the population, and the sampling distribution. Note that we use Greek letters to refer to both the sampling and the population distributions.

*The Population* We began with the *population distribution* of 20 individuals. This distribution actually exists. It is an empirical distribution that is usually unknown to us. We are interested in estimating the mean income for this population.

*The Sample* We drew a sample from that population. The *sample distribution* is an empirical distribution that is known to us and is used to help us estimate the mean of the population. We selected 50 samples of $N = 3$ and calculated the mean income. We usually use the sample mean $\bar{Y}$ as an estimate of the population mean $\mu_Y$.

**Figure 6.3**   Two Income Distributions

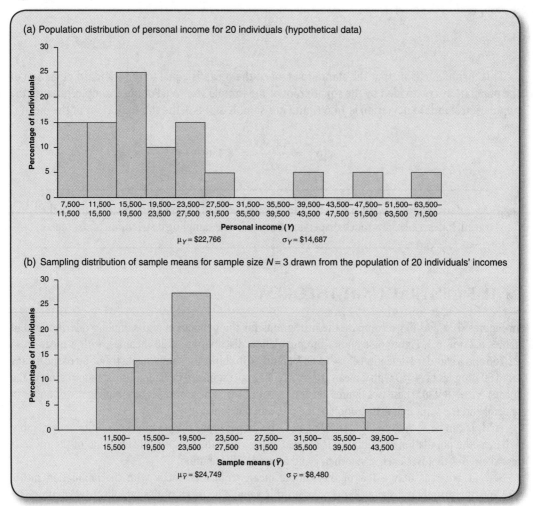

The Sampling Distribution of the Mean For illustration, we generated an approximation of the sampling distribution of the mean, consisting of 50 samples of $N = 3$. *The sampling distribution of the mean does not really exist. It is a theoretical distribution.*

Third, the variability of the sampling distribution is considerably smaller than the variability of the population distribution. Note that the standard deviation for the sampling distribution ($\sigma_{\bar{Y}} = 8{,}480$) is almost half that for the population ($\sigma_Y = 14{,}687$).

These properties of the sampling distribution of the mean are summarized more systematically in one of the most important statistical principles underlying statistical inference. It is called the **central limit theorem**, and it states that if all possible random samples of size $N$ are drawn from a population with a mean $\mu_Y$ and a standard deviation $\sigma_Y$, then as $N$ becomes larger,

| | *Mean* | *Standard Deviation* |
|---|:---:|:---:|
| Sample distribution | $\bar{Y}$ | $S_Y$ |
| Population distribution | $\mu_Y$ | $\sigma_Y$ |
| Sampling distribution of $\bar{Y}$ | $\mu_{\bar{Y}}$ | $\sigma_{\bar{Y}}$ |

the sampling distribution of sample means becomes approximately normal, with mean $\mu_{\bar{Y}}$ and standard deviation

$$\sigma_{\bar{Y}} = \frac{\sigma_Y}{\sqrt{N}}$$

---

***Central limit theorem*** If all possible random samples of size $N$ are drawn from a population with a mean $\mu_Y$ and a standard deviation $\sigma_Y$, then as $N$ becomes larger, the sampling distribution of sample means becomes approximately normal, with mean $\mu_{\bar{Y}}$ and standard deviation

$$\sigma_{\bar{Y}} = \frac{\sigma_Y}{\sqrt{N}}$$

---

The significance of the central limit theorem is that it tells us that with a *sufficient sample size* the sampling distribution of the mean will be normal regardless of the shape of the population distribution. Therefore, even when the population distribution is skewed, we can still assume that the sampling distribution of the mean is normal, given random samples of large enough size. Furthermore, the central limit theorem also assures us that (1) as the sample size gets larger, the mean of the sampling distribution becomes equal to the population mean and (2) as the sample size gets larger, the standard error of the mean (the standard deviation of the sampling distribution of the mean) decreases in size. The standard error of the mean tells how much variability in the sample estimates there is from sample to sample. The smaller the standard error of the mean, the closer (on average) the sample means will be to the population mean. Thus, the larger the sample, the more closely the sample statistic clusters around the population parameter.

*Make sure you understand the difference between the number of samples that can be drawn from a population and the sample size. Whereas the number of samples is infinite in theory, the sample size is under the control of the investigator.*

✓ *Learning Check*

## The Size of the Sample

Although there is no hard-and-fast rule, a general rule of thumb is that when $N$ is 50 or more, the sampling distribution of the mean will be approximately normal regardless of the shape of the distribution. However, we can assume that the sampling distribution will be normal even with samples as small as 30 if we know that the population distribution approximates normality.

✓ *Learning*
*Check*

*What is a normal population distribution? If you can't answer this question, go back to Chapter 5. You must understand the concept of a normal distribution before you can understand the techniques involved in inferential statistics.*

## The Significance of the Sampling Distribution and the Central Limit Theorem

To estimate the mean income of a population of 20 individuals, we drew a sample of three cases and calculated the mean income for that sample. Our sample mean, $\bar{Y} = 20,817$, differs from the actual population parameter, $\mu_Y = 22,766$. When we selected different samples, we found each time that the sample mean differed from the population mean. These discrepancies are due to sampling errors. Had we taken a number of additional samples, we probably would have found that the mean was different each time because every sample differs slightly. Few, if any, sample means would correspond exactly to the actual population mean. Usually we have only one sample statistic as our best estimate of the population parameter.

If sample estimates vary and if most result in some sort of sampling error, how much confidence can we place in the estimate? On what basis can we infer from the sample to the population?

The solution lies in the sampling distribution and its properties. Because the sampling distribution is a theoretical distribution that includes all possible sample outcomes, we can compare our sample outcome with it and estimate the likelihood of its occurrence.

Our knowledge is based on what the central limit theorem tells us about the properties of the sampling distribution of the mean. If our sample size is large enough (at least 50 cases), most sample means will be quite close to the true population mean. It is highly unlikely that our sample mean would deviate much from the actual population mean.

In Chapter 5, we saw that in all normal curves, a constant proportion of the area under the curve lies between the mean and any given distance from the mean when measured in standard deviation units, or $Z$ scores. We can find this proportion in the standard normal table (Appendix A).

Knowing that the sampling distribution of the means is approximately normal, with a mean $\mu_{\bar{Y}}$ and a standard deviation $\sigma_Y / \sqrt{N}$ (the standard error of the mean), we can use Appendix A to determine the probability that a sample mean will fall within a certain distance—measured in standard deviation units, or $Z$ scores—of $\mu_{\bar{Y}}$ or $\mu_Y$. For example, we can expect approximately 68% of all sample means to fall within ±1 standard error ($\sigma_{\bar{Y}} = \sigma_Y / \sqrt{N}$, or the standard deviation of the sampling distribution of the mean) of $\mu_{\bar{Y}}$ or $\mu_Y$. This information helps us evaluate the accuracy of our sample estimates.

✓ *Learning Check*

*Suppose a population distribution has a mean $\mu_Y = 150$ and a standard deviation $\sigma_Y = 30$, and you draw a simple random sample of $N = 100$ cases. What is the probability that the mean is between 147 and 153? What is the probability that the sample mean exceeds 153? Would you be surprised to find a mean score of 159? Why? (Hint: To answer these questions, you need to apply what you learned in Chapter 5 about Z scores and areas under the normal curve [Appendix A].) Remember, to translate a raw score into a Z score we used this formula:*

$$Z = \frac{Y - \bar{Y}}{S_Y}$$

*However, because here we are dealing with a sampling distribution, replace Y with the sample mean $\bar{Y}$, $\bar{Y}$ with the sampling distribution's mean $\mu_{\bar{Y}}$, and $\sigma_Y$ with the standard error of the mean $\sigma_Y / \sqrt{N}$.*

$$Z = \frac{\bar{Y} - \mu_{\bar{Y}}}{\sigma_Y / \sqrt{N}}$$

## ▣ ESTIMATION

In this section, we discuss the procedures involved in estimating population means and proportions. These procedures are based on the principles of sampling and statistical inference discussed in previous sections. Knowledge about the sampling distribution allows us to estimate population means and proportions from sample outcomes and to assess the accuracy of these estimates.

*Example:* Every other year, the National Opinion Research Center (NORC) conducts the General Social Survey (GSS) on a representative sample of about 1,500 respondents. The GSS, from which many of the examples in this book are selected, is designed to provide social science researchers with a readily accessible database of socially relevant attitudes, behaviors, and attributes of a cross section of the U.S. adult population. For example, in analyzing the responses to the 2008 GSS, researchers found that the average respondent's education was about 13.43 years. This average probably differs from the average of the population from which the GSS sample was drawn. However, we can establish that in most cases the sample mean (in this case, 13.43 years) is fairly close to the actual true average in the population.

The actual average level of education in the United States is a population parameter. The average level of education in the United States as calculated from the GSS is a sample estimate of a population parameter. Sample estimates are used to calculate population parameters; the mean number of years of education of 13.43 calculated from the GSS sample can be used to estimate the mean education of all adults in the United States.

This is an example of estimation. **Estimation** is a process whereby we select a random sample from a population and use a sample statistic to estimate a population parameter. We can use sample proportions as estimates of population proportions, sample means as estimates of population means, or sample variances as estimates of population variances.

*Estimation*    A process whereby we select a random sample from a population and use a sample statistic to estimate a population parameter.

## Reasons for Estimation

The goal of most research is to find the population parameter. Yet we hardly ever have enough resources to collect information about the entire population. We rarely know the value of the population parameter. On the other hand, we can learn a lot about a population by randomly selecting a sample from that population and obtaining an estimate of the population parameter. The major objective of sampling theory and statistical inference is to provide estimates of unknown population parameters from sample statistics. Refer to Figure 6.4.

**Figure 6.4**    Estimation as a Type of Inference

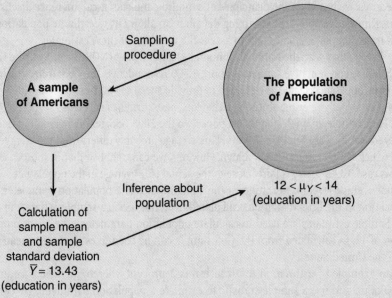

## Point and Interval Estimation

Estimates of population characteristics can be divided into two types: point estimates and interval estimates. **Point estimates** are sample statistics used to estimate the exact value of a population parameter. When we report the average level of education of the population of adult Americans to be exactly 13.43 years, we would be using a point estimate.

---

*Point estimate*   A sample statistic used to estimate the exact value of a population parameter.

---

The problem with point estimates is that sample estimates usually vary, and most result in some sort of sampling error. As a result, when we use a sample statistic to estimate the exact value of a population parameter, we never really know how accurate it is.

One method of increasing accuracy is to use an interval estimate rather than a point estimate. In interval estimation, we identify a range of values within which the population parameter may fall. This range of values is called a **confidence interval (CI)**. Instead of using a single value, 13.43 years, as an estimate of the mean education of adult Americans, we could say that the population mean is somewhere between 12 and 14 years.

---

*Confidence interval (CI)*   A range of values defined by the confidence level within which the population parameter is estimated to fall. Sometimes confidence intervals are referred to as margin of error.

---

When we use confidence intervals to estimate population parameters, such as mean educational levels, we can also evaluate the accuracy of this estimate by assessing the likelihood that any given interval will contain the mean. This likelihood, expressed as a percentage or a probability, is called a **confidence level**. Confidence intervals are defined in terms of confidence levels. Thus, by selecting a 95% confidence level, we are saying that there is a .95 probability—or 95 chances out of 100—that a specified interval will contain the population mean. Confidence intervals can be constructed for any level of confidence, but the most common ones are the 90%, 95%, and 99% levels. You should also know that confidence intervals are sometimes referred to in terms of **margin of error**. In short, margin of error is simply the radius of a confidence interval. If we select a 95% confidence level, we would have a margin of error of ±5 percentage points.

---

*Confidence level*   The likelihood, expressed as a percentage or a probability, that a specified interval will contain the population parameter.

*Margin of error*   The radius of a confidence interval.

---

*What is the difference between a point estimate and a confidence interval?*

✓ *Learning* *Check*

# ▣ PROCEDURES FOR ESTIMATING CONFIDENCE INTERVALS FOR MEANS

To illustrate the procedure for establishing confidence intervals for means, we'll reintroduce one of the research examples mentioned at the beginning of this chapter (in example 3)—assessing the needs of commuting students on our campus.

Recall that we have been given enough money to survey a random sample of 500 students. One of our tasks is to estimate the average commuting time of all 15,000 commuters on our campus—the population parameter. To obtain this estimate, we calculate the average commuting time for the sample. Suppose the sample average is $\bar{Y} = 7.5$ hr/week, and we want to use it as an estimate of the true average commuting time for the entire population of commuting students.

Because it is based on a sample, this estimate is subject to sampling error. We do not know how close it is to the true population mean. However, based on what the central limit theorem tells us about the properties of the sampling distribution of the mean, we know that with a large enough sample size, most sample means will tend to be close to the true population mean. Therefore, it is unlikely that our sample mean, $\bar{Y} = 7.5$ hr/week, deviates much from the true population mean.

We know that the sampling distribution of the mean is approximately normal with a mean $\mu_{\bar{Y}}$ equal to the population mean $\mu_Y$ and a standard error $\sigma_{\bar{Y}}$ (standard deviation of the sampling distribution) as follows:

$$\sigma_{\bar{Y}} = \frac{\sigma_Y}{\sqrt{N}} \qquad (6.1)$$

This information allows us to use the normal distribution to determine the probability that a sample mean will fall within a certain distance—measured in standard deviation (standard error) units or $Z$ scores—of $\mu_Y$ or $\mu_{\bar{Y}}$. We can make the following assumptions:

- A total of 68% of all random sample means will fall within ±1 standard error of the true population mean.
- A total of 95% of all random sample means will fall within ±1.96 standard errors of the true population mean.
- A total of 99% of all random sample means will fall within ±2.58 standard errors of the true population mean.

On the basis of these assumptions and the value of the standard error, we can establish a range of values—a confidence interval—that is likely to contain the actual population mean. We can also evaluate the accuracy of this estimate by assessing the likelihood that this range of values will actually contain the population mean.

The general formula for constructing a confidence interval (CI) for any level is

$$CI = \bar{Y} \pm Z(\sigma_{\bar{Y}}) \qquad (6.2)$$

Note that to calculate a confidence interval, we take the sample mean and add to or subtract from it the product of a $Z$ value and the standard error.

The $Z$ score we choose depends on the desired confidence level. For example, to obtain a 95% confidence interval we would choose a $Z$ of 1.96 because we know (from Appendix A) that 95% of the area under the curve lies between ±1.96. Similarly, for a 99% confidence level, we would choose a $Z$ of 2.58. The relationship between the confidence level and $Z$ is illustrated in Figure 6.5 for the 95% and 99% confidence levels.

> *To understand the relationship between the confidence level and Z, review the material in Chapter 5. What would be the appropriate Z value for a 98% confidence interval?*  ✓ *Learning Check*

**Figure 6.5**   Relationship Between Confidence Level and $Z$ for 95% and 99% Confidence Intervals

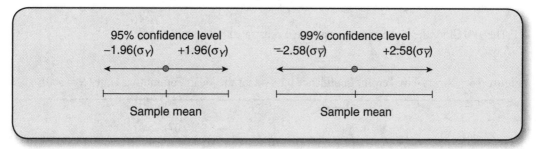

*Source:* From David Freedman, Robert Pisani, Roger Purves, and Ani Akhikari, *Statistics,* 2nd ed. (New York: W. W. Norton, 1991). Copyright ©1991 by W. W. Norton & Company, Inc. Used by permission of W. W. Norton and Company, Inc.

## Determining the Confidence Interval

To determine the confidence interval for means, follow these steps:

1. Calculate the standard error of the mean.

2. Decide on the level of confidence, and find the corresponding $Z$ value.

3. Calculate the confidence interval.

4. Interpret the results.

Let's return to the problem of estimating the mean commuting time of the population of students on our campus. How would you find the 95% confidence interval?

### Calculating the Standard Error of the Mean

Let's suppose that the standard deviation for our population of commuters is $\sigma_Y = 1.5$. We calculate the standard error for the sampling distribution of the mean:

$$\sigma_{\bar{Y}} = \frac{\sigma_Y}{\sqrt{N}} = \frac{1.5}{\sqrt{500}} = 0.07$$

### Deciding on the Level of Confidence
### and Finding the Corresponding Z Value

We decide on a 95% confidence level. The $Z$ value corresponding to a 95% confidence level is 1.96.

### Calculating the Confidence Interval

The confidence interval is calculated by adding and subtracting from the observed sample mean the product of the standard error and $Z$:

$$95\% \text{ CI} = 7.5 \pm 1.96(0.07)$$
$$= 7.5 \pm 0.14$$
$$= 7.36 \text{ to } 7.64$$

The 95% CI for the mean commuting time is illustrated in Figure 6.6.

**Figure 6.6**   Ninety-Five Percent Confidence Interval for the Mean Commuting Time ($N = 500$)

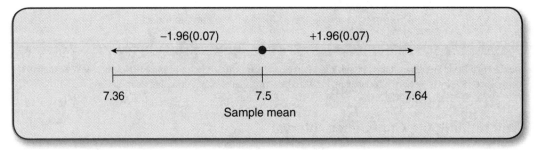

### Interpreting the Results

We can be 95% confident that the actual mean commuting time—the true population mean—is not less than 7.36 hr and not greater than 7.64 hr. In other words, if we collected a large number of samples ($N = 500$) from the population of commuting students, 95 times out of 100, the true population mean would be included within our computed interval. With a 95% confidence level, there is a 5% risk that we are wrong. Five times out of 100, the true population mean will not be included in the specified interval.

Remember that we can never be sure whether the population mean is actually contained within the confidence interval. Once the sample is selected and the confidence interval defined, the confidence interval either does or does not contain the population mean—but we will never be sure.

*✓ Learning Check*

*What is the 90% confidence interval for the mean commuting time? (Hint: First, find the Z value associated with a 90% confidence level.)*

To further illustrate the concept of confidence intervals, let's suppose that we draw 10 different samples ($N = 500$) from the population of commuting students. For each sample mean, we construct a 95% confidence interval. Figure 6.7 displays these confidence intervals. Each horizontal line represents a 95% confidence interval constructed around a sample mean (marked with a circle).

The vertical line represents the population mean. Note that the horizontal lines that intersect the vertical line are the intervals that contain the true population mean. Only 1 out of the 10 confidence intervals does not intersect the vertical line, meaning it does not contain the population mean. Drawing all possible samples, 95% of the intervals will include the true population mean and 5% will not.

**Figure 6.7**  Ninety-Five Percent Confidence Intervals for 10 Samples

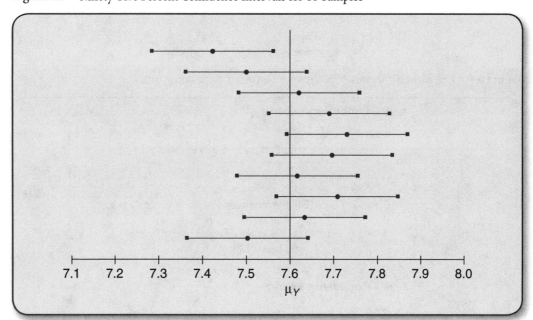

## Reducing Risk

One way to reduce the risk of being incorrect is by increasing the level of confidence. For instance, we can increase our confidence level from 95% to 99%. The 99% confidence interval for our commuting example is

$$99\% \ CI = 7.5 \pm 2.58(0.07)$$
$$= 7.5 \pm 0.18$$
$$= 7.32 \text{ to } 7.68$$

When using the 99% confidence interval, there is only a 1% risk that we are wrong and the specified interval does not contain the true population mean. We can be almost certain that the true population mean is included in the interval ranging from 7.32 to 7.68 hr/week. Note that by increasing the confidence level, we have also increased the width of the confidence interval from 0.28 (7.36–7.64) to 0.36 hr (7.32–7.68), thereby making our estimate less precise.

You can see in Figure 6.8 that there is a trade-off between achieving greater confidence in an estimate and the precision of that estimate. Although using a higher level of confidence (e.g., 99%) increases our confidence that the true population mean is included in our confidence interval, the estimate becomes less precise as the width of the interval increases.

**Table 6.4** Confidence Levels and Corresponding $Z$ Values

| Confidence Level | Z Value |
| --- | --- |
| 90% | 1.65 |
| 95% | 1.96 |
| 99% | 2.58 |

**Figure 6.8** Confidence Intervals, 95% Versus 99% (mean commuting time)

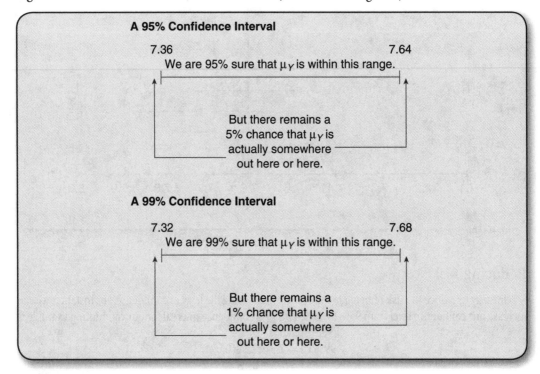

## Estimating Sigma

To calculate confidence intervals, we need to know the standard error of the sampling distribution, $\sigma_{\bar{Y}}$. The standard error is a function of the population standard deviation and the sample size:

$$\sigma_{\bar{Y}} = \frac{\sigma_Y}{\sqrt{N}}$$

In our commuting example, we have been using a hypothetical value, $\sigma_Y = 1.5$, for the population standard deviation. Typically, both the mean ($\mu_Y$) and the standard deviation ($\sigma_Y$) of the population are unknown to us. When $N \geq 50$, however, the sample standard deviation, $S_Y$, is a good estimate of $\sigma_Y$. The standard error is then calculated as follows:

$$S_{\bar{Y}} = \frac{S_Y}{\sqrt{N}} \qquad (6.3)$$

As an example, we'll estimate the mean hours per day that Americans spend watching television based on the 2008 GSS survey. The mean hours per day spent watching television for a sample of $N = 562$ is $\bar{Y} = 2.98$ hr, and the standard deviation $S_Y = 2.66$ hr. Let's determine the 95% confidence interval for these data.

### Calculating the Estimated Standard Error of the Mean

The estimated standard error for the sampling distribution of the mean is

$$S_{\bar{Y}} = \frac{S_Y}{\sqrt{N}} = \frac{2.66}{\sqrt{562}} = 0.11$$

### Deciding on the Level of Confidence and Finding the Corresponding Z Value

We decide on a 95% confidence level. The Z value corresponding to a 95% confidence level is 1.96.

### Calculating the Confidence Interval

The confidence interval is calculated by adding to and subtracting from the observed sample mean the product of the standard error and $Z$:

$$95\% \text{ CI} = 2.98 \pm 1.96(0.11)$$
$$= 2.98 \pm 0.22$$
$$= 2.76 \text{ to } 3.20$$

### Interpreting the Results

We can be 95% confident that the actual mean hours spent watching television by Americans from which the GSS sample was taken is not less than 2.76 hr and not greater than 3.20 hr. In other words, if we drew a large number of samples ($N = 562$) from this population, then 95 times out of 100, the true population mean would be included within our computed interval.

## Sample Size and Confidence Intervals

Researchers can increase the precision of their estimate by increasing the sample size. Earlier, we learned that larger samples result in smaller standard errors and, therefore, sampling distributions are more clustered around the population mean. A more tightly clustered sampling distribution means that our confidence intervals will be narrower and more precise. To illustrate the relationship between sample size and the standard error, and thus the confidence interval, let's calculate the 95% confidence interval for our GSS data with (1) a sample of $N = 195$ and (2) a sample of $N = 1,987$.

With a sample size $N = 195$, the estimated standard error for the sampling distribution is

$$S_{\bar{Y}} = \frac{S_Y}{\sqrt{N}} = \frac{2.66}{\sqrt{195}} = 0.19$$

and the 95% confidence interval is

$$95\% \text{ CI} = 2.98 \pm 1.96(0.19)$$
$$= 2.98 \pm 0.37$$
$$= 2.61 \text{ to } 3.35$$

With a sample size $N = 1,987$, the estimated standard error for the sampling distribution is

$$S_{\bar{Y}} = \frac{S_Y}{\sqrt{N}} = \frac{2.66}{\sqrt{1,987}} = 0.06$$

and the 95% confidence interval is

$$95\% \text{ CI} = 2.98 \pm 1.96(0.06)$$
$$= 2.98 \pm 0.12$$
$$= 2.86 \text{ to } 3.10$$

In Table 6.5, we summarize the 95% confidence intervals for the mean number of hours watching television for these three sample sizes: $N = 195$, $N = 562$, and $N = 1,987$.

**Table 6.5**   Ninety-Five Percent Confidence Interval and Width for Mean Number of Hours Watching Television for Three Different Sample Sizes

| Sample Size (N) | Confidence Interval | Interval Width | $S_Y$ | $S_{\bar{Y}}$ |
|---|---|---|---|---|
| 195 | 2.61–3.35 | 0.74 | 2.66 | 0.19 |
| 562 | 2.76–3.20 | 0.44 | 2.66 | 0.11 |
| 1,987 | 2.86–3.10 | 0.24 | 2.66 | 0.06 |

Note that there is an inverse relationship between sample size and the width of the confidence interval. The increase in sample size is linked with increased precision of the confidence interval. The 95% confidence interval for the GSS sample of 195 cases is 0.74 hr. But the interval widths decrease to 0.44 and 0.24 hr, respectively, as the sample sizes increase to $N = 562$ and then to $N = 1,987$. We had to nearly quadruple the size of the sample (from 562 to 1,987) to reduce the confidence interval by about one half (from 0.44 to 0.24 hr). Researchers have to balance increased precision with the additional costs for larger samples.

# ▣ CONFIDENCE INTERVALS FOR PROPORTIONS

Confidence intervals can also be computed for sample proportions or percentages to estimate population proportions or percentages. The procedures for estimating proportions and percentages are identical. We can obtain a confidence interval for a percentage by calculating the confidence interval for a proportion and then multiplying the result by 100.

Earlier, we saw that the sampling distribution of the means underlies the process of estimating population means from sample means. Similarly, the *sampling distribution of proportions* underlies the estimation of population proportions from sample proportions. Based on the central limit theorem, we know that with sufficient sample size the sampling distribution of proportions is approximately normal, with mean $\mu_p$ equal to the population proportion $\pi$ and with a standard error of proportions (the standard deviation of the sampling distribution of proportions) equal to

$$\sigma_p = \sqrt{\frac{(\pi)(1-\pi)}{N}} \tag{6.4}$$

where

$\sigma_p$ = the standard error of proportions

$\pi$ = the population proportion

$N$ = the population size

However, since the population proportion, $\pi$, is unknown to us (that is what we are trying to estimate), we can use the sample proportion, $p$, as an estimate of $\pi$. The estimated standard error then becomes

$$S_p = \sqrt{\frac{(p)(1-p)}{N}} \tag{6.5}$$

where

$S_p$ = the estimated standard error of proportions

$p$ = the sample proportion

$N$ = the sample size

As an example, let's calculate the estimated standard error for the survey by Gallup. Based on a random sample of 1,013 adults, the percentage who favors the new immigration law introduced by the state of Arizona in 2010 was estimated to be 51%. Based on Formula 6.5, with $p = 0.51$, $1 - p = (1 - 0.51) = 0.49$, and $N = 1,013$, the standard error is $S_p = \sqrt{(0.51)(1-0.51)/1,013} = 0.016$. We will have to consider two factors to meet the assumption of normality with the sampling distribution of proportions: (1) the sample size $N$ and (2) the sample proportions $p$ and $1 - p$. When $p$ and $1 - p$ are about 0.50, a sample size of at least 50 is sufficient. But when $p > 0.50$ (or $1 - p < 0.50$), a larger sample is required to meet the assumption of normality. Usually, a sample of 100 or more is adequate for any single estimate of a population proportion.

## Procedures for Estimating Proportions

Because the sampling distribution of proportions is approximately normal, we can use the normal distribution to establish confidence intervals for proportions in the same manner that we used the normal distribution to establish confidence intervals or means.

The general formula for constructing confidence intervals for proportions for any level of confidence is

$$CI = p \pm Z(S_p) \tag{6.6}$$

where

$CI$ = the confidence interval

$p$  = the observed sample proportion

$Z$  = the $Z$ corresponding to the confidence level

$S_p$ = the estimated standard error of proportions

To determine the confidence interval for a proportion, we follow the same steps that were used to find confidence intervals for means.

To illustrate these steps, we use the results of the Gallup survey on the percentage of Americans who favor the new immigration law introduced by the state of Arizona in 2010.

### Calculating the Estimated Standard Error of the Proportion

The standard error of the proportion 0.51 (51%) with a sample $N = 1,013$ is 0.016.

### Deciding on the Desired Level of Confidence and Finding the Corresponding Z Value

We choose the 95% confidence level. The $Z$ corresponding to a 95% confidence level is 1.96.

### Calculating the Confidence Interval

We calculate the confidence interval by adding to and subtracting from the observed sample proportion the product of the standard error and $Z$:

$$95\% \ CI = 0.51 \pm 1.96(0.016)$$
$$= 0.51 \pm 0.03$$
$$= 0.48 \text{ to } 0.54$$

### Interpreting the Results

We are 95% confident that the true population proportion is somewhere between 0.48 and 0.54. In other words, if we drew a large number of samples from the population of adults, then 95 times out of 100, the confidence interval we obtained would contain the true population proportion. We can also express this result in percentages and say that we are 95% confident that the true population percentage of Americans who favor the new immigration law introduced by the state of Arizona in 2010 is included somewhere within our computed interval of 48% to 54%.

Note that with a 95% confidence level, there is a 5% risk that we are wrong. If we continued to draw large samples from this population, in 5 out of 100 samples the true population proportion would not be included in the specified interval.

We can decrease our risk by increasing the confidence level from 95% to 99%.

$$99\% \text{ CI} = 0.51 \pm 2.58(0.016)$$
$$= 0.51 \pm 0.04$$
$$= 0.47 \text{ to } 0.55$$

When using the 99% confidence interval, we can be almost certain (99 times out of 100) that the true population proportion is included in the interval ranging from 0.47 (47%) to 0.55 (55%). However, as we saw earlier, there is a trade-off between achieving greater confidence in making an estimate and the precision of that estimate.[5,6]

## ▣ STATISTICS IN PRACTICE: HEALTH CARE REFORM

Poll or survey results may be limited to a single estimate of a parameter. For instance, political pollsters could estimate the percentage of Americans who approve of the new health care law. Often, however, separate estimates are reported for subgroups within the overall population of interest. In a report released on April 8–11, 2010, Gallup compared Americans' reaction to the passage of Healthcare Reform Bill by Party Identification. They were interested in exploring whether or not there were differences across groups with different political affiliations.[7]

When estimates are reported for subgroups, the confidence intervals are likely to vary from subgroup to subgroup. To illustrate this, let's calculate the 95% confidence intervals for the proportions of Democrats and Republicans who think it is a good thing that Congress passed the new health care legislation. Out of 470 Democrats in the sample, 0.81 (or 81%) indicated support for the legislation. In contrast, of the 459 Republicans surveyed, only 0.10 (or 10%) expressed support for the new health care legislation.

### Calculating the Estimated Standard Error of the Proportion

The estimated standard error for the proportion of Democrats is

$$S_p = \sqrt{\frac{(0.81)(1 - 0.81)}{470}} = 0.02$$

The estimated standard error for the proportion of Republicans is

$$S_p = \sqrt{\frac{(0.10)(1 - 0.10)}{459}} = 0.01$$

## Calculating the Confidence Interval

For Democrats,

$$95\% \, CI = 0.81 \pm 1.96(0.02)$$
$$= 0.81 \pm 0.04$$
$$= 0.77 \text{ to } 0.85$$

and for Republicans,

$$95\% \, CI = 0.10 \pm 1.96(0.01)$$
$$= 0.10 \pm 0.02$$
$$= 0.08 \text{ to } 0.12$$

The 95% confidence interval for the proportion of Democrats and Republicans surveyed who supported the new health care law is illustrated in Figure 6.9.

**Figure 6.9**   Ninety-Five Percent Confidence Interval for the Proportion of Democrats and Republicans Who Supported the New Health Care Law

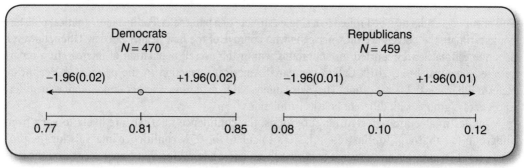

*Source:* Data from Frank Newport, "Americans' Views on Healthcare Law Remain Stable," *Gallup*, April 15, 2010.

## Interpreting the Results

We are 95% confident that the true population proportion supporting the new health care law in April 2010 was between 0.77 and 0.85 (or between 77% and 85%) for Democrats, and somewhere between 0.08 and 0.12 (or between 8% and 12%) for Republicans. Based on the sample, it is clear that among Americans there were partisan differences in support for the new health care law. Democrats were much more likely than Republicans to support the health care reform.

## ▣ STATISTICS IN PRACTICE: THE MARGIN OF ERROR

The most common application of estimation using confidence intervals (also called the margin of error) is demonstrated in opinion and election polls. For example, a 2009 *New York Times* survey[8] of 708 unemployed adults reported that 69% are more stressed than usual. Also reported was the poll's margin of error of plus or minus 4 percentage points. The margin of error in a poll tells us how well a randomly selected sample represents the population from which it was selected. The *New York Times* poll results indicate that we can be 95% confident that the true percentage of unemployed who feel stressed is somewhere between 65% (69% − 4%) and 73% (69% + 4%).

A margin of error of 4% means that with a sample of 708 individuals, we can be 95% confident that the true value of 69% is within ±4 percentage points of what it would be if the entire unemployed adult population had been polled.

It should be emphasized that the margin of error measures only the extent of random sampling errors in samples that were randomly selected. Unfortunately, there are several other possible sources of errors in all polls or surveys. They include refusals to be interviewed (nonresponse), question wording, or interviewer bias. Such errors may arise even when the poll's sampling design follows sound methodological principles. Such systematic error may affect the accuracy of polls, yet it is difficult or impossible to quantify the errors that may result from these factors.

A classic example of the failure to predict election results due to systematic and unquantifiable error is the 2008 New Hampshire presidential primaries, when then Senator Barack Obama ran against Senator Hillary Clinton, in a race where all the published polls predicted victory for Obama, forecasting that he would have a 10- to 15-point lead over Clinton. When the votes were counted, contrary to the polls' outcome, Clinton ended up beating Obama 39.4% to 36.8%.

According to several polling experts,[9] the main problem in the race was the division among voters along socioeconomic lines. Clinton beat Obama by 12 points (47% to 35%) among those with family income below $50,000. In contrast, Obama beat Clinton by five points (40% to 35%) among those earning more than $50,000. There was an education gap too. College graduates voted for Obama 39% to 34%. Clinton won among those who had never attended college, 43% to 35%. Another problem was the long-standing pattern in preelection polls for poor white voters to overstate their support for black candidates.[10]

## MAIN POINTS

- Through the process of sampling, researchers attempt to generalize the characteristics of a large group (the population) from a subset (sample) selected from that group. The term *parameter*, associated with the population, refers to the information we are interested in finding out. *Statistic* refers to a corresponding calculated sample statistic.

- A probability sample design allows us to estimate the extent to which the findings based on one sample are likely to differ from what we would find by studying the entire population.

- A simple random sample is chosen in such a way as to ensure that every member of the population and every combination of *N* members have an equal chance of being chosen.

- The sampling distribution is a theoretical probability distribution of all possible sample values for the statistic in which we are interested. The sampling distribution of the mean is a frequency distribution of all possible sample means of the same size that can be drawn from the population of interest.

- According to the central limit theorem, if all possible random samples of size *N* are drawn from a population with a mean $\mu_Y$ and a standard deviation $\sigma_Y$, then as *N* becomes larger, the sampling distribution of sample means becomes approximately normal, with mean $\mu_{\bar{Y}}$ and standard deviation $\sigma_Y/\sqrt{N}$.

- The central limit theorem tells us that with sufficient sample size, the sampling distribution of the mean will be normal regardless of the shape of the population distribution. Therefore, even when the population distribution is skewed, we can still assume that the sampling distribution of the mean is normal, given a large enough randomly selected sample size.

• The goal of most research is to find population parameters. The major objective of sampling theory and statistical inference is to provide estimates of unknown parameters from sample statistics.

• Researchers make point estimates and interval estimates. Point estimates are sample statistics used to estimate the exact value of a population parameter. Interval estimates are ranges of values within which the population parameter may fall.

• Confidence intervals can be used to estimate population parameters such as means or proportions. Their accuracy is defined with the confidence level. The most common confidence levels are 90%, 95%, and 99%.

• To establish a confidence interval for a mean or a proportion, add or subtract from the mean or the proportion the product of the standard error and the Z value corresponding to the confidence level.

## KEY TERMS

central limit theorem
confidence interval
    (interval estimate)
confidence level
estimation
margin of error

parameter
point estimate
population
probability sample
sample
sampling distribution

sampling distribution
    of the mean
sampling error
simple random sample
standard error of the mean
statistic

## ON YOUR OWN

Log on to the web-based student study site at **www.sagepub.com/ssdsessentials** for additional study questions, web quizzes, web resources, flashcards, codebooks and datasets, web exercises, appendices, and links to social science journal articles reflecting the statistics used in this chapter.

## CHAPTER EXERCISES

1. Can the standard error of a variable ever be larger than, or even equal in size to, the standard deviation for the same variable? Justify your answer by means of both a formula and a discussion of the relationship between these two concepts.

2. When taking a random sample from a very large population, how does the standard error of the mean change when
   a. the sample size is increased from 100 to 1,600?
   b. the sample size is decreased from 300 to 150?
   c. the sample size is multiplied by 4?

Exercises

3. The mean family income in Wisconsin, according to the 2008 data, is about $64,750, with a standard deviation (for the population) of $59,750.
   a. Imagine that you are taking a subsample of 200 state residents. What is the probability that your sample mean is between $61,000 and $64,750?
   b. For this same sample size, what is the probability that the sample mean exceeds $75,000?

4. A small population of $N = 10$ has values of 4, 7, 2, 11, 5, 3, 4, 6, 10, and 1.
   a. Calculate the mean and standard deviation for the population.
   b. Take 10 simple random samples of size 3, and calculate the mean for each.
   c. Calculate the mean and standard deviation of all these sample means. How closely does the mean of all sample means match the population mean? How is the standard deviation of the means related to the standard deviation for the population?

5. The following data summarize the 2007 external debt (in millions of dollars) for seven countries.

| Country | Debt (in millions of dollars) |
|---|---|
| Brazil | 237,500 |
| Chile | 58,600 |
| Colombia | 45,000 |
| India | 221,000 |
| Mexico | 178,100 |
| South Africa | 43,400 |
| Turkey | 251,500 |

*Source:* U.S. Census Bureau, *Statistical Abstract of the United States*, 2010, Table 1367.

   a. Assume that $\sigma_y = 93,500$. Calculate the standard error and interpret. (*Hint:* Consider the formula for the standard error.) Since you are provided with the population standard deviation, calculating the standard error requires only minor calculations.
   b. Write a report wherein you discuss the following: the standard error compared with the standard deviation of the population, the shape of the sampling distribution, and suggestions for reducing the standard error.

6. Use the data on education from Chapter 5, Exercise 4.

| | Mean | Standard Deviation | N |
|---|---|---|---|
| Lower class | 11.36 | 2.96 | 121 |
| Working class | 12.73 | 2.79 | 676 |
| Middle class | 14.40 | 3.04 | 636 |
| Upper class | 15.49 | 2.95 | 53 |

    a. Construct the 95% confidence interval for the mean number of years of education for lower-class and middle-class respondents.

    b. Construct the 99% confidence interval for the mean number of years of education for lower-class and middle-class respondents.

    c. As our confidence in the result increases, how does the size of the confidence interval change? Explain why this is so.

7. There has been a great deal of discussion about global warming in recent years. In 2006,[11] the Pew Research Center conducted a survey of 1,501 Americans to assess their opinion of global warming. The data show that 615 respondents of the 1,501 surveyed felt global warming is a very serious problem.

    a. Estimate the proportion of all adult Americans who felt global warming is a very serious problem at the 95% confidence interval.

    b. Estimate the proportion of all adult Americans who felt global warming is a very serious problem at the 99% confidence interval.

    c. If you were going to write a report on this poll result, would you prefer to use the 99% or 95% confidence interval? Explain why.

8. You have been doing research for your statistics class on how nervous the American adults are in general. You have decided to use HINTS 2007 data set that has a scale (going from 0 to 24) measuring the psychological distress of the respondents.

    a. According to HINTS 2007 data, the average psychological distress score, for this sample of size 1,390, is 4.75, with a standard deviation of 4.53. Construct the 95% confidence interval for the true average psychological distress score.

    b. One of your classmates, who claims to be good at statistics, complains about your confidence interval calculation. She or he asserts that the psychological distress scores are not normally distributed, which in turn makes the confidence interval calculation meaningless. Assume that she or he is correct about the distribution of psychological distress scores. Does that imply that the calculation of a confidence interval is not appropriate? Why or why not?

9. From the 2008 GSS subsample, we find that 78.6% of respondents believe in some form of life after death ($N = 1,303$).

    a. What is the 95% confidence interval for the percentage of the U.S. population who believe in life after death?

    b. Without doing any calculations, make an educated guess at the lower and upper bounds of 90% and 99% confidence intervals.

10. A social service agency plans to conduct a survey to determine the mean income of its clients. The director of the agency prefers that you measure the mean income very accurately, to within ±$500. From a sample taken 2 years ago, you estimate that the standard deviation of income for this population is about $5,000. Your job is to figure out the necessary sample size to reduce sampling error to ±$500.

    a. Do you need to have an estimate of the current mean income to answer this question? Why or why not?

    b. What sample size should be drawn to meet the director's requirement at the 95% level of confidence? (*Hint:* Use the formula for a confidence interval and solve for *N*, the sample size.)

    c. What sample size should be drawn to meet the director's requirement at the 99% level of confidence?

**Exercises**

11. Data from a 2008 GSS subsample show that the mean number of children per respondent was 1.94, with a standard deviation of 1.70. A total of 2,020 people answered this question. Estimate the population mean number of children per adult using a 90% confidence interval.

12. A subsample of the 2008 MTF survey suggests that adolescents are generally concerned with social issues. In fact, 79.9% of the 1,488 respondents who answered the question reported that they either sometimes or often/all the time think about social issues. Estimate at the 95% and 99% confidence levels the proportion of all adolescents who sometimes or often think about social issues.

13. According to a 2008 survey[12] by the Pew Research Center, two out of three Americans aged between 18 and 29 years use social networking websites such as MySpace and Facebook. Interestingly, 27% of the 225 18- to 29-year-olds surveyed say that they have received information about the 2008 presidential political candidates from social networking websites. What is the 95% confidence interval for the percentage of Americans aged between 18 and 29 years that use social networking websites to obtain information about political candidates?

14. According to a report[13] published by the Pew Research Center in February 2010, 61% of Millenials (Americans in their teens and 20s) think that their generation has a unique and distinctive identity ($N = 527$).
    a. Calculate the 95% confidence interval to estimate the percentage of Millenials who believe that their generation has a distinctive identity as compared with the other generations (the Generation X, Baby Boomers, or the Silent Generation).
    b. Calculate the 99% confidence interval.
    c. Are both these results compatible with the conclusion that the majority of Millenials believe that they have a unique identity that separates them from the previous generations?

Exercises

# Chapter 7

# Testing Hypotheses

## Chapter Learning Objectives

❖ Understanding the assumptions of statistical hypothesis testing

❖ Defining and applying the components in hypothesis testing: the research and null hypotheses, sampling distribution, and test statistic

❖ Understanding what it means to reject or fail to reject a null hypothesis

❖ Applying hypothesis testing to two sample cases, with means or proportions

In the past, the increase in the price of gasoline could be attributed to major national or global event, such as the Lebanon and Israeli war or Hurricane Katrina. However, in 2005, the price for a gallon of regular gasoline reached $3.00 and remained high for a long time afterward. The impact of unpredictable fuel prices is still felt across the nation, but the burden is greater among distinct social economic groups and geographic areas.

Lower-income Americans spend eight times more of their disposable income on gasoline than wealthier Americans do.[1] For example, in Wilcox, Alabama, individuals spend 12.72% of their income to fuel one vehicle, while in Hunterdon Co., New Jersey, people spend 1.52%. Nationally, Americans spend 3.8% of their income fueling one vehicle.

The first state to reach the $3.00-per-gallon milestone was California in 2005. California's drivers were especially hit hard by the rising price of gas, due in part to their reliance on automobiles, especially for work commuters. Analysts predicted that gas prices would continue to rise nationally. Declines in consumer spending and confidence in the economy have been attributed in part to the high (and rising) cost of gasoline.

In 2010, gasoline prices have remained higher for states along the West Coast, particularly in Alaska and California. Let's say we drew a random sample of California gas stations ($N = 100$) and calculated the mean price for a gallon of regular gas. Based on consumer information,[2] we also know that nationally the mean price of a gallon was $2.86, with a standard deviation of 0.17 for the same week. We can thus compare the mean price of gas in California with the mean price of all gas stations in April 2010. By comparing these means, we are asking whether it is reasonable to consider our random sample of California gas as representative of the population of gas stations in the United States. Actually, we expect to find that the average price of gas from a sample of California gas stations will be unrepresentative of the population of gas stations because we assume higher gas prices in the state.

The mean price for our sample is $3.11. This figure is higher than $2.86, the mean price per gallon across the nation. But is the observed gap of 25 cents ($3.11 − $2.86) large enough to convince us that the sample of California gas stations is not representative of the population?

There is no easy answer to this question. The sample mean of $3.11 is higher than the population mean, but it is an estimate based on a single sample. Thus, it could mean one of two things: (1) The average price of gas in California is indeed higher than the national average or (2) the average price of gas in California is about the same as the national average, and this sample happens to show a particularly high mean.

How can we decide which of these explanations makes more sense? Because most estimates are based on single samples and different samples may result in different estimates, sampling results cannot be used directly to make statements about a population. We need a procedure that allows us to evaluate hypotheses about population parameters based on sample statistics. In Chapter 6, we saw that population parameters can be estimated from sample statistics. In this chapter, we will learn how to use sample statistics to make decisions about population parameters. This procedure is called **statistical hypothesis testing**.

---

**Statistical hypothesis testing**   A procedure that allows us to evaluate hypotheses about population parameters based on sample statistics.

---

## ASSUMPTIONS OF STATISTICAL HYPOTHESIS TESTING

Statistical hypothesis testing requires several assumptions. These assumptions include considerations of the level of measurement of the variable, the method of sampling, the shape of the population distribution, and the sample size. The specific assumptions may vary, depending on the test or the conditions of testing. However, without exception, *all* statistical tests assume random sampling. Tests of hypotheses about means also assume interval-ratio level of measurement and require that the population under consideration be normally distributed or that the sample size be larger than 50.

Based on our data, we can test the hypothesis that the average price of gas in California is higher than the average national price of gas. The test we are considering meets these conditions:

1. The sample of California gas stations was randomly selected.

2. The variable *price per gallon* is measured at the interval-ratio level.

3. We cannot assume that the population is normally distributed. However, because our sample size is sufficiently large ($N > 50$), we know, based on the central limit theorem, that the sampling distribution of the mean will be approximately normal.

## STATING THE RESEARCH AND NULL HYPOTHESES

Hypotheses are usually defined in terms of interrelations between variables and are often based on a substantive theory. Earlier, we defined *hypotheses* as tentative answers to research questions. They are tentative because we can find evidence for them only after being empirically tested. The testing of hypotheses is an important step in this evidence-gathering process.

## The Research Hypothesis ($H_1$)

Our first step is to formally express the hypothesis in a way that makes it amenable to a statistical test. The substantive hypothesis is called the **research hypothesis** and is symbolized as $H_1$. Research hypotheses are always expressed in terms of population parameters because we are interested in making statements about population parameters based on our sample statistics.

---

*Research hypothesis* ($H_1$)   A statement reflecting the substantive hypothesis. It is always expressed in terms of population parameters, but its specific form varies from test to test.

---

In our research hypothesis ($H_1$), we state that the average price of gas in California is higher than the average price of gas nationally. Symbolically, we use $\mu_Y$ to represent the population mean; our hypothesis can be expressed as

$$H_1: \mu_Y > \$2.86$$

In general, the research hypothesis ($H_1$) specifies that the population parameter is one of the following:

1. Not equal to some specified value: $\mu_Y \neq$ some specified value

2. Greater than some specified value: $\mu_Y >$ some specified value

3. Less than some specified value: $\mu_Y <$ some specified value

## The Null Hypothesis ($H_0$)

Is it possible that in the population there is no real difference between the mean price of gas in California and the mean price of gas in the nation and that the observed difference of 0.25 is actually due to the fact that this particular sample happened to contain California gas stations with higher prices? Since statistical inference is based on probability theory, it is not possible to prove or disprove the research hypothesis directly. We can, at best, estimate the *likelihood* that it is true or false.

To assess this likelihood, statisticians set up a hypothesis that is counter to the research hypothesis. The **null hypothesis**, symbolized as $H_0$, contradicts the research hypothesis and usually states that there is no difference between the population mean and some specified value. It is also referred to as the hypothesis of "no difference." Our null hypothesis can be stated symbolically as

$$H_0: \mu_Y = \$2.86$$

Rather than directly testing the substantive hypothesis ($H_1$) that there is a difference between the mean price of gas in California and the mean price nationally, we test the null hypothesis ($H_0$) that there is no difference in prices. In hypothesis testing, we hope to reject the null hypothesis to provide support for the research hypothesis. Rejection of the null hypothesis will strengthen our belief in the research hypothesis and increase our confidence in the importance and utility of the broader theory from which the research hypothesis was derived.

---

*Null hypothesis ($H_0$)*   A statement of "no difference" that contradicts the research hypothesis and is always expressed in terms of population parameters.

---

## More About Research Hypotheses: One- and Two-Tailed Tests

In a **one-tailed test**, the research hypothesis is directional; that is, it specifies that a population mean is either less than (<) or greater than (>) some specified value. We can express our research hypothesis as either

$$H_1: \mu_Y < \text{some specified value}$$

or

$$H_1: \mu_Y > \text{some specified value}$$

The research hypothesis we've stated for the average price of a gallon of regular gas in California is a one-tailed test.

When a one-tailed test specifies that the population mean is *greater than* some specified value, we call it a **right-tailed test** because we will evaluate the outcome at the right tail of the sampling distribution. If the research hypothesis specifies that the population mean is *less than* some specified value, it is called a **left-tailed test** because the outcome will be evaluated at the left tail of the sampling distribution. Our example is a right-tailed test because the research hypothesis states that the mean gas prices in California are higher than $2.86. (Refer to Figure 7.1 on page 163.)

Sometimes, we have some theoretical basis to believe that there is a difference between groups, but we cannot anticipate the direction of that difference. For example, we may have reason to believe that the average price of California gas is *different* from that of the general population, but we may not have enough research or support to predict whether it is *higher* or *lower*. When we have no theoretical reason for specifying a direction in the research hypothesis, we conduct a **two-tailed test**. The research hypothesis specifies that the population mean is not equal to some specified value. For example, we can express the research hypothesis about the mean price of gas as

$$H_1: \mu_Y \neq \$2.86$$

With both one- and two-tailed tests, our null hypothesis of no difference remains the same. It can be expressed as

$$H_0: \mu_Y = \text{some specified value}$$

*One-tailed test*   A type of hypothesis test that involves a directional research hypothesis. It specifies that the values of one group are either larger or smaller than some specified population value.

*Right-tailed test*   A one-tailed test in which the sample outcome is hypothesized to be at the right tail of the sampling distribution.

*Left-tailed test*   A one-tailed test in which the sample outcome is hypothesized to be at the left tail of the sampling distribution.

*Two-tailed test*   A type of hypothesis test that involves a nondirectional research hypothesis. We are equally interested in whether the values are less than or greater than one another. The sample outcome may be located at both the lower and the higher ends of the sampling distribution.

## DETERMINING WHAT IS SUFFICIENTLY IMPROBABLE: PROBABILITY VALUES AND ALPHA

Now let's put all our information together. We're assuming that our null hypothesis ($\mu_Y = \$2.86$) is true, and we want to determine whether our sample evidence casts doubt on that assumption, suggesting that there is evidence for research hypothesis, $\mu_Y > \$2.86$. What are the chances that we would have randomly selected a sample of California gas stations such that the average price per gallon is higher than \$2.86, the average for the nation? We can determine the chances or probability because of what we know about the sampling distribution and its properties. We know, based on the central limit theorem, that if our sample size is larger than 50, the sampling distribution of the mean is approximately normal, with a mean $\mu_{\bar{Y}} = \mu_Y$ and a standard deviation (standard error) of

$$\sigma_{\bar{Y}} = \frac{\sigma_Y}{\sqrt{N}}$$

We are going to assume that the null hypothesis is true and then see if our sample evidence casts doubt on that assumption. We have a population mean $\mu_Y = \$2.86$ and a standard deviation $\sigma_Y = 0.17$. Our sample size is $N = 100$, and the sample mean is \$3.11. We can assume that the distribution of means of all possible samples of size $N = 100$ drawn from this distribution would be approximately normal, with a mean of \$2.86 and a standard deviation of

$$\sigma_{\bar{Y}} = \frac{\sigma_Y}{\sqrt{N}} = \frac{0.17}{\sqrt{100}} = 0.017$$

This sampling distribution is shown in Figure 7.1. Also shown in Figure 7.1 is the mean gas price we observed for our sample of California gas stations.

**Figure 7.1**   Sampling Distribution of Sample Means Assuming $H_0$ Is True for a Sample $N = 100$

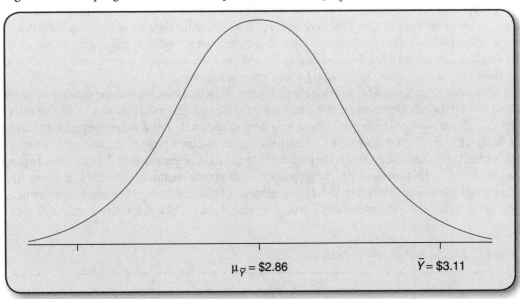

Because this distribution of sample means is normal, we can use Appendix A to determine the probability of drawing a sample mean of $3.11 or higher from this population. We will translate our sample mean into a $Z$ score so that we can determine its location relative to the population mean. In Chapter 5, we learned how to translate a raw score into a $Z$ score by using Formula 5.1:

$$Z = \frac{Y - \bar{Y}}{S_Y}$$

Because we are dealing with a sampling distribution in which our raw score is $\bar{Y}$, the mean, and the standard deviation (standard error) is $\sigma_Y / \sqrt{N}$, we need to modify the formula somewhat:

$$Z = \frac{\bar{Y} - \mu_Y}{\sigma_Y / \sqrt{N}} \tag{7.1}$$

Converting the sample mean to a $Z$-score equivalent is called computing the *test statistic*. The $Z$ value we obtain is called the **$Z$ statistic (obtained)**. The obtained $Z$ gives us the number of standard deviations (standard errors) that our sample is from the hypothesized value ($\mu_Y$ or $\mu_{\bar{Y}}$), assuming the null hypothesis is true. For our example, the obtained $Z$ is

$$Z = \frac{3.11 - 2.86}{0.17 / \sqrt{100}} = 14.70$$

---

**$Z$ statistic (obtained)**   The test statistic computed by converting a sample statistic (such as the mean) to a $Z$ score. The formula for obtaining $Z$ varies from test to test.

---

Before we determine the probability of our obtained $Z$ statistic, let's determine whether it is consistent with our research hypothesis. Recall that we defined our research hypothesis as a right-tailed test ($\mu_Y > \$2.86$), predicting that the difference would be assessed on the right tail of the sampling distribution. The positive value of our obtained $Z$ statistic confirms that we will be evaluating the difference on the right tail. (If we had a negative obtained $Z$, it would mean the difference would have to be evaluated at the left tail of the distribution, contrary to our research hypothesis.)

To determine the probability of observing a $Z$ value of 14.70, assuming that the null hypothesis is true, look up the value in Appendix A to find the area to the right of (above) the $Z$ of 14.70. Our calculated $Z$ value is not listed in Appendix A, so we'll need to rely on the last $Z$ value reported in the table, 4.00. Recall from Chapter 5, where we calculated $Z$ scores and their probability, that the $Z$ values are located in Column A. The $P$ value is the probability to the right of the obtained $Z$, or the "area beyond $Z$" in Column C. This area includes the proportion of all sample means that are \$3.11 or higher. The proportion is less than 0.0001 (Figure 7.2). This value is the probability of getting a result as extreme as the sample result if the null hypothesis is true; it is symbolized as $P$. Thus, for our example, $P \leq .0001$.

**Figure 7.2**    The Probability ($P$) Associated With $Z \geq 14.70$

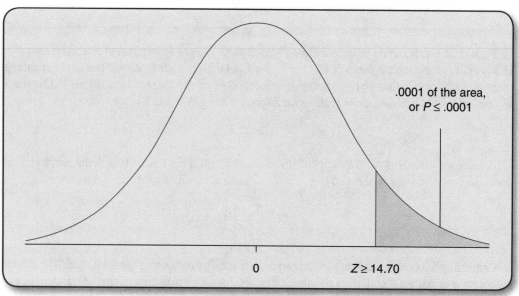

A **P value** can be defined as the actual probability associated with the obtained value of $Z$. It is a measure of how unusual or rare our obtained statistic is compared with what is stated in our null hypothesis. The smaller the $P$ value, the more evidence we have that the null hypothesis should be rejected in favor of the research hypothesis.

---

**P value**    The probability associated with the obtained value of $Z$.

---

Researchers usually define in advance what a sufficiently improbable $Z$ value is by specifying a cutoff point below which $P$ must fall to reject the null hypothesis. This cutoff point, called **alpha** and denoted by the Greek letter $\alpha$, is customarily set at the .05, .01, or .001 level. Let's say that we decide to reject the null hypothesis if $P \leq .05$. The value .05 is referred to as alpha ($\alpha$); it defines for us what result is sufficiently improbable to allow us to take the risk and reject the null hypothesis. An alpha ($\alpha$) of .05 means that even if the obtained $Z$ statistic is due to sampling error, so that the null hypothesis is true, we would allow a 5% risk of rejecting it. Alpha values of .01 and .001 are more cautionary levels of risk. The difference between $P$ and alpha is that $P$ is the *actual probability* associated with the obtained value of $Z$, whereas alpha is the level of probability *determined in advance* at which the null hypothesis is rejected. The null hypothesis is rejected when $P \leq \alpha$.

---

*alpha ($\alpha$)*　The level of probability at which the null hypothesis is rejected. It is customary to set alpha at the .05, .01, or .001 level.

---

We have already determined that our obtained $Z$ has a probability value less than .0001. Since our observed $P$ is less than .05 ($P = .0001 < \alpha = .05$), we reject the null hypothesis. The value of .0001 means that fewer than 1 out of 10,000 samples drawn from this population are likely to have a mean that is 14.70 $Z$ scores above the hypothesized mean of $2.86. Another way to say it is as follows: There is only 1 chance out of 10,000 (or .0001%) that we would draw a random sample with a $Z \geq 14.70$ if the mean price of California gas were equal to the national mean price.

Based on the $P$ value, we can also make a statement regarding the "significance" of the results. If the $P$ value is equal to or less than our alpha level, our obtained $Z$ statistic is considered *statistically significant*—that is to say, it is very unlikely to have occurred by random chance or sampling error. We can state that the difference between the average price of gas in California and nationally is significantly different at the .05 level, or we can specify the actual level of significance by saying that the level of significance is less than .0001.

Recall that our hypothesis was a one-tailed test ($\mu_Y > \$2.86$). In a two-tailed test, sample outcomes may be located at both the higher and the lower ends of the sampling distribution. Thus, the null hypothesis will be rejected if our sample outcome falls either at the left or right tail of the sampling distribution. For instance, a .05 alpha or $P$ level means that $H_0$ will be rejected if our sample outcome falls among either the lowest or the highest 5% of the sampling distribution.

Suppose we had expressed our research hypothesis about the mean price of gas as

$$H_1: \mu_Y \neq \$2.86$$

The null hypothesis to be directly tested still takes the form $H_0: \mu_Y = \$2.86$ and our obtained $Z$ is calculated using the same formula (7.1) as was used with a one-tailed test. To find $P$ for a two-tailed test, look up the area in Column C of Appendix A that corresponds to your obtained $Z$ (as we did earlier) and then multiply it by 2 to obtain the two-tailed probability. Thus, the two-tailed $P$ value for $Z = 14.70$ is $.0001 \times 2 = .0002$. This probability is less than our stated alpha (.05), and thus, we reject the null hypothesis.

## ▣ THE FIVE STEPS IN HYPOTHESIS TESTING: A SUMMARY

Regardless of the particular application or problem, statistical hypothesis testing can be organized into five basic steps. Let's summarize these steps:

1. Making assumptions

2. Stating the research and null hypotheses and selecting alpha

3. Selecting the sampling distribution and specifying the test statistic

4. Computing the test statistic

5. Making a decision and interpreting the results

*Making Assumptions.* Statistical hypothesis testing involves making several assumptions regarding the level of measurement of the variable, the method of sampling, the shape of the population distribution, and the sample size. In our example, we made the following assumptions:

1. A random sample was used.

2. The variable *price per gallon* is measured on an interval-ratio level of measurement.

3. Because $N > 50$, the assumption of normal population is not required.

*Stating the Research and Null Hypotheses and Selecting Alpha.* The substantive hypothesis is called the *research hypothesis* and is symbolized as $H_1$. Research hypotheses are always expressed in terms of population parameters because we are interested in making statements about population parameters based on sample statistics. Our research hypothesis was

$$H_1: \mu_Y > \$2.86$$

The *null hypothesis*, symbolized as $H_0$, contradicts the research hypothesis in a statement of no difference between the population mean and our hypothesized value. For our example, the null hypothesis was stated symbolically as

$$H_0: \mu_Y = \$2.86$$

We set alpha at .05, meaning that we would reject the null hypothesis if the probability of our obtained $Z$ was less than or equal to .05.

*Selecting the Sampling Distribution and Specifying the Test Statistic.* The normal distribution and the $Z$ statistic are used to test the null hypothesis.

*Computing the Test Statistic.* Based on Formula 7.1, our $Z$ statistic is 14.70.

*Making a Decision and Interpreting the Results.* We confirm that our obtained $Z$ is on the right tail of the distribution, consistent with our research hypothesis. Based on our obtained $Z$ statistic of 14.70, we determine that its *P* value is less than .0001, less than our .05 alpha levels. We have evidence to reject the null hypothesis of no difference between the mean price of California gas and the mean price of gas nationally. We thus conclude that the price of California gas is, on average, significantly higher than the national average.

## ▣ ERRORS IN HYPOTHESIS TESTING

We should emphasize that because our conclusion is based on sample data, we will never really know if the null hypothesis is true or false. In fact, as we have seen, there is a 0.01% chance that the null hypothesis is true and that we are making an error by rejecting it.

The null hypothesis can be either true or false, and in either case, it can be rejected or not rejected. If the null hypothesis is true and we reject it nonetheless, we are making an incorrect decision. This type of error is called a **Type I error.** Conversely, if the null hypothesis is false but we fail to reject it, this incorrect decision is a **Type II error.**

---

*Type I error*   The probability associated with rejecting a null hypothesis when it is true.

*Type II error*   The probability associated with failing to reject a null hypothesis when it is false.

---

In Table 7.1, we show the relationship between the two types of errors and the decisions we make regarding the null hypothesis. The probability of a Type I error—rejecting a true hypothesis—is equal to the chosen alpha level. For example, when we set alpha at the .05 level, we know that the probability that the null hypothesis is in fact true is .05 (or 5%).

**Table 7.1**   Type I and Type II Errors

| | True State of Affairs | |
|---|---|---|
| *Decision Made* | $H_0$ *Is True* | $H_0$ *Is False* |
| Reject $H_0$ | Type I error ($\alpha$) | Correct decision |
| Do not reject $H_0$ | Correct decision | Type II error |

We can control the risk of rejecting a true hypothesis by manipulating alpha. For example, by setting alpha at .01, we are reducing the risk of making a Type I error to 1%. Unfortunately, however, Type I and Type II errors are inversely related; thus, by reducing alpha and lowering the risk of making a Type I error, we are increasing the risk of making a Type II error (Table 7.1).

As long as we base our decisions on sample statistics and not population parameters, we have to accept a degree of uncertainty as part of the process of statistical inference.

*The implications of research findings are not created equal. For example, researchers might hypothesize that eating spinach increases the strength of weight lifters. Little harm will be done if the null hypothesis that eating spinach has no effect on the strength of weight lifters is rejected in error. The researchers would most likely be willing to risk a high probability of a Type I error, and all weight lifters would eat spinach. However, when the implications of research have important consequences, the balancing act between Type I and Type II errors becomes more important. Can you think of some examples where researchers would want to minimize Type I errors? When might they want to minimize Type II errors?*

## The *t* Statistic and Estimating the Standard Error

The *Z* statistic we have calculated (Formula 7.1) to test the hypothesis involving a sample of California gas stations assumes that the population standard deviation $\sigma_Y$ is known. The value of $\sigma_Y$ is required to calculate the standard error

$$\frac{\sigma_Y}{\sqrt{N}}$$

In most situations, $\sigma_Y$ will not be known, and we will need to estimate it using the sample standard deviation $S_Y$. We then use the *t* statistic instead of the *Z* statistic to test the null hypothesis. The formula for computing the *t* statistic is

$$t = \frac{\bar{Y} - \mu_Y}{S_Y/\sqrt{N}} \tag{7.2}$$

The *t* value we calculate is called the *t* statistic (obtained). The obtained *t* represents the number of standard deviation units (or standard error units) that our sample mean is from the hypothesized value of $\mu_Y$, assuming that the null hypothesis is true.

---

**t *statistic (obtained)***  The test statistic computed to test the null hypothesis about a population mean when the population standard deviation is unknown and is estimated using the sample standard deviation.

---

## The *t* Distribution and Degrees of Freedom

To understand the *t* statistic, we should first be familiar with its distribution. The **t distribution** is actually a family of curves, each determined by its *degrees of freedom*. The concept of degrees of freedom is used in calculating several statistics, including the *t* statistic. The **degrees of freedom** (*df*) represent the number of scores that are free to vary in calculating each statistic.

---

t *distribution*    A family of curves, each determined by its degrees of freedom *(df)*. It is used when the population standard deviation is unknown and the standard error is estimated from the sample standard deviation.

*Degrees of freedom* (**df***)*    The number of scores that are free to vary in calculating a statistic.

---

To calculate the degrees of freedom, we must know the sample size and whether there are any restrictions in calculating that statistic. The number of restrictions is then subtracted from the sample size to determine the degrees of freedom. When calculating the *t* statistic for a one-sample test, we start with the sample size *N* and lose 1 degree of freedom for the population standard deviation we estimate.[3] Note that the degrees of freedom will increase as the sample size increases. In the case of a single-sample mean, the *df* is calculated as follows:

$$df = N - 1 \qquad\qquad (7.3)$$

## Comparing the *t* and *Z* Statistics

Notice the similarities between the formulas for the *t* and *Z* statistics. The only apparent difference is in the denominator. The denominator of *Z* is the standard error based on the population standard deviation $\sigma_Y$. For the denominator of *t*, we replace $\sigma_Y / \sqrt{N}$ with $S_Y / \sqrt{N}$, the estimated standard error based on the sample standard deviation.

However, there is another important difference between the *Z* and *t* statistics: Because it is estimated from sample data, the denominator of the *t* statistic is subject to sampling error. The sampling distribution of the test statistic is not normal, and the standard normal distribution cannot be used to determine probabilities associated with it.

In Figure 7.3, we present the *t* distribution for several *df*s. Like the standard normal distribution, the *t* distribution is bell shaped. The *t* statistic, similar to the *Z* statistic, can have positive and negative values. A positive *t* statistic corresponds to the right tail of the distribution; a negative value corresponds to the left tail. Note that when the *df* is small, the *t* distribution is much flatter than the normal curve. But as the degrees of freedom increase, the shape of the *t* distribution gets closer to the normal distribution, until the two are almost identical when *df* is greater than 120.

Appendix B summarizes the *t* distribution. We have reproduced a small part of this appendix in Table 7.2. Note that the *t* table differs from the normal (*Z*) table in several ways. First, the column on the left side of the table shows the degrees of freedom. The *t* statistic will vary depending on the degrees of freedom, which must first be computed (*df* = *N* − 1). Second, the probabilities or alpha, denoted as significance levels, are arrayed across the top of the table in two rows, the first for a one-tailed and the second for a two-tailed test. Finally, the values of *t*, listed as the entries of this table, are a function of (1) the degrees of freedom, (2) the level of significance (or probability), and (3) whether the test is a one- or a two-tailed test.

**Figure 7.3**    The Normal Distribution and $t$ Distributions for 1, 5, 20, and $\infty$ Degrees of Freedom

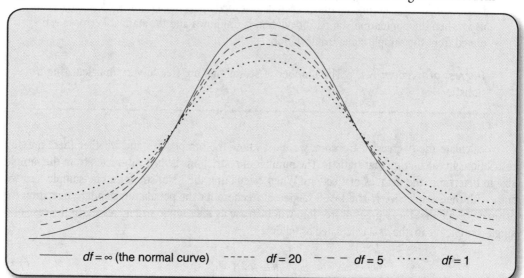

To illustrate the use of this table, let's determine the probability of observing a $t$ value of 2.021 with 40 degrees of freedom and a two-tailed test. Locating the proper row ($df = 40$) and column (two-tailed test), we find the $t$ statistic of 2.021 corresponding to the .05 level of significance. Restated, we can say that the probability of obtaining a $t$ statistic of 2.021 is .05, or that there are less than 5 chances out of 100 that we would have drawn a random sample with an obtained $t$ of 2.021 if the null hypothesis were correct.

## Statistics in Practice: The Earnings of White Women

We drew a 2002 General Social Survey (GSS) sample ($N = 371$) of white females who worked full-time. We found their mean earnings to be $28,889, with a standard deviation $S_Y = \$21,071$. Based on the Current Population Survey,[4] we also know that the 2002 mean earnings nationally for all women was $\mu_Y = \$24,146$. However, we do not know the value of the population standard deviation. We want to determine whether the sample of white women was representative of the population of all full-time women workers in 2002. Although we suspect that white American women experienced a relative advantage in earnings, we are not sure enough to predict that their earnings were indeed higher than the earnings of all women nationally. Therefore, the statistical test is two-tailed.

Let's apply the five-step model to test the hypothesis that the average earnings of white women differed from the average earnings of all women working full-time in the United States in 2002.

*Making Assumptions.* Our assumptions are as follows:

1. A random sample is selected.
2. Because $N > 50$, the assumption of normal population is not required.
3. The level of measurement of the variable *income* is interval ratio.

**Table 7.2** Values of the *t* Distribution

| df | .20 | .10 | .05 | .02 | .01 | .001 |
|---|---|---|---|---|---|---|
| | **Level of Significance for One-Tailed Test** | | | | | |
| | .10 | .05 | .025 | .01 | .005 | .0005 |
| | **Level of Significance for Two-Tailed Test** | | | | | |
| 1 | 3.078 | 6.314 | 12.706 | 31.821 | 63.657 | 636.619 |
| 2 | 1.886 | 2.920 | 4.303 | 6.965 | 9.925 | 31.598 |
| 3 | 1.638 | 2.353 | 3.182 | 4.541 | 5.841 | 12.941 |
| 4 | 1.533 | 2.132 | 2.776 | 3.747 | 4.604 | 8.610 |
| 5 | 1.476 | 2.015 | 2.571 | 3.365 | 4.032 | 6.859 |
| 10 | 1.372 | 1.812 | 2.228 | 2.764 | 3.169 | 4.587 |
| 15 | 1.341 | 1.753 | 2.131 | 2.602 | 2.947 | 4.073 |
| 20 | 1.325 | 1.725 | 2.086 | 2.528 | 2.845 | 3.850 |
| 25 | 1.316 | 1.708 | 2.060 | 2.485 | 2.787 | 3.725 |
| 30 | 1.310 | 1.697 | 2.042 | 2.457 | 2.750 | 3.646 |
| 40 | 1.303 | 1.684 | 2.021 | 2.423 | 2.704 | 3.551 |
| 60 | 1.296 | 1.671 | 2.000 | 2.390 | 2.660 | 3.460 |
| 80 | 1.289 | 1.658 | 1.980 | 2.358 | 2.617 | 3.373 |
| ∞ | 1.282 | 1.645 | 1.960 | 2.326 | 2.576 | 3.291 |

*Source:* Abridged from R. A. Fisher and F. Yates, *Statistical Tables for Biological, Agricultural and Medical Research,* Table 111. Copyright © R. A. Fisher and F. Yates, 1963. Published by Pearson Education Limited.

*Stating the Research and the Null Hypotheses and Selecting Alpha.* The research hypothesis is

$$H_1: \mu_Y \neq \$24,146$$

and the null hypothesis is

$$H_0: \mu_Y = \$24,146$$

We'll set alpha at .05, meaning that we will reject the null hypothesis if the probability of our obtained statistic is less than or equal to .05.

*Selecting the Sampling Distribution and Specifying the Test Statistic.* We use the *t* distribution and the *t* statistic to test the null hypothesis.

*Computing the Test Statistic.* We first calculate the *df* associated with our test:

$$df = (N-1) = (371-1) = 370$$

We need to calculate the obtained $t$ statistic by using Formula 7.2:

$$t = \frac{\bar{Y} - \mu_Y}{S_Y/\sqrt{N}} = \frac{28,889 - 24,146}{21,071/\sqrt{371}} = 4.33$$

*Making a Decision and Interpreting the Results.* Given our research hypothesis, we will conduct a two-tailed test. To determine the probability of observing a $t$ value of 4.33 with 370 degrees of freedom, let's refer to Table 7.2. From the first column, we can see that 370 degrees of freedom is not listed, so we'll have to use the last row, $df = \infty$, to locate our obtained $t$ statistic.

Though our obtained $t$ statistic of 4.33 is not listed in the last row of $t$ statistics, in fact, it is greater than the last value listed in the row, 3.291. The $t$ statistic of 3.291 corresponds to the .001 level of significance for two-tailed tests. Restated, we can say that our obtained $t$ statistic of 4.33 is greater than 3.291 or the probability of obtaining a $t$ statistic of 4.33 is less than .001 ($P < .001$). This $P$ value is below our .05 alpha level. The probability of obtaining the difference of $4,743 ($28,889 − $24,146) between the income of white women and the national average for all women, if the null hypothesis were true, is extremely low. We have sufficient evidence to reject the null hypothesis and conclude that the average earnings of white women in 2002 were significantly different from the average earnings of all women. The difference of $4,743 is significant at the .05 level. We can also say that the level of significance is less than .001.

## ▣ TESTING HYPOTHESES ABOUT TWO SAMPLES

In practice, social scientists are often more interested in situations involving two (sample) parameters than those involving one. For example, we may be interested in finding out whether the average years of education for one racial/ethnic group is the same, lower, or higher than another group.

U.S. data on educational attainment reveal that Asians and Pacific Islanders have more years of education than any other racial/ethnic groups; this includes the percentage of those earning a high school degree or higher or a college degree or higher. Though years of education have steadily increased for blacks and Hispanics since 1990, their numbers remain behind Asians and Pacific Islanders and whites.

Using data from the 2008 GSS, we examine the difference in black and Hispanic educational attainment. From the GSS sample, black respondents reported an average of 12.80 years of education and Hispanics an average of 10.63 years as shown in Table 7.3. These sample averages could mean either (1) the average number of years of education for blacks is higher than the average for Hispanics or (2) the average for blacks is actually about the same as for Hispanics, but our sample just happens to indicate a higher average for blacks. What we are applying here is a bivariate analysis (for more information, refer to Chapter 8), a method to detect and describe the relationship between two variables—race/ethnicity and educational attainment.

The statistical procedures discussed in the following sections allow us to test whether the differences that we observe between two samples are large enough for us to conclude that the populations from which these samples are drawn are different as well. We present tests for the significance of the differences between two groups. Primarily, we consider differences between sample means and differences between sample proportions.

**Table 7.3**  Years of Education for Black and Hispanic Men and
Women, GSS 2008

|  | *Blacks (Sample 1)* | *Hispanics (Sample 2)* |
|---|---|---|
| Mean | 12.80 | 10.63 |
| Standard deviation | 2.55 | 3.76 |
| Variance | 6.50 | 14.14 |
| *N* | 279 | 82 |

## The Assumption of Independent Samples

With a two-sample case, we assume that the samples are independent of each other. The choice of sample members from one population has no effect on the choice of sample members from the second population. In our comparison of blacks and Hispanics, we are assuming that the selection of blacks is independent of the selection of Hispanics. (The requirement of independence is also satisfied by selecting one sample randomly, then dividing the sample into appropriate subgroups. For example, we could randomly select a sample and then divide it into groups based on gender, religion, income, or any other attribute that we are interested in.)

## Stating the Research and Null Hypotheses

With two-sample tests, we compare two population parameters.

Our research hypothesis ($H_1$) is that the average years of education for blacks is not equal to the average years of education for Hispanic respondents. We are stating a hypothesis about the relationship between race/ethnicity and education in the general population by comparing the mean educational attainment of blacks with the mean educational attainment of Hispanics. Symbolically, we use $\mu$ to represent the population mean; the subscript 1 refers to our first sample (blacks) and subscript 2 to our second sample (Hispanics). Our research hypothesis can then be expressed as

$$H_1: \mu_1 \neq \mu_2$$

Because $H_1$ specifies that the mean education for blacks is not equal to the mean education for Hispanics, it is a nondirectional hypothesis. Thus, our test will be a two-tailed test. Alternatively, if there were sufficient basis for deciding which population mean score is larger (or smaller), the research hypothesis for our test would be a one-tailed test:

$$H_1: \mu_1 < \mu_2 \text{ or } H_1: \mu_1 > \mu_2$$

In either case, the null hypothesis states that there are no differences between the two population means:

$$H_0: \mu_1 = \mu_2$$

We are interested in finding evidence to reject the null hypothesis of no difference so that we have sufficient support for our research hypothesis.

✓ *Learning*
*Check*

> *For the following research situations, state your research and null hypotheses:*
>
> - *There is a difference between the mean statistics grades of social science majors and the mean statistics grades of business majors.*
> - *The average number of children in two-parent black families is lower than the average number of children in two-parent nonblack families.*
> - *Grade point averages are higher among girls who participate in organized sports than among girls who do not.*

## ▣ THE SAMPLING DISTRIBUTION OF THE DIFFERENCE BETWEEN MEANS

The sampling distribution allows us to compare our sample results with all possible sample outcomes and estimate the likelihood of their occurrence. Tests about differences between two sample means are based on the **sampling distribution of the difference between means**. The sampling distribution of the difference between two sample means is a theoretical probability distribution that would be obtained by calculating all the possible mean differences ($\bar{Y}_1 - \bar{Y}_2$) by drawing all possible independent random samples of size $N_1$ and $N_2$ from two populations.

---

*Sampling distribution of the difference between means*   A theoretical probability distribution that would be obtained by calculating all the possible mean differences ($\bar{Y}_1 - \bar{Y}_2$) that would be obtained by drawing all the possible independent random samples of size $N_1$ and $N_2$ from two populations where $N_1$ and $N_2$ are both greater than 50.

---

The properties of the sampling distribution of the difference between two sample means are determined by a corollary to the central limit theorem. This theorem assumes that our samples are independently drawn from normal populations, but that with sufficient sample size ($N_1 > 50, N_2 > 50$) the sampling distribution of the difference between means will be approximately normal, even if the original populations are not normal. This sampling distribution has a mean $\mu_{\bar{Y}_1} - \mu_{\bar{Y}_2}$ and a standard deviation (standard error)

$$\sigma_{\bar{Y}_1 - \bar{Y}_2} = \sqrt{\frac{\sigma^2_{Y_1}}{N_1} + \frac{\sigma^2_{Y_2}}{N_2}} \tag{7.4}$$

which is based on the variances in each of the two populations ($\sigma^2_{Y_1}$ and $\sigma^2_{Y_2}$).

## Estimating the Standard Error

Formula 7.4 assumes that the population variances are known and that we can calculate the standard error $\sigma_{\bar{Y}_1 - \bar{Y}_2}$ (the standard deviation of the sampling distribution). However, in most situations, the only data we have are based on sample data, and we do not know the true value of the population variances, $\sigma^2_{Y_1}$ and $\sigma^2_{Y_2}$. Thus, we need to estimate the standard error from the sample variances, $S^2_{Y_1}$ and $S^2_{Y_2}$. The estimated standard error of the difference between means is symbolized as $S_{\bar{Y}_1 - \bar{Y}_2}$ (instead of $\sigma_{\bar{Y}_1 - \bar{Y}_2}$).

## Calculating the Estimated Standard Error

When we can assume that the two population variances are equal, we combine information from the two sample variances to calculate the estimated standard error.

$$S_{\bar{Y}_1 - \bar{Y}_2} = \sqrt{\frac{(N_1 - 1)S^2_{Y_1} + (N_2 - 1)S^2_{Y_2}}{(N_1 + N_2) - 2}} \sqrt{\frac{N_1 + N_2}{N_1 N_2}} \tag{7.5}$$

where $S_{\bar{Y}_1 - \bar{Y}_2}$ is the estimated standard error of the difference between means, and $S^2_{Y_1}$ and $S^2_{Y_2}$ are the variances of the two samples. As a rule of thumb, when either sample variance is more than *twice* as large as the other, we can no longer assume that the two population variances are equal and would need to use Formula 7.8.

## The *t* Statistic

As with single sample means, we use the *t* distribution and the *t* statistic whenever we estimate the standard error for a difference between means test. The *t* value we calculate is the obtained *t*. It represents the number of standard deviation units (or standard error units) that our mean difference $(\bar{Y}_1 - \bar{Y}_2)$ is from the hypothesized value of $\mu_1 - \mu_2$, assuming that the null hypothesis is true.

The formula for computing the *t* statistic for a difference between means test is

$$t = \frac{\bar{Y}_1 - \bar{Y}_2}{S_{\bar{Y}_1 - \bar{Y}_2}} \tag{7.6}$$

where $S_{\bar{Y}_1 - \bar{Y}_2}$ is the estimated standard error.

## Calculating the Degrees of Freedom for a Difference Between Means Test

To use the *t* distribution for testing the difference between two sample means, we need to calculate the degrees of freedom. As we saw earlier, the degrees of freedom (*df*) represent the number of scores that are free to vary in calculating each statistic. When calculating the *t* statistic for the two-sample test, we lose 2 degrees of freedom, one for every population variance we estimate. When population variances are assumed to be equal or if the size of both samples is greater than 50, the *df* is calculated as follows:

$$df = (N_1 + N_2) - 2 \tag{7.7}$$

When we cannot assume that the population variances are equal and when the size of one or both samples is equal to or less than 50, we use Formula 7.9 to calculate the degrees of freedom.

## ▣ POPULATION VARIANCES ARE ASSUMED TO BE UNEQUAL

If the variances of the two samples ($S^2_{Y_1}$ and $S^2_{Y_2}$) are very different (one variance is twice as large as the other), the formula for the estimated standard error becomes

$$S_{\bar{Y}_1 - \bar{Y}_2} = \sqrt{\frac{S^2_{Y_1}}{N_1} + \frac{S^2_{Y_2}}{N_2}} \tag{7.8}$$

When the population variances are unequal and the size of one or both samples is equal to or less than 50, we use another formula to calculate the degrees of freedom associated with the $t$ statistic:[5]

$$df = \frac{(S^2_{Y_1}/N_1 + S^2_{Y_2}/N_2)^2}{(S^2_{Y_1}/N_1)^2/(N_1 - 1) + (S^2_{Y_2}/N_2)^2/(N_2 - 1)} \tag{7.9}$$

## ▣ THE FIVE STEPS IN HYPOTHESIS TESTING ABOUT DIFFERENCE BETWEEN MEANS: A SUMMARY

As with single-sample tests, statistical hypothesis testing involving two sample means can be organized into five basic steps. Let's summarize these steps:

1. Making assumptions
2. Stating the research and null hypotheses and selecting alpha
3. Selecting the sampling distribution and specifying the test statistic
4. Computing the test statistic
5. Making a decision and interpreting the results

*Making Assumptions.* In our example, we made the following assumptions:

1. Independent random samples are used.
2. The variable *years of education* is measured at an interval-ratio level of measurement.
3. Because $N_1 > 50$ and $N_2 > 50$, the assumption of normal population is not required.
4. The population variances are assumed to be equal.

*Stating the Research and Null Hypotheses and Selecting Alpha.* Our research hypothesis is that the mean education of blacks is different from the mean education of Hispanics, indicating a two-tailed test. Symbolically, the research hypothesis is expressed as

$$H_1: \mu_1 \neq \mu_2$$

with $\mu_1$ representing the mean education of blacks and $\mu_2$ the mean education of Hispanics.

The null hypothesis states that there are no differences between the two population means, or

$$H_0: \mu_1 = \mu_2$$

We are interested in finding evidence to reject the null hypothesis of no difference so that we have sufficient support for our research hypothesis. We will reject the null hypothesis if the probability of $t$ (obtained) is less than or equal to .05 (our alpha value).

*Selecting the Sampling Distribution and Specifying the Test Statistic.* The $t$ distribution and the $t$ statistic are used to test the significance of the difference between the two sample means.

*Computing the Test Statistic.* To test the null hypothesis about the differences between the mean education of blacks and Hispanics, we need to translate the ratio of the observed differences to its standard error into a $t$ statistic (based on data presented in Table 7.3). The obtained $t$ statistic is calculated using Formula 7.6:

$$t = \frac{\bar{Y}_1 - \bar{Y}_2}{S_{\bar{Y}_1 - \bar{Y}_2}}$$

where $S_{\bar{Y}_1 - \bar{Y}_2}$ is the estimated standard error of the sampling distribution. Because the population variances are assumed to be equal, $df$ is $(N_1 + N_2) - 2 = (279 + 82) - 2 = 359$ and we can combine information from the two sample variances to estimate the standard error (Formula 7.5):

$$S_{\bar{Y}_1 - \bar{Y}_2} = \sqrt{\frac{(279 - 1)2.55^2 + (82 - 1)3.76^2}{(279 + 82) - 2}} \sqrt{\frac{279 + 82}{279(82)}} = (2.87)(0.13) = 0.37$$

We substitute this value into the denominator for the $t$ statistic (Formula 7.6):

$$t = \frac{12.80 - 10.63}{0.37} = 5.86$$

*Making a Decision and Interpreting the Results.* We confirm that our obtained $t$ is on the right tail of the distribution. Since our obtained $t$ statistic of 5.86 is greater than $t = 3.291$ ($df = \infty$, two-tailed; see Appendix B), we can state that its probability is less than .001. This is less than our .05 alpha level, and we can reject the null hypothesis of no difference between the educational attainment of blacks and Hispanics. We conclude that black men and women, on average, have significantly higher years of education than Hispanic men and women do.

## ▣ TESTING THE SIGNIFICANCE OF THE DIFFERENCE BETWEEN TWO SAMPLE PROPORTIONS

Numerous variables in the social sciences are measured at a nominal or an ordinal level. These variables are often described in terms of proportions. For example, we might be interested in comparing the proportion of those who support immigrant policy reform among Hispanics and non-Hispanics or the proportion of men and women who supported the Democratic candidate during the last presidential election. In this section, we present statistical inference techniques to test for significant differences between two sample proportions.

Hypothesis testing with two sample proportions follows the same structure as the statistical tests presented earlier: The assumptions of the test are stated, the research and null hypotheses are formulated, the sampling distribution and the test statistic are specified, the test statistic is calculated, and a decision is made whether or not to reject the null hypothesis.

## Statistics in Practice: Political Party Affiliation and Confidence in the Executive Branch

Pollsters often collect data to measure public opinions on current social and political issues. Differences are often categorized by income, race/ethnicity, gender, region, or political groups. For example, do Republicans and Democrats have the same confidence in the executive branch of government? We can use data from the 2008 GSS to test the null hypothesis that the proportion of Republicans and Democrats who have a "great deal" of confidence in the executive branch of the government is equal. The proportion of Republicans who reported a great deal of confidence was 0.21 $(p_1)$; the proportion of Democrats with the same response was lower at 0.06 $(p_2)$. A total of 141 Republicans $(N_1)$ and 267 Democratic respondents $(N_2)$ answered this question.

*Making Assumptions.* Our assumptions are as follows:

1. Independent random samples of $N_1 > 50$ and $N_2 > 50$ are used.

2. The level of measurement of the variable is nominal.

*Stating the Research and Null Hypotheses and Selecting Alpha.* We propose a two-tailed test that the population proportions for Republicans and Democrats are not equal.

$$H_1: \pi_1 \neq \pi_2$$
$$H_0: \pi_1 = \pi_2$$

We decide to set alpha at .05.

*Selecting the Sampling Distribution and Specifying the Test Statistic.* The population distributions of dichotomies are not normal. However, based on the central limit theorem, we know that the sampling distribution of the difference between sample proportions is normally distributed when the sample size is large (when $N_1 > 50$ and $N_2 > 50$), with mean $\mu_{p_1-p_2}$ and the estimated standard error $S_{p_1-p_2}$. Therefore, we can use the normal distribution as the sampling distribution and we can calculate $Z$ as the test statistic.[6]

The formula for computing the $Z$ statistic for a difference between proportions test is

$$Z = \frac{p_1 - p_2}{S_{p_1 - p_2}} \tag{7.10}$$

where $p_1$ and $p_2$ are the sample proportions for Republicans and Democrats, and $S_{p_1 - p_2}$ is the estimated standard error of the sampling distribution of the difference between sample proportions.

The estimated standard error is calculated using the following formula:

$$S_{p_1 - p_2} = \sqrt{\frac{p_1(1 - p_1)}{N_1} + \frac{p_2(1 - p_2)}{N_2}} \qquad (7.11)$$

*Calculating the Test Statistic.* We calculate the standard error using Formula 7.11:

$$S_{p_1 - p_2} = \sqrt{\frac{0.21(1 - 0.21)}{141} + \frac{0.06(1 - 0.06)}{267}} = .037 = 0.04$$

Substituting this value into the denominator of Formula 7.10, we get

$$Z = \frac{0.21 - 0.06}{0.04} = 3.75$$

*Making a Decision and Interpreting the Results.* Our obtained $Z$ of 3.75 indicates that the difference between the two proportions will be evaluated at the right tail of the $Z$ distribution. To determine the probability of observing a $Z$ value of 3.75 if the null hypothesis is true, look up the value in Appendix A (Column C) to find the area to the right of (above) the obtained $Z$.

The $P$ value corresponding to a $Z$ score of 3.75 is .0001. However, for a two-tailed test, we'll have to multiply $P$ by 2 (.0001 × 2 = .0002). The probability of 3.75 for a two-tailed test is less than our alpha level of .05 (.0002 < .05).

Thus, we reject the null hypothesis of no difference and conclude that there is a significant political party difference in confidence in the executive branch of the government. Republicans are more likely than Democrats to report a great deal of confidence in the executive branch.

## ▣ IS THERE A SIGNIFICANT DIFFERENCE?

The news media made note of a 2010 Centers for Disease Control (CDC) study that examined the difference in length of marriage between couples in 2005 who first cohabited before marriage and couples who did not cohabit before marriage. Several news services released stories noting the "troubles" associated with living together. As reported, the percentage of marriages surviving to the 10th anniversary, among those who cohabited before marriage, was lower than those who did not cohabit before their first marriage. A closer look at the report reveals important (overlooked) details.

CDC researchers Goodwin, Mosher, and Chandra (2010) reported that previous cohabitation experience was significantly associated with marriage survival probabilities for men. On the other hand, though the probability that a woman's marriage would last at least 10 years was lower for those who cohabited before marriage (60%) than for women who did not (66%), the researchers wrote, "however, in the 2002 data, the difference was not significant at the 5% level" (p. 7).[7]

Throughout this chapter, we've assessed the difference between two means and two proportions, attempting to determine whether the difference between them is due to real effects in the population or due to sampling error. A significant difference is one that confirms that effects of the independent variable, such as cohabiting before marriage, are real. As in the case of marriage survival rate, cohabitation before marriage makes a significant difference in marital outcomes for men, but not for women in the CDC sample. Take caution in accepting comparative statements that fail to mention significance. There may be a difference, but you have to ask, is it a significant difference?

# 🔲 READING THE RESEARCH LITERATURE: REPORTING THE RESULTS OF STATISTICAL HYPOTHESIS TESTING

Robert Emmet Jones and Shirley A. Rainey (2006) examined the relationship between race, environmental attitudes, and perceptions about environmental health and justice.[8] Researchers have documented how people of color and the poor are more likely than whites and more affluent groups to live in areas with poor environmental quality and protection, exposing them to greater health risks. Yet little is known about how this disproportional exposure and risk are perceived by those affected. Jones and Rainey studied black and white residents from the Red River community in Tennessee, collecting data from interviews and a mail survey during 2001 to 2003.

They created a series of index scales measuring residents' attitudes pertaining to environmental problems and issues. The Environmental Concern (EC) Index measures public concern for specific environmental problems in the neighborhood. It includes questions on drinking water quality, landfills, loss of trees, lead paint and poisoning, the condition of green areas, and stream and river conditions. EC-II measures public concern (very unconcerned to very concerned) for the overall environmental quality in the neighborhood. EC-III measures the seriousness (not serious at all to very serious) of environmental problems in the neighborhood. Higher scores on all EC indicators indicate greater concern for environmental problems in their neighborhood. The Environmental Health (EH) Index measures public perceptions of certain physical side effects, such as headaches, nervous disorders, significant weight loss or gain, skin rashes, and breathing problems. The EH Index measures the likelihood (very unlikely to very likely) that the person believes that he or she or a household member experienced health problems due to exposure to environmental contaminants in his or her neighborhood. Higher EH scores reflect a greater likelihood that respondents believe that they have experienced health problems from exposure to environmental contaminants. Finally, the Environmental Justice (EJ) Index measures public perceptions about environmental justice, measuring the extent to which they agreed (or disagreed) that public officials had informed residents about environmental problems, enforced environmental laws, or held meetings to address residents' concerns. A higher mean EJ score indicates a greater likelihood that respondents think public officials failed to deal with environmental problems in their neighborhood. Index score comparisons between black and white respondents are presented in Table 7.4.

Let's examine the table carefully. Each row represents a single index measurement, reporting means and standard deviations separately for black and white residents. Obtained *t*-test statistics are reported in the second to last column. The probability of each *t* test is reported in the last column ($P < .001$), indicating a significant difference in responses between the two groups. All index score comparisons are significant at the .001 level.

While not referring to specific differences in index scores or to *t*-test results, Jones and Rainey use data from this table to summarize the differences between black and white residents on the three environmental index measurements:

The results presented [in Table 1] suggest that as a group, Blacks are significantly more concerned than Whites about local environmental conditions (EC Index).... The results ... also indicate that

**Table 7.4**   Environmental Concern (EC), Environmental Health (EH), and Environmental Justice (EJ)

| Indicator | Group | Mean | Standard Deviation | t | Significance (one-tailed) |
|-----------|-------|------|--------------------|---|---------------------------|
| EC Index | Blacks | 56.2 | 13.7 | 6.2 | <0.001 |
|          | Whites | 42.6 | 15.5 | | |
| EC-II | Blacks | 4.4 | 1.0 | 5.6 | <0.001 |
|       | Whites | 3.5 | 1.3 | | |
| EC-III | Blacks | 3.4 | 1.1 | 6.7 | <0.001 |
|        | Whites | 2.3 | 1.0 | | |
| EH Index | Blacks | 23.0 | 10.5 | 5.1 | <0.001 |
|          | Whites | 16.0 | 7.3 | | |
| EJ Index | Blacks | 31.0 | 7.3 | 3.8 | <0.001 |
|          | Whites | 27.2 | 6.3 | | |

*Source:* Robert E. Jones and Shirley A. Rainey, "Examining Linkages Between Race, Environmental Concern, Health and Justice in a Highly Polluted Community of Color," *Journal of Black Studies* 36, no. 4 (2006): 473–496.

*Note: N* = 78 blacks, 113 whites.

as a group, Blacks believe they have suffered more health problems from exposure to poor environmental conditions in their neighborhood than Whites (EH Index). . . . [T]here is greater likelihood that Blacks feel local public agencies and officials failed to deal with environmental problems in their neighborhood in a fair, just, and effective manner (EJ Index). (p. 485)

## MAIN POINTS

- Statistical hypothesis testing is a decision-making process that enables us to determine whether a particular sample result falls within a range that can occur by an acceptable level of chance. The process of statistical hypothesis testing consists of five steps: (1) making assumptions, (2) stating the research and null hypotheses and selecting alpha, (3) selecting a sampling distribution and a test statistic, (4) computing the test statistic, and (5) making a decision and interpreting the results.

- Statistical hypothesis testing may involve a comparison between a sample mean and a population mean or a comparison between two sample means. If we know the population variance(s) when testing for differences between means, we can use the $Z$ statistic and the normal distribution. However, in practice, we are unlikely to have this information.

- When testing for differences between means when the population variance(s) are unknown, we use the $t$ statistic and the $t$ distribution.

- Tests involving differences between proportions follow the same procedure as tests for differences between means when population variances are known. The test statistic is $Z$, and the sampling distribution is approximated by the normal distribution.

## KEY TERMS

alpha ($\alpha$)

degrees of freedom ($df$)

left-tailed test

null hypothesis ($H_0$)

one-tailed test

$P$ value

research hypothesis ($H_1$)

right-tailed test

sampling distribution of the
difference between means

statistical hypothesis testing

$t$ distribution

$t$ statistic (obtained)

two-tailed test

Type I error

Type II error

$Z$ statistic (obtained)

## ON YOUR OWN

Log on to the web-based student study site at **www.sagepub.com/ssdsessentials** for additional study questions, web quizzes, web resources, flashcards, codebooks and datasets, web exercises, appendices, and links to social science journal articles reflecting the statistics used in this chapter.

## CHAPTER EXERCISES

1. It is known that, nationally, doctors working for health maintenance organizations (HMOs) average 13.5 years of experience in their specialties, with a standard deviation of 7.6 years. The executive director of an HMO in a western state is interested in determining whether or not its doctors have less experience than the national average. A random sample of 150 doctors from HMOs shows a mean of only 10.9 years of experience.
   a. State the research and the null hypotheses to test whether or not doctors in this HMO have less experience than the national average.
   b. Using an alpha level of .01, make this test.

2. Consider the problem facing security personnel at a military facility in the Southwest. Their job is to detect infiltrators (spies trying to break in). The facility has an alarm system to assist the security officers. However, sometimes the alarm doesn't work properly, and sometimes the officers don't notice a real alarm. In general, the security personnel must decide between these two alternatives at any given time:

   $H_0$: Everything is fine; no one is attempting an illegal entry.

   $H_1$: There are problems; someone is trying to break into the facility.

   Based on this information, fill in the blanks in these statements:
   a. A "missed alarm" is a Type ____ error, and its probability of occurrence is denoted as ____.
   b. A "false alarm" is a Type ____ error.

3. For each of the following situations determine whether a one- or a two-tailed test is appropriate. Also, state the research and the null hypotheses.
   a. You are interested in finding out if the average household income of residents in your state is different from the national average household. According to the U.S. Census, for 2010, the national average household income is $50,303.
   b. You believe that students in small liberal arts colleges attend more parties per month than students nationwide. It is known that, nationally, undergraduate students attend an average of 3.2 parties per month. The average number of parties per month will be calculated from a random sample of students from small liberal arts colleges.
   c. A sociologist believes that the average income of elderly women is lower than the average income of elderly men.

d. Is there a difference in the amount of study time on-campus and off-campus students devote to their schoolwork during an average week? You prepare a survey to determine the average number of study hours for each group of students.

e. Reading scores for a group of third graders enrolled in an accelerated reading program are predicted to be higher than the scores for nonenrolled third graders.

f. Stress (measured on an ordinal scale) is predicted to be lower for adults who own dogs (or other pets) than for non–pet owners.

4. a. For each situation in Exercise 3, describe the Type I and Type II errors that could occur.
   b. What are the general implications of making a Type I error? Of making a Type II error?
   c. When would you want to minimize Type I error? Type II error?

5. One way to check on how representative a survey is of the population from which it was drawn is to compare various characteristics of the sample with the population characteristics. A typical variable used for this purpose is age. The 2008 GSS of the American adult population found a mean age of 47.71 and a standard deviation of 17.35 for its sample of 2,013 adults. Assume that we know from census data that the mean age of all American adults is 37.7. Use this information to answer these questions.
   a. State the research and the null hypotheses for a two-tailed test.
   b. Calculate the $t$ statistic and test the null hypothesis at the .001 significance level. What did you find?
   c. What is your decision about the null hypothesis? What does this tell us about how representative the sample is of the American adult population?

6. Using data on average school grade for first-time cigarette use from MTF 2008, use the $t$ test to conduct a one-tailed test of the null hypothesis, assuming that the average grade is higher for males than for females. Set alpha at .05. What can you conclude? Would your conclusions have been different if you had used a two-tailed test?

|  | *Males* | *Females* |
|---|---|---|
| Grade first tried cigarettes | Mean = 4.90 | Mean = 4.76 |
|  | $S_Y = 1.74$ | $S_Y = 1.73$ |
|  | $N = 150$ | $N = 169$ |

*Source:* Monitoring the Future, 2008.

*Note:* Only valid responses are included in the table. Students who indicated that they have never used the drug are not included.

7. In this exercise, we will examine the attitudes of liberals and conservatives toward affirmative action policies in the workplace. Data from the 2008 GSS reveal that 10% of conservatives ($N = 424$) and 28% of liberals ($N = 336$) indicate that they "strongly support" or "support" affirmative action policies for African Americans in the workplace.
   a. What is the appropriate test statistic? Why?
   b. Test the null hypothesis with a one-tailed test (conservatives are less likely to support affirmative action policies than liberals); $\alpha = .05$. What do you conclude about the difference in attitudes between conservatives and liberals?
   c. If you conducted a two-tailed test with $\alpha = .05$, would your decision have been different?

8. Let's continue our analysis of liberals and conservatives, taking a look this time at differences in their educational attainment. We obtain the following information from the 2008 GSS—the average educational attainment for liberals is 13.90 years ($S_Y = 3.27$) and the average educational attainment for conservatives is 13.55 years ($S_Y = 2.82$). Data are based on 187 liberals and 227 conservative responses.
   a. Test the research hypothesis that there is a difference in level of education between liberals and conservatives; set alpha at .01.
   b. Would your decision have been different if alpha were set at .05?

9. During the 2008 Democratic presidential campaign, gender was considered more of an issue for Hillary Clinton's campaign than for her opponent, Barack Obama. During the campaign, pollsters consistently reported how Clinton's supporters were mostly (older) women, with men less likely to support her candidacy. In surveys conducted during December 2007 and January 2008, the Pew Research Center reported that among 240 men, 41% indicated that Senator Clinton was their first-choice candidate. Among 381 women, 49% reported the same. Do these differences reflect a gender gap among Clinton supporters?
   a. If you wanted to test the research hypothesis that the proportion of male voters identifying Senator Clinton as their first-choice candidate is less than female voters, would you conduct a one- or a two-tailed test?
   b. Test the null hypothesis at the .05 alpha level. What do you conclude?
   c. If alpha were changed to .01, would your decision remain the same?

10. Data from the MTF 2008 survey reveal that 75.7% (493 out of 651) of males and 70.4% (501 out of 712) of females reported trying alcohol. You wonder whether there is any difference between males and females in the population trying alcohol (this variable does not measure regular use, only if the student had ever tried alcohol). Use a test of the difference between proportions when answering these questions.
   a. What is the research hypothesis? Should you conduct a one- or a two-tailed test? Why?
   b. Test your hypothesis at the .05 level. What do you conclude?

11. Marcelline Fusilier, Subhash Durlabhji, Alain Cucchi, and Michael Collins (2005) examined Internet usage among college students in the United States and in India.[9] Usage was divided into two categories, for personal use and course-related use. We compare average hours between the U.S. and Indian students in the following table.

**Internet Use Means and Standard Deviations for U.S. and Indian Students Sample**

|  | Course Work Hours | Personal Hours |
|---|---|---|
| U.S. students | $\bar{Y} = 1.76$ | $\bar{Y} = 2.08$ |
| $N = 149$ | $S_Y = 1.52$ | $S_Y = 1.91$ |
| Indian students | $\bar{Y} = 0.73$ | $\bar{Y} = 0.87$ |
| $N = 306$ | $S_Y = 0.79$ | $S_Y = 0.78$ |

Source: Marcelline Fusilier, Subhash Durlabhji, Alain Cucchi, and Michael Collins, "A four country investigation of factors facilitating student Internet use," CyberPsychology and Behavior 8(5): 454–464. Copyright © Mary Ann Liebert, Inc. Publishers.

Exercises

a. Determine whether U.S. students have significantly higher Internet use for course work than the Indian students. Test at the .05 alpha level.

b. Test whether there is a significant difference in Internet use for personal use between the U.S. and Indian students. Test at the .01 alpha level.

12. Do men and women have different beliefs on the ideal number of children in a family? Based on the following GSS 2008 data and obtained $t$ statistic, what would you conclude? (*Hint:* Assume a two-tailed test; $\alpha = .05$.)

|  | *Men* | *Women* |
|---|---|---|
| Mean ideal number of children | 3.06 | 3.22 |
| Standard deviation | 1.92 | 1.99 |
| $N$ | 604 | 678 |
| Obtained $t$ statistic | −1.450 | |

13. Does access to cocaine vary by sex? Data from the MTF 2008 survey indicate that among male students ($N = 670$) 18% say that it would be "very easy" to obtain cocaine compared with 14% of female students ($N = 723$). Is there a significant difference in proportion of easy cocaine access between the two student groups? Set alpha at .01. What can you conclude?

14. We recalculated our comparison of ideal number of children, this time only for women, separating them into two groups, those who indicated that they were "very happy" or "not too happy" in general. Results, based on the GSS 2008, are presented below.

|  | *Very Happy* | *Not Too Happy* |
|---|---|---|
| Mean ideal number of children | 3.49 | 2.73 |
| Standard deviation | 2.24 | 1.35 |
| $N$ | 228 | 90 |
| Obtained $t$ statistic | 2.98 | |

Based on a two-tailed test, $\alpha = .05$. What do you conclude?

# Chapter 8

# Relationships Between Two Variables

*Cross-Tabulation*

---

**Chapter Learning Objectives**

- ❖ Constructing a bivariate table
- ❖ Determining the properties of a bivariate relationship: existence, strength, and direction
- ❖ Understanding hypothesis testing with chi-square
- ❖ Recognizing the limitations of chi-square test—sample size and statistical significance
- ❖ Understanding the concept of PRE (proportional reduction of error) and how to interpret measures of association
- ❖ Calculating and interpreting lambda: a measure of association for nominal variables
- ❖ Calculating and interpreting gamma and Kendall's tau-*b*: measures of association for ordinal variables
- ❖ Interpreting Cramer's *V*: a chi-square-related measure of association

---

One of the main objectives of social science is to make sense out of human and social experience by uncovering regular patterns among events. Therefore, the language of *relationships* is at the heart of social science inquiry. Consider the following examples from articles and research reports:

*Example 1:* Americans 45 years and older are more likely to support the state of Arizona's 2010 immigration reform law than younger Americans. The new law permits Arizona police officers to detain individuals suspected of entering the country illegally.[1] (This example indicates a relationship between age and immigration reform.)

*Example 2:* Contrary to the stereotype, whites use government safety net programs more than blacks or Latinos, and they are more likely than minorities to be lifted out of poverty by the tax-payer money that they get.[2] (This example indicates a relationship between race and receipt of government aid.)

*Example 3:* On average, blacks and Latinos have a lower likelihood of access to health care than whites.[3] (This example indicates a relationship between race and access to health care.)

In each of these examples, a relationship means that certain values of one variable tend to "go together" with certain values of the other variable. For example, in Example 1, age "goes together" with immigration reform and being older is associated with support of Arizona's immigration reform law.

In this chapter, we introduce one of the most common techniques used in the analysis of relationships between two variables: *cross-tabulation*. **Cross-tabulation** is a technique for analyzing the relationship between two variables that have been organized in a table. A cross-tabulation is a type of **bivariate analysis**, a method designed to detect and describe the relationship between two nominal or ordinal variables. We have already applied bivariate analysis in Chapter 7, using *t* tests and *Z* tests to determine the difference between two means or proportions.[4]

---

*Cross-tabulation*   A technique for analyzing the relationship between two nominal or ordinal variables that have been organized in a table.

*Bivariate analysis*   A statistical method designed to detect and describe the relationship between two nominal or ordinal variables.

---

## ▣ INDEPENDENT AND DEPENDENT VARIABLES

In the social sciences, an important aspect in research design and statistics is the distinction between the *independent variable* and the *dependent variable*. These terms, first introduced in Chapter 1, are used throughout this chapter as well as in the following chapters, and therefore, it is important that you understand the distinction between them.

In each of the illustrations given, there are two variables: an independent and a dependent variable. In Example 1, the purpose of the research is to explain support for immigration reform laws. One of the variables hypothesized as being connected to support for immigration reform is age. Therefore, support for immigration reform is the dependent variable, and age is the independent variable. In Example 2, the object of the investigation is to examine the common stereotype that people of color use government aid more than white Americans. The investigator is trying to explain differences in utilization of government aid using race as an explanatory variable. Therefore, utilization of government aid is the dependent variable, and race is the independent variable. Similarly, in Example 3, access to health care is the dependent variable because it is the variable to be explained, whereas race, the explanatory variable, is the independent variable.

The statistical techniques discussed in this chapter will help the researcher decide the strength of the relationship between the independent and dependent variables.

> *For some variables, whether it is the independent or dependent variable depends on the research question. If you are still having trouble distinguishing between an independent and a dependent variable, go back to Chapter 1 for a detailed discussion.*

## ▣ HOW TO CONSTRUCT A BIVARIATE TABLE: RACE AND HOME OWNERSHIP

A **bivariate table** displays the distribution of one variable across the categories of another variable. It is obtained by classifying cases based on their joint scores on two nominal or ordinal variables. It can be thought of as a series of frequency distributions joined to make one table. The data in Table 8.1 represent a sample of General Social Survey (GSS) respondents by race and whether they own or rent their home (in this case, both variables are nominal-level measurements).

**Table 8.1**   Race and Home Ownership for 17 GSS Respondents

| Respondent | Race | Home Ownership |
|---|---|---|
| 1 | Black | Own |
| 2 | Black | Own |
| 3 | White | Rent |
| 4 | White | Rent |
| 5 | White | Own |
| 6 | White | Own |
| 7 | White | Own |
| 8 | Black | Rent |
| 9 | Black | Rent |
| 10 | Black | Rent |
| 11 | White | Own |
| 12 | White | Own |
| 13 | White | Rent |
| 14 | White | Own |
| 15 | Black | Rent |
| 16 | White | Own |
| 17 | Black | Rent |

To make sense out of these data, we must first construct the table in which these individual scores will be classified. In Table 8.2, the 17 respondents have been classified according to joint scores on race and home ownership.

**Table 8.2**    Home Ownership by Race (absolute frequencies), GSS 2006

| HOME OWNERSHIP | RACE | | | |
| --- | --- | --- | --- | --- |
| | Black | White | | |
| Own | 2 | 7 | 9 | Row Marginals (Row Total) |
| Rent | 5 | 3 | 8 | |
| | 7 | 10 | 17 | Total Cases (*N*) |
| | Column Marginals (Column Total) | | | |

The table has the following features typical of most bivariate tables:

1. The table's title is descriptive, identifying its content in terms of the two variables.
2. It has two dimensions, one for race and one for home ownership. The variable *home ownership* is represented in the rows of the table, with one row for owners and another for renters. The variable *race* makes up the columns of the table, with one column for each racial group. A table may have more columns and more rows, depending on how many categories the variables represent. Usually, the independent variable is the **column variable** and the dependent variable is the **row variable**.
3. The intersection of a row and a column is called a **cell**. For example, the two individuals represented in the upper left cell are blacks who are also homeowners.
4. The column and row totals are the frequency distribution for each variable, respectively. The column total is the frequency distribution for *race,* the row total for *home ownership.* Row and column totals are sometimes called **marginals**. The total number of cases (*N*) is the number reported at the intersection of the row and column totals. (These elements are all labeled in the table.)
5. The table is a 2 × 2 table because it has two rows and two columns (not counting the marginals). We usually refer to this as an *r* × *c* table, in which *r* represents the number of rows and *c* the number of columns. Thus, a table in which the row variable has three categories and the column variable has two categories would be designated as a 3 × 2 table.
6. The source of the data should also be clearly noted in a source note to the table. This is consistent with what we reviewed in Chapter 2, The Organization and Graphic Presentation of Data.

---

*Bivariate table*    A table that displays the distribution of one variable across the categories of another variable.

*Column variable*    A variable whose categories are the columns of a bivariate table.

*Row variable*    A variable whose categories are the rows of a bivariate table.

*Cell*    The intersection of a row and a column in a bivariate table.

*Marginals*    The row and column totals in a bivariate table.

---

✓ *Learning*
*Check*

> *Examine Table 8.2. Make sure you can identify all the parts just described and that you understand how the numbers were obtained. Can you identify the independent and dependent variables in the table? You will need to know this to convert the frequencies to percentages.*

## ▣ HOW TO COMPUTE PERCENTAGES IN A BIVARIATE TABLE

To compare home ownership status for blacks and whites, we need to convert the raw frequencies to percentages because the column totals are not equal. Recall from Chapter 2 that percentages are especially useful for comparing two or more groups that differ in size. There are two basic rules for computing and analyzing percentages in a bivariate table:

1. Calculate percentages within each category of the independent variable.

2. Interpret the table by comparing the percentage point difference for different categories of the independent variable.

### Calculating Percentages Within Each Category of the Independent Variable

The first rule means that we have to calculate percentages within each category of the variable that the investigator defines as the independent variable. When the independent variable is arrayed in the *columns*, we compute percentages within each column separately. The frequencies within each cell and the row marginals are divided by the total of the column in which they are located, and the column totals should sum to 100%. When the independent variable is arrayed in the *rows*, we compute percentages within each row separately. The frequencies within each cell and the column marginals are divided by the total of the row in which they are located, and the row totals should sum to 100%.

In our example, we are interested in *race* as the independent variable and in its relationship with *home ownership*. Therefore, we are going to calculate percentages by using the column total of each racial group as the base of the percentage. For example, the percentage of black respondents who own their homes is obtained by dividing the number of black homeowners by the total number of blacks in the sample:

$$(100)\frac{2}{7} = 28.6\%$$

Table 8.3 presents percentages based on the data in Table 8.2. Notice that the percentages in each column add up to 100%, including the total column percentages. Always show the *N*s that are used to compute the percentages—in this case, the column totals.

**Table 8.3** Home Ownership by Race (in percentages)

| | Race | | |
| --- | --- | --- | --- |
| Home Ownership | Black | White | Total |
| Own | 28.6% | 70.0% | 52.9% |
| Rent | 71.4% | 30.0% | 47.1% |
| Total | 100% | 100% | 100% |
| (*N*) | (7) | (10) | (17) |

## Comparing the Percentages Across Different Categories of the Independent Variable

The second rule tells us to compare how home ownership varies between blacks and whites. Comparisons are made by examining differences between percentage points across different categories of the independent variable. Some researchers limit their comparisons to categories with at least a 10 percentage point difference. In our comparison, we can see that there is a 41.4 percentage point difference between the percentage of white homeowners (70%) and black homeowners (28.6%). In other words, in this group, whites are more likely to be homeowners than blacks.[5] Therefore, we can conclude that one's race appears to be associated with the likelihood of being a homeowner.

Note that the same conclusion would be drawn had we compared the percentage of black and white renters. However, since the percentages of homeowners and renters within each racial group sum to 100%, we need to make only one comparison. In fact, for any 2 × 2 table, only one comparison needs to be made to interpret the table. For a larger table, more than one comparison can be made and used in interpretation.

## ▣ HOW TO DEAL WITH AMBIGUOUS RELATIONSHIPS BETWEEN VARIABLES

Sometimes it isn't apparent which variable is independent or dependent; sometimes the data can be viewed either way. In this case, you might compute both row and column percentages. For example, Table 8.4 presents three sets of figures for the variables SPANKING and FEFAM for a sample of 127 GSS 2008 respondents: (a) the absolute frequencies, (b) the column percentages, and (c) the row percentages. SPANKING is measured with the survey question "Do you favor spanking to discipline a child?" The variable FEFAM measures whether the respondent agrees or disagrees with the statement "a man should work and a woman should stay at home." Table 8.4b shows that respondents who strongly disagree with spanking a child are less likely to agree to the FEFAM statement than those who strongly agree with spanking (10% compared with 50%). Table 8.4c shows that individuals who strongly agree that a man should work and a woman should stay at home are more likely to agree to spanking than those who disagree with the statement on men's and women's roles (94% compared with 63%).

Thus, percentaging within each *column* (Table 8.4b) allows us to examine the hypothesis that spanking (the independent variable) is associated with agreement to the FEFAM statement (dependent variable). When we percentage within each *row* (Table 8.4c), the hypothesis is that agreement or disagreement with the FEFAM statement (the independent variable) may be related to SPANKING (the dependent variable).[6]

**Table 8.4**   The Different Ways Percentages Can Be Computed: SPANKING by FEFAM

| | SPANKING | | |
| --- | --- | --- | --- |
| **FEFAM** | **Strongly Agree** | **Strongly Disagree** | **Row Total** |
| a. Absolute frequencies | | | |
|   Strongly agree | 48 | 3 | 51 |
|   Strongly disagree | 48 | 28 | 76 |
|   Column total | 96 | 31 | 127 |
| b. Column percentages (column totals as base) | | | |
|   Strongly agree | 50% | 10% | 40% |
|   Strongly disagree | 50% | 90% | 60% |
|   Column total | 100% | 100% | 100% |
|   | (96) | (31) | (127) |
| c. Row percentages (row totals as base) | | | |
|   Strongly agree | 94% | 6% | 100% |
|   | | | (51) |
|   Strongly disagree | 63% | 37% | 100% |
|   | | | (76) |
|   Column total | 76% | 24% | 100% |
|   | | | (127) |

*Source:* General Social Survey, 2008.

Finally, it is important to understand that ultimately what guides the construction and interpretation of bivariate tables is the theoretical question posed by the researcher. Although the particular example in Table 8.4 makes sense if interpreted using row or column percentages, not all data can be interpreted this way. For example, a table comparing women's and men's attitudes toward sexual harassment in the workplace could provide a sensible explanation in only one direction. Gender might influence a person's attitude toward sexual harassment; however, a person's attitude toward sexual harassment certainly couldn't influence her or his gender. Therefore, either row or column percentages are appropriate, depending on the way the variables are arrayed, but not both.

## 🔲 READING THE RESEARCH LITERATURE: PLACE OF DEATH IN AMERICA

The guidelines for constructing and interpreting bivariate tables discussed in this chapter are not always strictly followed. Most bivariate tables presented in the professional literature are a good deal more complex than those we have just been describing. Let's conclude this section with a typical example of how bivariate tables are presented in social science literature. The following example is drawn from a 2007 study by Andrea Gruneir, Vincent Mor, Sherry Weitzen, Rachael Truchill, Joan Teno, and Jason Roy on understanding variations on the sites of death in America.

According to the researchers, driven in part by the increasing aging of the U.S. population, there has been an increase in the level of public and professional concern about the quality of end-of-life care. Though public surveys confirm that most would prefer to die at home, the majority of Americans die in an institutional setting, such as a hospital or care facility. Acute care hospitals are still the number one site of death for people with chronic illnesses. In this study, Gruneir and her colleagues explored the likelihood of home versus hospital or nursing home death, explaining that

> where individuals spend their last days of life is influenced by individual demographic and clinical characteristics as well as to the degree which their community has an interest in and sufficient wealth to invest in service resources such as hospitals, nursing homes, or home and hospice care services. (p. 359)

The study examines differences in the place of death by gender, age, marital status, race/ethnicity, education, and cause of death. Researchers relied on data from the National Vital Statistics System (NVSS) for 1997. The data set includes death certificates for 1,402,167 deaths, identifying place of death as either acute care hospital, nursing care facility, home, or other.

Table 8.5 shows the results of the survey.

Follow these steps in examining the researchers' findings.

1. Identify the dependent variable and the type of unit of analysis it describes (such as individual, city, or child). Here the dependent variable is *place of death in 1997*. The categories for this variable are "hospital," "home," or "nursing home." The type of unit used in this table is individual.

2. Identify the independent variables included in the table and the categories of each. There are five independent variables: gender, age, marital status, race/ethnicity, and education. For the first variable, gender, the categories include "male" and "female." Review the table to determine the categories for the remaining variables.

3. Clarify the structure of the table. Note that the independent variables are arrayed in the rows of the table and the dependent variable, *place of death,* is arrayed in the columns. The table is divided into five panels, one for each independent variable. There are actually five bivariate tables here—one for each independent variable.

   Since the independent variables are arrayed in the rows, percentages are calculated within each row separately, with the row totals (not shown) serving as the base for the percentages. From the table, we know that the largest number of 1997 deaths occurred in a hospital—740,405 died in a hospital compared with 330,447 who died at home and 331,315 who died in a nursing home. Combining categories for hospital and nursing home, as Gruneir and her colleagues explained, the majority of Americans died in an institutional setting rather than at home. Though not calculated, the row percentages should total 100%.

4. Using Table 8.5, we can make a number of comparisons, depending on which independent variable we are examining. For example, to determine the relationship between gender and place of death, compare the percentages between males and females. For example, we know that the

Table 8.5    Place of Death, 1997, by Gender, Age, Marital Status, Race/
Ethnicity, and Education (percentages reported)

| | Place of Death, 1997 | | |
| | Hospital (N = 740,405) | Home (N = 330,447) | Nursing home (N = 331,315) |
|---|---|---|---|
| **Gender** | | | |
| Male | 57.3 | 25.5 | 17.2 |
| Female | 48.7 | 21.8 | 29.5 |
| **Age** | | | |
| <65 | 64.4 | 29.2 | 6.4 |
| 65–74 | 59.3 | 28.1 | 12.6 |
| 75–84 | 52.7 | 22.7 | 24.6 |
| 85–94 | 40.8 | 16.9 | 42.3 |
| 95+ | 28.0 | 14.5 | 57.6 |
| **Marital status** | | | |
| Never married | 55.9 | 21.5 | 22.7 |
| Married | 59.0 | 26.9 | 14.1 |
| Widowed | 45.2 | 19.9 | 34.9 |
| Divorced | 54.6 | 26.1 | 19.2 |
| Not stated | 57.1 | 24.9 | 18.0 |
| **Race/ethnicity** | | | |
| White | 49.7 | 24.2 | 26.1 |
| Black | 66.4 | 20.2 | 13.5 |
| Hispanic | 65.2 | 22.7 | 12.1 |
| Other/unknown | 63.4 | 21.7 | 14.9 |
| **Education (years)** | | | |
| <9 | 50.8 | 20.3 | 29.0 |
| 9–11 | 54.7 | 23.0 | 22.3 |
| 12 | 53.5 | 23.8 | 22.7 |
| 13–15 | 51.9 | 26.4 | 21.7 |
| 16+ | 50.8 | 27.7 | 21.6 |
| Unknown | 56.8 | 18.8 | 24.4 |

Source: Andrea Gruneir, Vincent Mor, Sherry Weitzen, Rachael Truchil, Joan Teno, and Jason Roy, "Where People Die: A Multilevel Approach to Understanding Influences on Site of Death in America," *Medical Care Research Review 64* (2007): 351–378.

largest percentages for both male and females are in the category "died in hospital." Yet looking at each place of death category, we see that the percentages are about the same for those who died at home (25.5 males, 21.8 females), but greater differences exist between the percentage of those who died in a nursing home (17.2 males, 29.5 females) and in hospitals (57.3 males, 48.7 females). A higher percentage of females than males were reported dying in a nursing home, while a higher percentage of males than females were reported dying in a hospital.

You can make similar comparisons to determine the association between age, marital status, race/ethnicity, and education.

5. Finally, what conclusions can you draw about variations in place of death? The researchers offer this interpretation of the findings presented in the table.

> The frequency of nursing home death increased with age and among the oldest adults, nursing homes were the most common site of death. A greater percentage of women than men died in the nursing home (29.5% vs. 17.2%) but the converse was seen in other sites of death. Married and divorced decedents showed the greatest frequency of home death (26.9% and 26.1%, respectively) while widowed decedents showed the greatest frequency of nursing home death (34.9%). Approximately half of all white decedents died in hospital but well over 60% of each other racial/ethnic group died in hospital. Of those who died outside the hospital, white decedents were equivalently split between home and nursing home while other groups more frequently died at home than in nursing home. (p. 363)

✓ *Learning Check*

*Use Table 8.5 to verify each of the following conclusions drawn by the researchers about the place of death: (1) Among the oldest adults, nursing homes were the most common site of death. (2) A greater percentage of women died in nursing homes than men. (3) Marital status is related to home death. (4) Approximately half of all white respondents died at a hospital, while more than 60% of all other racial/ethnic groups died in a hospital. Can you explain these patterns? What other questions do these patterns raise about place of death?*

## ◉ THE PROPERTIES OF A BIVARIATE RELATIONSHIP

So far, we have looked at the general principles of a bivariate relationship as well as the more specific "mechanics" involved in examining bivariate tables. In this section, we present some detailed observations that we may want to make about the "properties" of a bivariate association. These properties can be expressed as three questions to ask when examining a bivariate relationship:[7]

1. Does there appear to be a relationship?

2. How strong is it?

3. What is the direction of the relationship?

### The Existence of the Relationship

We have seen earlier in this chapter that calculating percentages and comparing them are the two operations necessary to analyze a bivariate table. Based on Table 8.6, we want to examine whether the frequency of church attendance by respondents had an effect on their support for abortion. Support for

Table 8.6   Support for Abortion by Church Attendance

|  | Church Attendance | | | |
| Abortion | Never | Infrequently | Frequently | Total |
| --- | --- | --- | --- | --- |
| Yes | 55% | 50% | 26% | 43% |
| No | 45% | 50% | 74% | 57% |
| Total | 100% | 100% | 100% | 100% |
| (N) | (111) | (212) | (157) | (480) |

*Source:* General Social Survey, 2002.

abortion was measured with the following question: "Please tell me whether or not you think it should be possible for a pregnant woman to obtain a legal abortion if the woman wants it for any reason." Frequency of church attendance was determined by asking respondents to indicate how often they attend religious services.[8]

Let's hypothesize that those who attend church frequently are more likely to be pro-life. We are not suggesting that church attendance necessarily "causes" pro-life attitudes, but that perhaps there is an indirect connection between the two. For example, perhaps those who attend church less frequently are more likely to want decisions about the body to be made on an individual basis through the right to choose an abortion.

In this formulation, church attendance is said to "influence" attitudes toward abortion, so it is the independent variable; therefore, percentages are calculated within each category of church attendance (church attendance is the column variable).

A relationship is said to exist between two variables in a bivariate table if the percentage distributions vary across the different categories of the independent variable, in this case church attendance. We can easily see that the percentage that supports abortion changes across the different levels of church attendance. Of those who never attend church, 55% are pro-choice; of those who infrequently attend church, 50% are pro-choice; and of those who frequently attend church, 26% are pro-choice.

Table 8.6 indicates that church attendance and support for abortion are associated as hypothesized.

If church attendance were unrelated to attitudes toward abortion among GSS respondents, then we would expect to find equal percentages of respondents who are pro-choice (or anti-choice) regardless of the level of church attendance. Table 8.7 is a fictional representation of a strictly hypothetical pattern of no association between abortion attitudes and church attendance. The percentage of respondents who are pro-choice in each category of church attendance is equal to the overall percentage of respondents in the sample who are pro-choice (43%).

## The Strength of the Relationship

In the preceding section, we saw how to establish whether an association exists in a bivariate table. If it does, how do we determine the strength of the association between the two variables? A quick method is to examine the percentage difference across the different categories of the independent variable. The larger the percentage difference across the categories, the stronger the association.

Table 8.7   Support for Abortion by Church Attendance
(a hypothetical illustration of no relationship)

| Abortion | Church Attendance | | | Total |
| --- | --- | --- | --- | --- |
| | *Never* | *Infrequently* | *Frequently* | |
| Yes | 43% | 43% | 43% | 43% |
| No | 57% | 57% | 57% | 57% |
| Total | 100% | 100% | 100% | 100% |
| (*N*) | (111) | (212) | (157) | (480) |

In the hypothetical example of no relationship between church attendance and attitude toward abortion (Table 8.7), there is a 0% difference between the columns. At the other extreme, if all respondents who never attended church were pro-choice and none of the respondents who frequently attended church were pro-choice, a perfect relationship would be manifested in a 100% difference. Most relationships, however, will be somewhere in between these two extremes. In fact, we rarely see a situation with either a 0% or a 100% difference. Going back to the observed percentages in Table 8.6, we find the largest percentage difference between respondents who never attend and who frequently attend church (55% − 26% = 29%). The difference between respondents who infrequently attend and who frequently attend church (50% − 26% = 24%), though not as large, is nonetheless substantial, indicating a moderate relationship between church attendance and attitudes toward abortion.

Percentage differences are a rough indicator of the strength of a relationship between two variables. Later in this chapter, we discuss measures of association that provide a more standardized indicator of the strength of an association.

## The Direction of the Relationship

When both the independent and dependent variables in a bivariate table are measured at the ordinal level or the interval-ratio level, we can talk about the relationship between the variables as being either positive or negative. A **positive** bivariate relationship exists when the variables vary in the same direction. Higher values of one variable "go together" with higher values of the other variable. In a **negative** bivariate relationship, the variables vary in opposite directions: Higher values of one variable "go together" with lower values of the other variable (and the lower values of one go together with the higher values of the other).

---

*Positive relationship*   A bivariate relationship between two variables measured at the ordinal level or higher in which the variables vary in the same direction.

*Negative relationship*   A bivariate relationship between two variables measured at the ordinal level or higher in which the variables vary in opposite directions.

---

**Table 8.8**  Willingness to Pay Higher Taxes by Willingness to Pay Higher Prices: A Positive Relationship

| Willingness to Pay Higher Taxes | Willingness to Pay Higher Prices | | |
| --- | --- | --- | --- |
| | Unwilling | Indifferent | Willing |
| Unwilling | 91.5% | 36.4% | 23.6% |
| Indifferent | 5.1% | 55.1% | 18.6% |
| Willing | 3.4% | 8.5% | 57.8% |
| Total | 100% | 100% | 100% |
| (N) | (529) | (352) | (532) |

Source: International Social Survey Programme, 2000.

Table 8.8, from the 2000 International Social Survey Programme, displays a positive relationship between willingness to pay higher taxes and willingness to pay higher prices. Examine each category separately. For respondents who are unwilling to pay higher prices, an unwillingness to pay higher taxes is most typical (91.5%). For respondents who are indifferent to paying higher prices, the most common response is to be indifferent to paying higher taxes (55.1%); and finally, for respondents who are willing to pay higher prices, a willingness to pay higher taxes is most typical (57.8%). This is a positive relationship, with a willingness to pay higher prices associated with a willingness to pay higher taxes and an unwillingness to pay higher prices associated with an unwillingness to pay higher taxes.

Table 8.9, also from the International Social Survey Programme, shows a negative association between educational level and attendance of religious services for a sample of about 400 international respondents.[9] Individuals with no education typically attended religious services two to three times per month or more (66.2%). Individuals with a secondary degree (i.e., roughly, the U.S. equivalent to high school) typically attended religious services infrequently, ranging from monthly to several times a year (35.0%); and for individuals who had completed work at a university, the most common category was "never," meaning they never attend religious services (37.3%). The relationship is a negative one because as educational level increases, the frequency of attendance of religious services decreases.

**Table 8.9**  Support for Attendance of Religious Services by Educational Level: A Negative Relationship

| Attendance of Religious Services | Educational Level | | |
| --- | --- | --- | --- |
| | None | Secondary Degree | University Degree |
| Never | 5.2% | 32.5% | 37.3% |
| Infrequently | 28.6% | 35.0% | 34.9% |
| 2 to 3 Times per Month or More | 66.2% | 32.5% | 27.8% |
| Total | 100% | 100% | 100% |
| (N) | (77) | (237) | (126) |

Source: International Social Survey Programme, 2000.

In the Health Information National Trends Survey (HINTS) 2007, the respondents were asked if they had ever looked for information about cancer from any source. Table 8.10 shows the distribution of responses to this question by age groups:

**Table 8.10**   Ever Looked for Information About Cancer by Age

| | Age Groups | | |
|---|---|---|---|
| *Ever Looked for Information About Cancer* | *18–34* | *35–44* | *45+* |
| No | 62.9% | 59.3% | 51.9% |
| Yes | 37.1% | 40.7% | 48.1% |
| Total | 100% | 100% | 100% |
| (*N*) | (224) | (194) | (1,068) |

*Source:* HINTS, 2007.

Although the majority of responses was "No" in all age groups, if we look at each category of the dependent variable (ever looked for information about cancer) across the categories of the independent variable (age), we see that the percentage of respondents who selected the lower category of the dependent variable (i.e., those who never looked for information about cancer) among the lowest age group (62.9%) is more than the comparable percentages in the older groups (59.3% and 51.9%, respectively). Similarly, the percentage of respondents who selected the higher category of the dependent variable (i.e., those who looked for information about cancer) among the highest age group (48.1%) is more than the comparable percentages in the younger groups (40.7% and 37.1%, respectively). In other words, the lower categories of each variable are associated with each other, and the same is true for the higher categories. Hence, we can conclude that there is a positive relationship between these two variables.

The data presented in Table 8.11 are also from HINTS 2007, showing the relationship between respondents' distress score (measured on a scale between 0 and 24) and age groups. The variable DISTRESSSCORE, created based on 5-point scale questions matching key elements of general depressive disorders, captures a respondent's current psychological distress/depressive symptomatology. We recoded the variable through a median split to come up with a binary (i.e., two-category)

**Table 8.11**   Distress Score by Age

| | Age Groups | | |
|---|---|---|---|
| *Distress Score* | *18–34* | *35–44* | *45+* |
| 1–3 | 32.0% | 35.5% | 45.8% |
| 4+ | 68.0% | 64.5% | 54.2% |
| Total | 100% | 100% | 100% |
| (*N*) | (197) | (166) | (825) |

*Source:* HINTS, 2007.

variable, the median score being the cutoff point. The median distress score was 3: About half of the respondents had a distress score of 3 or less, and the distress score of the other half was 4 or more. Hence, in our recoded variable, respondents with a distress score of 1, 2, or 3 fall into the first—lower—category, while the second category includes the respondents with higher distress scores (4 or more).

In this example, there is a negative relationship between distress score and age. If we look at the first row, we see that respondents in the highest age group are more likely to be associated with a lower distress score than the respondents in the younger age groups (45.8% vs. 35.5% and 32.0%, respectively). In a similar way, we can say that the percentage of those with a higher level of distress is more among the respondents in the lowest age group as compared with their older counterparts (68.0% vs. 64.5% and 54.2%, respectively). In other words, as respondent's age increases, her or his level of distress decreases, and vice versa, which indicates that age and distress are negatively associated.

Later in this chapter, we will see that measures of relationship for ordinal or interval-ratio variables take on a positive or a negative value, depending on the direction of the relationship.

✓ *Learning Check*

*Based on Table 8.9, collapse the attendance of religious services categories into two: never versus sometimes. Recalculate the bivariate table, estimating the percentages. Compare your results with Table 8.9. What can you say about the changes in the relationship between educational level and attendance of religious services?*

## ◨ ELABORATION

In the preceding sections, we have looked at relationships between two variables—an independent and a dependent variable. The examination of a possible relationship between two variables, however, is only a first step in data analysis. Having established through bivariate analysis that the independent and dependent variables are associated, we may seek to interpret and understand the nature of this relationship through a procedure called elaboration. **Elaboration** is a process designed to further explore a bivariate relationship, involving the introduction of additional variables, called **control variables.** By adding a control variable to our analysis, we are considering or "controlling" for the variable's effect on the bivariate relationship. Each potential control variable represents an alternative explanation for the bivariate relationship under consideration. The details of elaboration will not be considered in this text.

---

*Elaboration*   A process designed to further explore a bivariate relationship; it involves the introduction of control variables.

*Control variable*   An additional variable considered in a bivariate relationship. The variable is controlled for when we take into account its effect on the variables in the bivariate relationship.

---

# ▣ HYPOTHESIS TESTING AND BIVARIATE TABLES

We shift our examination to the educational experience by focusing on first-generation college students—that is, students whose parents never completed a postsecondary education. Most first-generation students begin college at 2-year programs or at community colleges. According to W. Elliot Inman and Larry Mayes (1999), since first-generation college students represent a large segment of the community college population, they bring with them a set of distinct goals and constraints. Inman and Mayes set out to examine first-generation college students' experiences, but they began first by determining who was most likely to be a first-generation college student.

Data from Inman and Mayes's study are presented in Table 8.12, a bivariate table, which includes *gender* and *first-generation college status*. From the table, we know that a higher percentage of women than men reported being first-generation college students, 46.6% versus 35.4%.

**Table 8.12**  Percentage of Men and Women Who Are First-Generation College Students

| First Generation | Men | Women | Total |
|---|---|---|---|
| Firsts | 35.4% | 46.6% | 41.9% |
| | (691) | (1,245) | (1,936) |
| Nonfirsts | 64.6% | 53.4% | 58.1% |
| | (1,259) | (1,425) | (2,684) |
| Total (*N*) | 100.0% | 100.0% | 100.0% |
| | (1,950) | (2,670) | (4,620) |

*Source:* Adapted from W. Elliot Inman and Larry Mayes, "The Importance of Being First: Unique Characteristics of First Generation Community College Students," *Community College Review* 26, no. 3 (1999): 8. Copyright © North Carolina State University. Published by SAGE.

The percentage differences between males and females in first-generation college status, shown in Table 8.12, suggest that there is a relationship. In inferential statistics, we base our statements about the larger population on what we observe in our sample. How do we know whether the gender differences in Table 8.12 reflect a real difference in first-generation college status among the larger population? How can we be sure that these differences are not just a quirk of sampling? If we took another sample, would these differences be wiped out or be even reversed?

Let's assume that men and women are equally likely to be first-generation college students—that in the population from which this sample was drawn, there are no real differences between them. What would be the expected percentages of men and women who are first-generation college students versus those who are not?

If gender and first-generation college status were not associated, we would expect the same percentage of men and women to be first-generation college students. Similarly, we would expect to see the same percentage of men and women who are nonfirsts. These percentages should be equal to the percentage of "firsts" and "nonfirsts" respondents in the sample as a whole (categories used by Inman and Mayes). The last column of

Table 8.12—the row marginals—displays these percentages: 41.9% of all respondents were first-generation students, whereas 58.1% were nonfirsts. Therefore, if there were no association between gender and first-generation college status, we would expect to see 41.9% of the men and 41.9% of the women in the sample as first-generation students. Similarly, 58.1% of the men and 58.1% of the women would not be.

Table 8.13 shows these hypothetical expected percentages. Because the percentage distributions of the variable *first-generation college status* are identical for men and women, we can say that Table 8.13 demonstrates a perfect model of "no association" between the variable *first-generation college status* and the variable *gender.*

**Table 8.13**   Percentage of Men and Women Who Are First-Generation College Students: Hypothetical Data Showing No Association

| First Generation | Men | Women | Total |
|---|---|---|---|
| Firsts | 41.9% | 41.9% | 41.9% (1,936) |
| Nonfirsts | 58.1% | 58.1% | 58.1% (2,684) |
| Total (N) | 100.0% (1,950) | 100.0% (2,670) | 100.0% (4,620) |

If there is an association between gender and first-generation college status, then at least some of the observed percentages in Table 8.12 should differ from the hypothetical expected percentages shown in Table 8.13. On the other hand, if gender and first-generation college status are not associated, the observed percentages should approximate the expected percentages shown in Table 8.13. In a cell-by-cell comparison of Tables 8.12 and 8.13, you can see that there is quite a disparity between the observed percentages and the hypothetical percentages. For example, in Table 8.12, 35.4% of the men reported that they were first-generation college students, whereas the corresponding cell for Table 8.13 shows that 41.9% of the men reported the same. The remaining three cells reveal similar discrepancies.

Are the disparities between the observed and expected percentages large enough to convince us that there is a genuine pattern in the population? The *chi-square* statistic helps us answer this question. It is obtained by comparing the actual observed frequencies in a bivariate table with the frequencies that are generated under an assumption that the two variables in the cross-tabulation are not associated with each other. If the observed and expected values are very close, the chi-square statistic will be small. If the disparities between the observed and expected values are large, the chi-square statistic will be large.

## ▣ THE CONCEPT OF CHI-SQUARE AS A STATISTICAL TEST

The **chi-square test** (pronounced kai-square and written as $\chi^2$) is an inferential statistical technique designed to test for significant relationships between two variables organized in a bivariate table. The test has a variety of research applications and is one of the most widely used tests in the social sciences. Chi-square requires no assumptions about the shape of the population distribution from which a sample is drawn. It can be applied to nominally or ordinally measured variables (including grouped interval-level data).

---

*Chi-square test*  An inferential statistical technique designed to test for significant relationships between two nominal or ordinal variables organized in a bivariate table.

---

The chi-square test can also be applied to the distribution of scores for a single variable. Also referred to as the goodness-of-fit test, the chi-square can compare the actual distribution of a variable with a set of expected frequencies. This application is not presented in this chapter.

## ▣ THE CONCEPT OF STATISTICAL INDEPENDENCE

When two variables are not associated (as in Table 8.13), one can say that they are **statistically independent**. That is, an individual's score on one variable is independent of his or her score on the second variable. We identify statistical independence in a bivariate table by comparing the distribution of the dependent variable in each category of the independent variable. When two variables are statistically independent, the percentage distributions of the dependent variable within each category of the independent variable are identical. The hypothetical data presented in Table 8.13 illustrate the notion of statistical independence. Based on Table 8.13, we would say that first-generation college status is independent of one's gender.[10]

---

*Statistical independence*  The absence of association between two cross-tabulated variables. The percentage distributions of the dependent variable within each category of the independent variable are identical.

---

✓ *Learning* *Check*

*The data we will use to practice calculating chi-square are also from Inman and Mayes's research. We will examine the relationship between age (independent variable) and first-generation college status (the dependent variable), as shown in the following bivariate table:*

*Age and First-Generation College Status*

| First-Generation Status | Years of Age | | Total |
| | 19 Years or Younger | 20 Years or Older | |
| --- | --- | --- | --- |
| Firsts | 916 (33.7%) | 1,018 (53.6%) | 1,934 (41.9%) |
| Nonfirsts | 1,802 (66.3%) | 881 (46.4%) | 2,683 (58.1%) |
| Total (*N*) | 2,718 (100.0%) | 1,899 (100.0%) | 4,617 (100.0%) |

*Source:* Adapted from W. Elliot Inman and Larry Mayes, "The Importance of Being First: Unique Characteristics of First Generation Community College Students," *Community College Review* 26, no. 3 (1999): 8.

*Construct a bivariate table (in percentages) showing no association between age and first-generation college status.*

## ▣ THE STRUCTURE OF HYPOTHESIS TESTING WITH CHI-SQUARE

The chi-square test follows the same five basic steps as the statistical tests presented in Chapter 7: (1) making assumptions, (2) stating the research and null hypotheses and selecting alpha, (3) selecting the sampling distribution and specifying the test statistic, (4) computing the test statistic, and (5) making a decision and interpreting the results. Before we apply the five-step model to a specific example, let's discuss some of the elements that are specific to the chi-square test.

### The Assumptions

The chi-square test requires no assumptions about the shape of the population distribution from which the sample was drawn. However, like all inferential techniques, it assumes random sampling. It can be applied to variables measured at a nominal and/or an ordinal level of measurement.

### Stating the Research and the Null Hypotheses

The research hypothesis ($H_1$) proposes that the two variables are related in the population.

$H_1$: The two variables are related in the population. (Gender and first-generation college status are statistically dependent.)

Like all other tests of statistical significance, the chi-square is a test of the null hypothesis. The null hypothesis ($H_0$) states that no association exists between two cross-tabulated variables in the population, and therefore, the variables are statistically independent.

$H_0$: There is no association between the two variables in the population. (Gender and first-generation college status are statistically independent.)

✓ *Learning Check*

> *Refer to the data in the previous Learning Check. Are the variables age and first-generation college status statistically independent? Write out the research and the null hypotheses for your practice data.*

### The Concept of Expected Frequencies

Assuming that the null hypothesis is true, we compute the cell frequencies that we would expect to find if the variables are statistically independent. These frequencies are called **expected frequencies** (and are symbolized as $f_e$). The chi-square test is based on cell-by-cell comparisons between the expected frequencies ($f_e$) and the frequencies actually observed (**observed frequencies** are symbolized as $f_o$).

---

*Expected frequencies (f$_e$)* The cell frequencies that would be expected in a bivariate table if the two variables were statistically independent.

*Observed frequencies (f₀)*   The cell frequencies actually observed in a bivariate table.

---

## Calculating the Expected Frequencies

The difference between $f_o$ and $f_e$ will determine the likelihood that the null hypothesis is true and that the variables are, in fact, statistically independent. When there is a large difference between $f_o$ and $f_e$, it is unlikely that the two variables are independent, and we will probably reject the null hypothesis. On the other hand, if there is little difference between $f_o$ and $f_e$, the variables are probably independent of each other, as stated by the null hypothesis (and therefore, we will not reject the null hypothesis).

The most important element in using chi-square to test for the statistical significance of cross-tabulated data is the determination of the expected frequencies. Because chi-square is computed on actual frequencies instead of on percentages, we need to calculate the expected frequencies based on the null hypothesis.

In practice, the expected frequencies are more easily computed directly from the row and column frequencies than from the percentages. We can calculate the expected frequencies using this formula:

$$f_e = \frac{(\text{Column marginal})(\text{Row marginal})}{N} \tag{8.1}$$

To obtain the expected frequencies for any cell in any cross-tabulation in which the two variables are assumed independent, multiply the row and column totals for that cell and divide the product by the total number of cases in the table.

Let's use this formula to recalculate the expected frequencies for our data on gender and first-generation college status as displayed in Table 8.12. Consider the men who were first-generation college students (the upper left cell). The expected frequency for this cell is the product of the column total (1,950) and the row total (1,936) divided by all the cases in the table (4,620):

$$f_e = \frac{1{,}936 \times 1{,}950}{4{,}620} = 817.14$$

For men who are nonfirsts (the lower left cell), the expected frequency is

$$f_e = \frac{2{,}684 \times 1{,}950}{4{,}620} = 1{,}132.86$$

Next, let's compute the expected frequencies for women who are first-generation college students (the upper right cell):

$$f_e = \frac{1{,}936 \times 2{,}670}{4{,}620} = 1{,}118.86$$

Finally, the expected frequency for women who are nonfirsts (the lower right cell) is

$$f_e = \frac{2{,}684 \times 2{,}670}{4{,}620} = 1{,}551.14$$

These expected frequencies are displayed in Table 8.14.

Table 8.14   Expected Frequencies of Men and Women and First-
Generation College Status

| First Generation | Men | Women | Total |
|---|---|---|---|
| Firsts | 817.14 | 1,118.86 | 1,936 |
| Nonfirsts | 1,132.86 | 1,551.14 | 2,684 |
| Total (N) | 1,950 | 2,670 | 4,620 |

Note that the table of expected frequencies contains identical row and column marginals as the original table (Table 8.12). Although the expected frequencies usually differ from the observed frequencies (depending on the degree of relationship between the variables), the row and column marginals must always be identical with the marginals in the original table.

✓ Learning
Check

*Refer to the data in the Learning Check on page 203. Calculate the expected frequencies for age and first-generation college status and construct a bivariate table. Are your column and row marginals the same as in the original table?*

## Calculating the Obtained Chi-Square

The next step in calculating chi-square is to compare the differences between the expected and observed frequencies across all cells in the table. In Table 8.15, the expected frequencies are shown next to the corresponding observed frequencies. Note that the difference between the observed and expected frequencies in each cell is quite large. Is it large enough to be significant? The way we decide is by calculating the **obtained chi-square** statistic:

$$\chi^2 = \sum \frac{(f_o - f_e)^2}{f_e} \qquad (8.2)$$

where

$f_o$ = observed frequencies

$f_e$ = expected frequencies

---

*Chi-square (obtained)*   The test statistic that summarizes the differences between the observed ($f_o$) and the expected ($f_e$) frequencies in a bivariate table.

---

According to this formula, for each cell, subtract the expected frequency from the observed frequency, square the difference, and divide by the expected frequency. After performing this operation for every cell, sum the results to obtain the chi-square statistic.

Let's follow these procedures using the observed and expected frequencies from Table 8.15. Our calculations are displayed in Table 8.16. The obtained chi-square statistic, 57.99, summarizes the differences between the observed frequencies and the frequencies that we would expect to see if the

**Table 8.15** Observed and Expected Frequencies of Men and
Women Who Are First-Generation College Students

| First Generation | Men $f_o$ | Men $f_e$ | Women $f_o$ | Women $f_e$ | Total |
|---|---|---|---|---|---|
| Firsts | 691 | 817.14 | 1,245 | 1,118.86 | 1,936 |
| Nonfirsts | 1,259 | 1,132.86 | 1,425 | 1,551.14 | 2,684 |
| Total ($N$) | 1,950 | 2,670 | 4,620 | | |

**Table 8.16** Calculating Chi-Square

| Gender and First-Generation College Status | $f_o$ | $f_e$ | $f_o - f_e$ | $(f_o - f_e)^2$ | $\dfrac{(f_o - f_e)^2}{f_e}$ |
|---|---|---|---|---|---|
| Men/firsts | 691 | 817.14 | −126.14 | 15,911.2996 | 19.47 |
| Men/nonfirsts | 1,259 | 1,132.86 | 126.14 | 15,911.2996 | 14.04 |
| Women/firsts | 1,245 | 1,118.86 | 126.14 | 15,911.2996 | 14.22 |
| Women/nonfirsts | 1,425 | 1,551.14 | −126.14 | 15,911.2996 | 10.26 |

$$\chi^2 = \sum \frac{(f_o - f_e)^2}{f_e} = 57.99$$

null hypothesis were true and the variables—gender and first-generation college status—were not associated. Next, we need to interpret our obtained chi-square statistic and decide whether it is large enough to allow us to reject the null hypothesis.

> *Using the format of Table 8.16, construct a table to calculate chi-square for age and educational attainment.*

✓ Learning
Check

## The Sampling Distribution of Chi-Square

Like other sampling distributions, the chi-square sampling distributions depend on the degrees of freedom. In fact, the chi-square sampling distribution is not one distribution, but—like the *t* distribution—is a family of distributions. The shape of a particular chi-square distribution depends on the number of degrees of freedom. This is illustrated in Figure 8.1, which shows chi-square distributions for 1, 5, and 9 degrees of freedom. Here are some of the main properties of the chi-square distributions that can be observed in this figure:

- The distributions are positively skewed. The research hypothesis for the chi-square is always a one-tailed test.
- Chi-square values are always positive. The minimum possible value is zero, with no upper limit to its maximum value. A chi-square of zero means that the variables are completely independent and the observed frequencies in every cell are equal to the corresponding expected frequencies.

- As the number of degrees of freedom increases, the chi-square distribution becomes more symmetrical and, with degrees of freedom greater than 30, begins to resemble the normal curve.

**Figure 8.1** Chi-Square Distributions for 1, 5, and 9 Degrees of Freedom

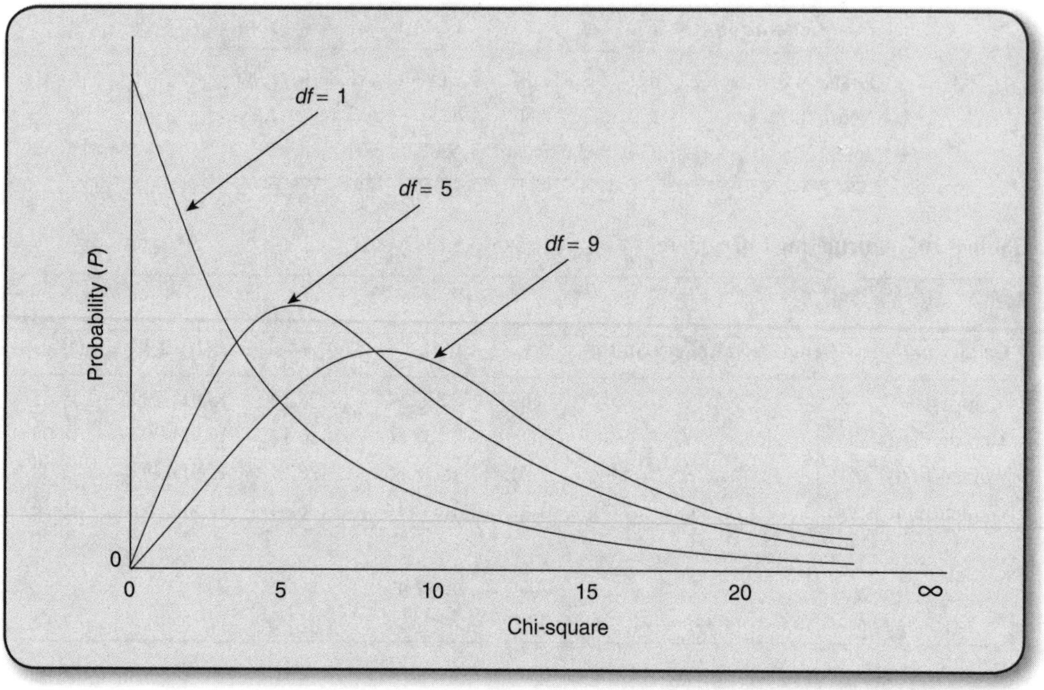

## Determining the Degrees of Freedom

With cross-tabulation data, we find the degrees of freedom by using the following formula:

$$df = (r-1)(c-1) \tag{8.3}$$

where

$r$ = the number of rows

$c$ = the number of columns

Thus, Table 8.12 with 2 rows and 2 columns has $(2-1)(2-1)$ or 1 degree of freedom. If the table had 3 rows and 2 columns, it would have $(3-1)(2-1)$ or 2 degrees of freedom.

Appendix C shows values of the chi-square distribution for various degrees of freedom. Notice how the table is arranged with the degrees of freedom listed down the first column and the level of significance (or $P$ values) arrayed across the top. For example, with 5 degrees of freedom, the probability associated with a chi-square as large as 15.086 is .01. An obtained chi-square as large as 15.086 would occur only once in 100 samples.

The degrees of freedom in a bivariate table can be interpreted as the number of cells in the table for which the expected frequencies are free to vary, given that the marginal totals are already set. Based on our data in Table 8.14, suppose we first calculate the expected frequencies for men who are first-generation college students ($f_e = 817.14$). Because the sum of the expected frequencies in the first column is set at 1,950, the expected frequency of men who are nonfirsts has to be 1,132.86 (1,950 − 817.14). Similarly, all other cells are predetermined by the marginal totals and are not free to vary. Therefore, this table has only 1 degree of freedom.

Data in a bivariate table can be distorted if by chance one cell is over- or undersampled and may influence the chi-square calculation. Calculation of the degrees of freedom compensates for this, but in the case of a 2 × 2 table with just 1 degree of freedom, the value of chi-square should be adjusted by applying the Yates's correction for continuity. The formula reduces the absolute value of each ($f_o - f_e$) by .5, then the difference is squared and then divided by the expected frequency for each cell. The formula for the Yates's correction for continuity is as follows:

$$\chi_c^2 = \sum \frac{(|f_o - f_e| - 0.5)^2}{f_e} \qquad (8.4)$$

✓ Learning Check

Based on Appendix C, identify the probability for each chi-square value (df in parentheses):

- 12.307 (15)
- 20.337 (21)
- 54.052 (24)

## Making a Final Decision

With the Yates's correction, the corrected chi-square is 57.54. Refer to Table 8.17 for calculations.

We can see that 57.54 does not appear on the first row ($df = 1$); in fact, it exceeds the largest chi-square value of 10.827 ($P = .001$). We can establish that the probability of obtaining a chi-square of 57.54 is less than .001 if the null hypothesis were true. If our alpha was preset at .05, the probability of 10.827 would be well below this. Therefore, we can reject the null hypothesis that gender and first-generation college status are not associated in the population from which our sample was drawn. Remember, the larger the chi-square statistic, the smaller the $P$ value providing us with more evidence to reject the null hypothesis. We can be very confident of our conclusion that there is a relationship between gender and first-generation college status in the population because the probability of this result occurring owing to sampling error is less than .001, a very rare occurrence.

✓ Learning Check

*What decision can you make about the association between age and first-generation college status? Should you reject the null hypothesis at the .05 alpha level or at the .01 level?*

**Table 8.17**   Calculating Yates's Correction

| Gender and First-Generation College Status | $\|f_o - f_e\|$ | $(\|f_o - f_e\| - .50)^2$ | $f_e$ | $\dfrac{(\|f_o - f_e\| - .5)^2}{f_e}$ |
|---|---|---|---|---|
| Men firsts | −126.14 | $(125.64)^2 = 15{,}785.41$ | 817.14 | 19.32 |
| Men nonfirsts | 126.14 | $(125.64)^2 = 15{,}785.41$ | 1,132.86 | 13.93 |
| Women firsts | 126.14 | $(125.64)^2 = 15{,}785.41$ | 1,118.86 | 14.11 |
| Women nonfirsts | −126.14 | $(125.64)^2 = 15{,}785.41$ | 1,551.14 | 10.18 |
| Total | | | | 57.54 |

## ▣ SAMPLE SIZE AND STATISTICAL SIGNIFICANCE FOR CHI-SQUARE

Although we found the relationship between gender and first-generation college status to be statistically significant, this in itself does not give us much information about the *strength* of the relationship or its *substantive significance* in the population. Statistical significance only helps us evaluate whether the argument (the null hypothesis) that the observed relationship occurred by chance is reasonable. It does not tell us anything about the relationship's theoretical importance or even if it is worth further investigation.

The distinction between statistical and substantive significance is important in applying any of the statistical tests discussed in Chapter 7. However, this distinction is of particular relevance for the chi-square test because of its sensitivity to sample size. The size of the calculated chi-square is directly proportional to the size of the sample, independent of the strength of the relationship between the variables.

For instance, suppose that we cut the observed frequencies for every cell in Table 8.12 exactly into half—which is equivalent to reducing the sample size by one half. This change will not affect the percentage distribution of firsts among men and women; therefore, the size of the percentage difference and the strength of the association between gender and first-generation college status will remain the same. However, reducing the observed frequencies by half will cut down our calculated chi-square by exactly half, from 57.54 to 28.77. (Can you verify this calculation?) Conversely, had we doubled the frequencies in each cell, the size of the calculated chi-square would have doubled, thereby making it easier to reject the null hypothesis.

This sensitivity of the chi-square test to the size of the sample means that a relatively strong association between the variables may not be significant when the sample size is small. Similarly, even when the association between variables is very weak, a large sample may result in a statistically significant relationship. However, just because the calculated chi-square is large and we are able to reject the null hypothesis by a large margin does not imply that the relationship between the variables is strong and substantively important.

Another limitation of the chi-square test is that it is sensitive to small expected frequencies in one or more of the cells in the table. Generally, when the expected frequency in one or more of the cells is below 5, the chi-square statistic may be unstable and lead to erroneous conclusions. There is no hard-and-fast rule regarding the size of the expected frequencies. Most researchers limit the use of chi-square to tables that either have no $f_e$ values below 5 or have no more than 20% of the $f_e$ values below 5.

Testing the statistical significance of a bivariate relationship is only a small step, although an important one, in examining a relationship between two variables. A significant chi-square suggests that a relationship, weak or strong, probably exists in the population and is not due to sampling fluctuation. However, to establish the strength of the association, we need to employ measures of association such as gamma, lambda, or Pearson's *r* (see Chapter 9). Used in conjunction, statistical tests of significance and measures of association can help determine the importance of the relationship and whether it is worth additional investigation.

## ◉ READING THE RESEARCH LITERATURE: VIOLENT OFFENSE ONSET BY GENDER, RACE, AND AGE

Paul Mazerolle, Alex Piquero, and Robert Brame (2010)[11] investigate whether violent onset offenders have distinct career dimensions from offenders whose initial offending involves nonviolence. In Table 8.18, the researchers examine the relationship between gender, race, and age and nonviolent versus violent onset using chi-square analysis. Their data are based on 1,503 juvenile offenders in Queensland, Australia.

Note that the obtained chi-square and its probability are reported for each pair of variables. There are two significant models, violent offense by gender and violent offense by age. For each cell, the *N* and percentage are reported.

According to the researchers' report, first, while males, in absolute terms, contribute many more violent onset youth to the overall total, in proportional terms, females exhibit a higher prevalence rate. The findings show that 18.32% of females exhibit violent onset compared with 11.71% of males.

Moreover, and perhaps unexpectedly, no differences were observed in the prevalence of violent onset across indigenous (11.83%) and nonindigenous groups (12.83%). However, in proportional terms, a higher prevalence of violent onset was observed for late-onset (14.74%) as opposed to early-onset offenders (9.67%), which was significant at $P < .01$.[12]

**Table 8.18**   Violent Offense Onset by Gender, Race, and Age

| | Nonviolent Onset N (%) | Violent Onset N (%) | Total N (%) |
|---|---|---|---|
| **Gender** | | | |
| Male | 1,146 (88.29) | 152 (11.71) | 1,298 (100) |
| Female | 156 (81.68) | 35 (18.32) | 191 (100) |
| Total | 1,302 | 187 | 1,489 |
| | | | $\chi^2 = 6.331$** |
| **Indigenous status** | | | |
| Nonindigenous | 815 (87.17) | 120 (12.83) | 935 (100) |
| Indigenous | 477 (88.17) | 64 (11.83) | 541 (100) |
| | | | $\chi^2 = 0.317$ |
| **Age at first offense** | | | |
| Less than 14 years | 579 (90.33) | 62 (9.67) | 641 (100) |
| 14 years and older | 723 (85.26) | 125 (14.74) | 848 (100) |
| | | | $\chi^2 = 8.539$** |

*Note:* **$P < .01$.

✓ *Learning*
*Check*

> *For the bivariate table with age and first-generation college status, the value of the obtained chi-square is 181.15 with 1 degree of freedom. Based on Appendix C, we determine that its probability is less than .001. This probability is less than our alpha level of .05. We reject the null hypothesis of no relationship between age and first-generation college status. If we reduce our sample size by half, the obtained chi-square is 90.58. Determine the P value for 90.58. What decision can you make about the null hypothesis?*

## ▣ PROPORTIONAL REDUCTION OF ERROR: A BRIEF INTRODUCTION

Earlier, we introduced cross-tabulation where the relationship between two variables was analyzed by making a number of percentage comparisons and incorporated chi-square into our discussion as an inferential test to examine whether two variables are statistically related. Now, we review special **measures of association** for nominal and ordinal variables. These measures enable us to use a single summarizing measure or number for analyzing the pattern of relationship between two variables. Unlike chi-square, measures of association reflect the strength of the relationship and, at times, its direction (whether it is positive or negative). They also indicate the usefulness of predicting the dependent variable from the independent variable.

In this section, we discuss four measures of association: lambda (measures of association for nominal variables), gamma and Kendall's tau-*b* (measures of association between ordinal variables), and Cramer's *V* (a chi-square-related measure of association). In Chapter 9, we introduce Pearson's correlation coefficient, which is used for measuring bivariate association between interval-ratio variables.

---

*Measure of association*   A single summarizing number that reflects the strength of a relationship, indicates the usefulness of predicting the dependent variable from the independent variable, and often shows the direction of the relationship.

---

Except Cramer's *V*, all the measures of association discussed here and Chapter 9 are based on the concept of the **proportional reduction of error**, often abbreviated as **PRE**. According to the concept of PRE, two variables are associated when information about one can help us improve our prediction of the other.

---

*Proportional reduction of error (PRE)*   The concept that underlies the definition and interpretation of several measures of association. PRE measures are derived by comparing the errors made in predicting the dependent variable while ignoring the independent variable with errors made when making predictions that use information about the independent variable.

---

Table 8.19 may help us grasp intuitively the general concept of PRE. Using General Social Survey (GSS) 2008 data, Table 8.19 shows a moderate relationship between the independent variable, *educational*

**Table 8.19**    Support for Abortion by Degree

| Support for Abortion | Degree | | Total |
|---|---|---|---|
| | High School or Less | More Than High School | |
| No | 399 | 156 | 555 |
| | 64.3% | 44.4% | 57.1% |
| Yes | 222 | 195 | 417 |
| | 35.7% | 55.6% | 42.9% |
| Total | 621 | 351 | 972 |
| | 100% | 100% | 100% |

*Source:* General Social Survey, 2008.

*attainment,* and the dependent variable, *support for abortion if the woman is poor and can't afford any more children.* The table shows that 64.3% of the respondents who had a high school degree or less were antiabortion, compared with only 44.4% of the respondents who had more than a high school degree.

The conceptual formula for all[13] PRE measures of association is

$$PRE = \frac{E_1 - E_2}{E_1} \tag{8.5}$$

where

$E_1$ = errors of prediction made when the independent variable is ignored (Prediction 1)

$E_2$ = errors of prediction made when the prediction is based on the independent variable (Prediction 2)

All PRE measures are based on comparing predictive error levels that result from each of the two methods of prediction. Let's say that we want to predict a respondent's position on abortion, but we do not know anything about the degree he or she has. Based on the row totals in Table 8.19, we could predict that every respondent in the sample is antiabortion because this is the modal category of the variable *abortion position.* With this prediction, we would make 417 errors because in fact 555 respondents in this group are antiabortion but 417 respondents are pro-choice. (See the row marginals in Table 8.19.) Thus,

$$E_1 = 972 - 555 = 417$$

How can we improve this prediction by using the information we have on each respondent's educational attainment? For our new prediction, we will use the following rule: If a respondent has a high school degree or less, we predict that he or she will be antiabortion; if a respondent has more than a high school degree, we predict that he or she is pro-choice. It makes sense to use this rule because we know, based on Table 8.19, that respondents with a lower educational attainment are more likely to be antiabortion, while respondents who have more than a high school degree are more likely to be pro-choice. Using this prediction rule, we will make 378 errors (instead of 417) because 156 of the respondents who have more than a high school degree are actually antiabortion, whereas 222 of the respondents who have a high school degree or less are pro-choice (156 + 222 = 378). Thus,

$$E_2 = 156 + 222 = 378$$

Our first prediction method, ignoring the independent variable (educational attainment), resulted in 417 errors. Our second prediction method, using information we have about the independent variable (educational attainment), resulted in 378 errors. If the variables are *associated*, the second method will result in fewer errors of prediction than the first method. The stronger the relationship is between the variables, the larger will be the reduction in the number of errors of prediction.

Let's calculate the proportional reduction of error for Table 8.19 using Formula 8.5. The proportional reduction of error resulting from using educational attainment to predict position on abortion is

$$\text{PRE} = \frac{417 - 378}{417} = 0.09$$

The more you work with various measures of association, the better feel you will have for what particular values mean. Until you develop this instinct, here are some guidelines regarding what is generally considered a strong relationship and what is considered a weak relationship.

PRE measures of association can range from 0.0 to ±1.0. A PRE of zero indicates that the two variables are not associated; information about the independent variable will not improve predictions about the dependent variable. A PRE of ±1.0 indicates a perfect positive or negative association between the variables; we can predict the dependent variable without error using information about the independent variable. Intermediate values of PRE will reflect the strength of the association between the two variables and therefore the utility of using one to predict the other. The more the measure of association departs from 0.00 in either direction, the stronger the association. PRE measures of association can be multiplied by 100 to indicate the percentage improvement in prediction.

**Figure 8.2**    A Guide to Interpreting Strength of Association

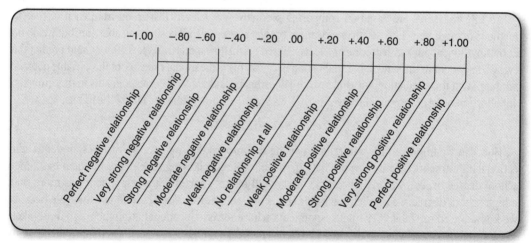

A PRE of 0.09 indicates that there is a weak relationship between respondents' educational attainment and their position on abortion. (Refer to Figure 8.2 for an approximation of the strength of a relationship.) A PRE of 0.09 means that we have improved our prediction of respondents' position on abortion by just 9% (0.09 × 100 = 9.0%) by using information on their educational attainment.

# ▣ LAMBDA: A MEASURE OF ASSOCIATION FOR NOMINAL VARIABLES

In this section, we examine the relationship between a voter's race, the independent variable, and the preferred Democratic candidate, the dependent variable, for the 2008 presidential nomination. Table 8.20 displays results from Gallup, Inc. based on a nationally represented survey of Democratic voters during the week of February 1, 2008. We can see 2,932 Democratic voters classified by their race and their preferred Democratic presidential candidate. We will consider a voter's race to be the independent variable and preferred Democratic candidate, which may be explained or predicted by the independent variable, to be the dependent variable.

Because race and Democratic vote are nominal variables, we need to apply a measure of association suitable for calculating relationships between nominal variables. Such a measure will help us determine how strongly associated race is with the 2008 preferred presidential Democratic candidate. **Lambda** is such a PRE measure.

**Table 8.20**  2008 Preferred Democratic Presidential Candidate by Race

| *2008 Preferred Democratic Presidential Candidate* | Race | | |
|---|---|---|---|
| | *Black* | *White* | *Row Total* |
| Barack Obama | 330 | 966 | 1,296 |
| Hillary Clinton | 117 | 1,519 | 1,636 |
| Total | 447 | 2,485 | 2,932 |

*Source: Gallup Poll Daily*, February 1–9, 2008.

---

*Lambda*  An asymmetrical measure of association, lambda is suitable for use with nominal variables and may range from 0.0 to 1.0. It provides us with an indication of the strength of an association between the independent and dependent variables.

---

## A Method for Calculating Lambda

Take a look at Table 8.20 and examine the row totals, which show the distribution of the variable *2008 preferred Democratic presidential candidate*. If we had to predict which candidate Democrats prefer, our best bet would be to guess the mode, which is that everyone voted for Senator Hillary Clinton. This prediction will result in the smallest possible error. The number of wrong predictions we make using this method is actually 1,296, since only 1,636 (the mode) out of 2,932 indicated their preference for Clinton (2,932 − 1,636 = 1,296).

Now take another look at Table 8.20, but this time let's consider race when we predict 2008 preferred Democratic presidential candidate. Again, we can use the mode of 2008 preferred Democratic presidential candidate, but this time we apply it separately for black and white Democrats. The mode for black Democrats is "Obama" (330 voters); therefore, we can predict that all black Democrats are Obama supporters. With this method of prediction, we make 117 errors, since 117 out of 447 black Democrats prefer Clinton (447 − 330 = 117). Next, we look at the group of white Democrats. The mode for this group is "Clinton"; this will be our prediction for this group. This method of prediction results in 966 errors (2,485 − 1,519 = 966). The total number of errors is thus 117 + 966 or 1,083 errors.

Let's now put it all together and state the procedure for calculating lambda in more general terms.

1. Find $E_1$, the errors of prediction made when the independent variable is ignored. To find $E_1$, find the mode of the dependent variable and subtract its frequency from $N$. For Table 8.20,

$$E_1 = N - \text{Modal frequency}$$

$$E_1 = 2,932 - 1,636 = 1,296$$

2. Find $E_2$, the errors made when the prediction is based on the independent variable. To find $E_2$, find the modal frequency for each category of the independent variable, subtract it from the category total to find the number of errors, and then add up all the errors. For Table 8.20,

$$\text{Black Democrats} = 447 - 330 = 117$$

$$\text{White Democrats} = 2,485 - 1,519 = 966$$

$$E_2 = 117 + 966 = 1,083$$

3. Calculate lambda using Formula 8.5:

$$\text{Lambda} = \frac{E_1 - E_2}{E_1} = \frac{1,296 - 1,083}{1,296} = 0.16$$

Lambda may range in value from 0.0 to 1.0. Zero indicates that there is nothing to be gained by using the independent variable to predict the dependent variable. A lambda of 1.0 indicates that by using the independent variable as a predictor, we are able to predict the dependent variable without any error. In our case, a lambda of 0.16 is less than one quarter of the distance between 0.0 and 1.0, indicating that for this sample of respondents, race and 2008 preferred Democratic presidential candidate are only slightly associated.

The proportional reduction of error indicated by lambda, when multiplied by 100, can be interpreted as follows: By using information on respondent's race to predict the 2008 preferred Democratic presidential candidate, we have reduced our error of prediction by 16% ($0.16 \times 100 = 16\%$). In other words, if we rely on respondent's race to predict preferred Democratic presidential candidate, we would reduce our error of prediction by 16 out of 100 or 16% ($0.16 \times 100$).

✓ *Learning*
*Check*

*Explain why lambda would not assume negative values.*

## Some Guidelines for Calculating Lambda

Lambda is an **asymmetrical measure of association**. This means that lambda will vary depending on which variable is considered the independent variable and which the dependent variable. In our example, we considered the 2008 preferred Democratic presidential candidate as the dependent variable and race as the independent variable, and not vice versa. Had we considered, instead, the 2008 preferred Democratic presidential candidate as the independent variable and race as the dependent variable, we would have obtained a slightly different lambda value.

---

*Asymmetrical measure of association*   A measure whose value may vary depending on which variable is considered the independent variable and which the dependent variable.

---

The method of calculation follows the same guidelines even when the variables are switched. However, exercise caution in calculating lambda, especially when the independent variable is arrayed in the rows rather than in the columns. To avoid confusion, it is safer to switch the variables and follow the convention of arraying the independent variable in the columns; then follow the exact guidelines suggested for calculating lambda. Remember, however, that although lambda can be calculated either way, ultimately what guides the decision of which variables to consider as independent or dependent is the theoretical question posed by the researcher.

Lambda is always zero in situations in which the mode for each category of the independent variable falls into the same category of the dependent variable. A problem with interpreting lambda arises in situations in which lambda is zero, but other measures of association indicate that the variables are associated. To avoid this potential problem, examine the percentage differences in the table whenever lambda is exactly equal to zero. If the percentage differences are very small (usually 5% or less), lambda is an appropriate measure of association for the table. However, if the percentage differences are larger, indicating that the two variables may be associated, lambda will be a poor choice as a measure of association. In such cases, we may want to discuss the association in terms of the percentage differences or select an alternative measure of association.

*Recalculate the lambda for Table 8.20, this time assuming the 2008 preferred Democratic presidential candidate as the independent variable and race as the dependent variable. How has the value of lambda changed?*   ✓ *Learning Check*

## ▣ CRAMER'S *V*: A CHI-SQUARE-RELATED MEASURE OF ASSOCIATION FOR NOMINAL VARIABLES

Cramer's *V* is an alternative measure of association that can be used for nominal variables. It is based on the value of chi-square and ranges between 0 to 1, with 0 indicating no association and 1 indicating perfect association. Because it cannot take negative values, it is considered a nondirectional measure. Unfortunately, Cramer's *V* is somewhat limited because the results cannot be interpreted using the PRE framework. It is calculated using the following formula:

$$\text{Cramer's } V = \sqrt{\frac{\chi^2}{N \times m}} \tag{8.6}$$

where $m$ = smaller of $(r - 1)$ or $(c - 1)$.

Hypothetically, say we tested the hypothesis that education and health are related in the population and our analysis yielded a chi-square value of 53.96, leading us to reject the null hypothesis that there are no differences in health among different educational groups. We would conclude that in the population from which our sample was drawn, health does vary by educational attainment.

We can use Cramer's $V$ to measure the relative strength of the association between health assessment and level of education using Formula 8.6 above.

$$\text{Cramer's } V = \sqrt{\frac{\chi^2}{N \times m}} = \sqrt{\frac{53.96}{844 \times 2}} = \sqrt{0.032} = 0.18$$

A Cramer's $V$ of 0.18 tells us that there is a weak association between health assessment and level of education.

## ▣ GAMMA AND KENDALL'S tau-*b*: ORDINAL MEASURES OF ASSOCIATION

In this section, we discuss a way to measure and interpret an association between two *ordinal* variables. If there is an association between the two variables, knowledge of one variable will enable us to make better predictions of the other variable.

Let's look at a research example in which the association between two ordinal variables is considered. We want to examine the hypothesis that the higher one's educational level, the more satisfied he or she is with one's financial situation. To examine this hypothesis, we selected two variables from the 2008 GSS. The variable education (EDUC) was recoded into a dichotomous variable, with those indicating that they had 11 or fewer years of education into one category and those reporting that they had 12 or more years of education into a second category. The first category was recoded to represent those without a high school diploma, while the second category was recoded to represent those with at least a high school diploma. The variable satisfaction with financial situation (SATFIN) was also recoded into a dichotomous variable, with response options "satisfied" and "more or less satisfied" coded as "satisfied." The response option "not at all satisfied" was left to represent those that reported that they were unsatisfied with their financial situation.

Table 8.21 displays the cross-tabulation of these two variables, with education as the independent variable and satisfaction with financial situation as the dependent variable. We find that 43.2% of those with less than a high school degree are unsatisfied with their financial situation, as compared with 28.5% of those with at least a high school degree. The percentage difference (43.2% − 28.5% = 14.7%) suggests that the variables are related. If we examine the percentage difference across those who are satisfied with their financial situation (71.5% − 56.8% = 14.7%), we find additional evidence that education and financial satisfaction are associated.

**Table 8.21**   Financial Satisfaction by Education

| Financial Satisfaction (Y) | Education (X) | | Total |
| --- | --- | --- | --- |
| | High School or More | Less Than High School | |
| Satisfied | 71.5% | 56.8% | 69% |
| | (1,194) | (193) | (1,387) |
| Unsatisfied | 28.5% | 43.2% | 31% |
| | (477) | (147) | (624) |
| Total | 100% | 100% | 100% |
| (N) | (1,671) | (340) | (2,011) |

*Source:* General Social Survey, 2008.

## ▣ CALCULATING GAMMA AND KENDALL'S tau-*b*

**Gamma** and **Kendall's tau-*b*** are symmetrical measures of association suitable for use with ordinal variables or with dichotomous nominal variables. This means that their value will be the same regardless of which variable is the independent variable or the dependent variable. Thus, if we had wanted to predict education from financial satisfaction rather than the opposite, we would have obtained the same gamma.

Both gamma and Kendall's tau-*b* can vary from 0.0 to ±1.0 and provide us with an indication of the strength and direction of the association between the variables. Gamma and Kendall's tau-*b* can be positive or negative. A gamma or Kendall's tau-*b* of 1.0 indicates that the relationship between the variables is positive and that the dependent variable can be predicted without any error based on the independent variable. A gamma of −1.0 indicates a perfect and a negative association between the variables.

A gamma or Kendall's tau-*b* of zero reflects no association between the two variables; hence, there is nothing to be gained by using the independent variable to predict the dependent variable. Because the calculations for these measures are quite laborious, we are using SPSS to calculate them. Earlier in the chapter, we used the Crosstabs procedure in SPSS to create bivariate tables. The same procedure is used to request measures of association. To analyze the relationship between Financial Satisfaction and Education, click on *Analyze, Descriptive Statistics,* then *Crosstabs* to get to the Crosstabs dialog box. Put Financial Satisfaction (SATFIN) in the *Row(s)* box and Education (EDUC) in the *Column(s)* box. Then click on the *Statistics* button. The Statistics dialog box (see Figure 8.3) has about a dozen statistics from which to choose. Note that four statistics are listed in separate categories for "Nominal" and "Ordinal" data. Lambda is listed in the former, and gamma and Kendall's tau-*b* in the latter. Cramer's *V* can be easily obtained by checking the Phi and Cramer's *V* box. The other measures of association, such as Somer's *d*, and phi, will not be discussed in this textbook. The chi-square statistic was discussed earlier.

Statistics are presented in Figure 8.4. A gamma of 0.347 indicates that there is a moderate positive association between education and financial satisfaction. We can conclude that using information on respondents' education helps us improve the prediction of their financial satisfaction by 34.7%.

**Figure 8.3**   The Statistics Dialog Box

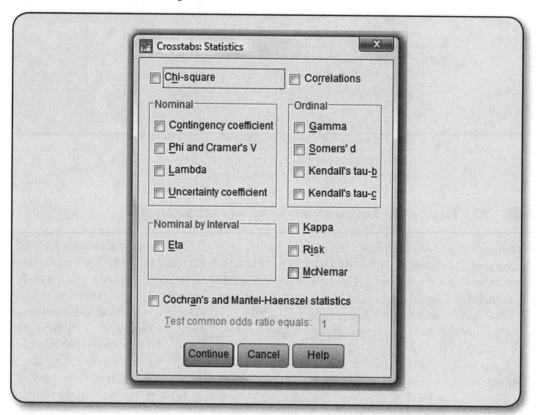

**Figure 8.4**   Output of Gamma and Kendall's tau-*b*

**Symmetric Measures**

|  |  | Value | Asymp. Std. Error[a] | Approx. T[b] | Approx. Sig. |
|---|---|---|---|---|---|
| Ordinal by Ordinal | Kendall's tau-b | .226 | .027 | 8.382 | .000 |
|  | Gamma | .347 | .040 | 8.382 | .000 |
| N of Valid Cases |  | 950 |  |  |  |

a. Not assuming the null hypothesis.

b. Using the asymptotic standard error assuming the null hypothesis.

*Gamma*    A symmetrical measure of association suitable for use with ordinal variables or with dichotomous nominal variables. It can vary from 0.0 to ±1.0 and provides us with an indication of the strength and direction of the association between the variables.

*Symmetrical measure of association*    A measure whose value will be the same when either variable is considered the independent variable or the dependent variable.

---

*Kendall's tau-b*    A symmetrical measure of association suitable for use with ordinal variables. Unlike gamma, it accounts for pairs tied on the independent and dependent variable. It can vary from 0.0 to ±1.0. It provides an indication of the strength and direction of the association between the variables.

---

A Kendall's tau-$b$ value of 0.226 indicates a weak positive association between education and financial satisfaction. By using information about respondent's education to predict financial satisfaction, we have reduced our prediction error by 22.6%. It's important to note that Kendall's tau-$b$ will always be lower than gamma because it accounts for variables tied on the independent and dependent variable.

## ▣ STATISTICS IN PRACTICE: EDUCATION AND APPROVAL OF HOMOSEXUALITY

Is there a relationship between educational level and approval of homosexuality? If so, what is the direction of the relationship? How strong is it? Let's use 2008 GSS data to examine these questions. As you can see, educational level and approval have been dichotomized to simplify our discussion. We will treat educational level as the independent variable and approval of homosexuality as the dependent variable. Table 8.22 illustrates the cross-tabulation. We can see that there appears to be a relationship between educational level and approval of homosexuality, as 41% of those with at least a high school degree approved of homosexuality as compared with 24.5% of those without a high school diploma. Gamma and Kendall's tau-$b$ support this finding. Using SPSS, we obtained a gamma value of 0.36, indicating a moderate positive association between educational level and approval of homosexuality. The higher one's education, the more likely he or she would be to approve of homosexuality. This gamma value also indicates that by using educational level to predict approval of homosexuality, we have reduced our prediction error by 36%. Kendall's tau-$b$ lends further support to this. SPSS generated a Kendall's tau-$b$ value of 0.13, which indicated a weak positive relationship between educational level and approval of homosexuality. Note that while direction of the relationship remains positive with both the gamma value and the Kendall's tau-$b$ value, the strength is different.

Table 8.22   Approval of Homosexuality by Education

| Approval of Homosexuality (Y) | Education (X) | | Total |
| | High School or More | Less Than High School | |
|---|---|---|---|
| Approve | 41.0% | 24.5% | 38.2% |
| | (431) | (52) | (483) |
| Disapprove | 59.0% | 75.5% | 61.8% |
| | (621) | (160) | (781) |
| Total | 100% | 100% | 100% |
| (N) | (1,052) | (212) | (1,264) |

*Source:* General Social Survey, 2008.

*Note:* Gamma = 0.36 and Kendall's tau-$b$ = 0.13.

## MAIN POINTS

- Bivariate analysis is a statistical technique designed to detect and describe the relationship between two variables. A relationship is said to exist when certain values of one variable tend to "go together" with certain values of the other variable.

- A bivariate table displays the distribution of one variable across the categories of another variable. It is obtained by classifying cases based on their joint scores for two variables.

- Percentaging bivariate tables are used to examine the relationship between two variables that have been organized in a bivariate table. The percentages are always calculated within each category of the independent variable.

- Bivariate tables are interpreted by comparing percentages across different categories of the independent variable. A relationship is said to exist if the percentage distributions vary across the categories of the independent variable.

- Variables measured at the ordinal or interval-ratio levels may be positively or negatively associated. With a positive association, higher values of one variable correspond to higher values of the other variable. When there is a negative association between variables, higher values of one variable correspond to lower values of the other variable.

- The chi-square test is an inferential statistical technique designed to test for a significant relationship between nominal or ordinal variables organized in a bivariate table. The test is conducted by testing the null hypothesis that no association exists between two cross-tabulated variables in the population, and therefore, the variables are statistically independent.

- The obtained chi-square ($\chi^2$) statistic summarizes the differences between the observed frequencies ($f_o$) and the expected frequencies ($f_e$)—the frequencies we would have expected to see if the null hypothesis were true and the variables were not associated. The Yates's correction for continuity is applied to all $2 \times 2$ tables.

- The sampling distribution of chi-square tells the probability of getting values of chi-square, assuming no relationship exists in the population. The shape of a particular chi-square sampling distribution depends on the number of degrees of freedom.

- Measures of association are single summarizing numbers that reflect the strength of the relationship between variables, indicate the usefulness of predicting the dependent variable from the independent variable, and often show the direction of the relationship.

- Proportional reduction of error (PRE) underlies the definition and interpretation of several measures of association. PRE measures are derived by comparing the errors made in predicting the dependent variable while ignoring the independent variable with errors made when making predictions that use information about the independent variable.

- PRE measures may range from 0.0 to ±1.0. A PRE of 0.0 indicates that the two variables are not associated and that information about the independent variable will not improve predictions about the dependent variable. A PRE of ±1.0 means that there is a perfect (positive or negative) association between the variables and that information about the independent variable results in a perfect (without any error) prediction of the dependent variable.

- Measures of association may be symmetrical or asymmetrical. When the measure is symmetrical, its value will be the same regardless of which of the two variables is considered the independent or dependent variable. In contrast, the value of asymmetrical measures of association may vary depending on which variable is considered the independent variable and which the dependent variable.

- Lambda is an asymmetrical measure of association suitable for use with nominal variables. It can range from 0.0 to 1.0 and gives an indication of the strength of an association between the independent and the dependent variables.

- Gamma is a symmetrical measure of association suitable for ordinal variables or for dichotomous nominal variables. It can vary from 0.0 to ±1.0 and reflects both the strength and direction of the association between two variables.

- Kendall's tau-*b* is a symmetrical measure of association suitable for use with ordinal variables. Unlike gamma, it accounts for pairs tied on the independent and dependent variable. It can vary from 0.0 to ±1.0. It provides an indication of the strength and direction of the association between two variables.

- Cramer's *V* is a measure of association for nominal variables. It is based on the value of chi-square and ranges between 0.0 to 1.0. Because it cannot take negative values, it is considered a nondirectional measure.

## KEY TERMS

| | | |
|---|---|---|
| asymmetrical measure of association | cross-tabulation | observed frequencies ($f_o$) |
| bivariate analysis | direct causal relationship | positive relationship |
| bivariate table | elaboration | proportional reduction |
| cell | expected frequencies ($f_e$) | of error (PRE) |
| chi-square (obtained) | gamma | row variable |
| chi-square test | Kendall's tau-*b* | statistical independence |
| column variable | lambda | symmetrical measure |
| control variable | marginals | of association |
| Cramer's *V* | measure of association | |
| | negative relationship | |

## ON YOUR OWN

Log on to the web-based student study site at **www.sagepub.com/ssdsessentials** for additional study questions, web quizzes, web resources, flashcards, codebooks and datasets, web exercises, appendices, and links to social science journal articles reflecting the statistics used in this chapter.

## CHAPTER EXERCISES

1. Use the following data on fear, race, and home ownership for this exercise. Variables measure respondent's race, whether the respondent fears walking alone at night, and his or her home ownership.

| Respondent | Race | Fear of Walking Alone | Rent/Own |
|:---:|:---:|:---:|:---:|
| 1 | W | N | R |
| 2 | B | N | R |
| 3 | W | Y | R |
| 4 | B | N | R |
| 5 | W | N | R |
| 6 | B | Y | O |
| 7 | W | Y | R |
| 8 | W | Y | R |
| 9 | W | N | O |
| 10 | W | N | O |
| 11 | W | Y | R |
| 12 | W | N | R |
| 13 | B | Y | O |
| 14 | W | N | R |
| 15 | B | N | O |
| 16 | B | N | R |
| 17 | W | N | O |
| 18 | W | N | O |
| 19 | B | N | R |
| 20 | W | N | O |
| 21 | B | Y | R |

*Source:* Data based on the General Social Survey, 2002.

*Notes:* Race: B = black, W = white; Fear: Y = yes, N = no; Rent/Own: R = rent, O = own.

a. Construct a bivariate table of frequencies for race and fear of walking alone at night. Which is the independent variable?

b. Calculate percentages for the table based on the independent variable. Describe the relationship between race and fear of walking alone using the table. What sampling issues are involved here?

c. Use the data to construct a bivariate table to compare fear of walking alone at night between people who own their homes and those who rent. Use percentages to show whether there is a difference between homeowners and renters in fear of walking alone.

2. The issue of how much should be spent to solve particular U.S. social problems is a complex matter, and people have diverse and conflicting ideas on these issues. Race and social class have an impact on how people perceive the extent of government spending. The 2006 GSS contains several questions on these topics. For this exercise, we present race and the variable NATFARE, which asked whether we were spending too much, too little, or the right amount of money to address welfare.

| | Race of Respondent | | | |
| Spending on Welfare | White | Black | Hispanic | Total |
|---|---|---|---|---|
| Too little | 128 | 41 | 21 | 190 |
| About right | 197 | 32 | 21 | 250 |
| Too much | 185 | 31 | 19 | 235 |
| Total | 510 | 104 | 61 | 675 |

a. How many degrees of freedom for the table?
b. Test whether race and NATFAE are independent ($\alpha = .05$). What do you conclude?

3. Is there a relationship between the race of violent offenders and their victims? Data from the U.S. Department of Justice (Expanded Homicide Data Table 5, 2007) are presented below.

| | Characteristics of Offender | | |
| Characteristics of Victim | White | Black | Other |
|---|---|---|---|
| White | 2,918 | 566 | 51 |
| Black | 245 | 2,905 | 11 |
| Other | 50 | 25 | 95 |

a. Let's treat race of offenders as the independent variable and race of victims as the dependent variable. If we first ignore the independent variable and try to predict race of victim, how many errors will we make?
b. If we now take into account the independent variable, how many errors of prediction will we make for those offenders who are white? Black offenders? Other offenders?
c. Combine the answers in (a) and (b) to calculate the proportional reduction in error for this table based on the independent variable. How does this statistic improve our understanding of the relationship between the two variables?

4. We continue our examination of attitudes regarding homosexuality. Suppose that a classmate of yours suggests that views about homosexual relations can be explained by the frequency of church attendance. Your classmate shows you the following table taken from the 2008 GSS sample. (Frequencies are shown on the following page.)

**Exercises**

|  | Church Attendance | | | |
| Homosexual Relations | Never | Several Times a Year | Every Week | Total |
| --- | --- | --- | --- | --- |
| Always wrong | 62 | 40 | 109 | 211 |
| Not wrong at all | 114 | 50 | 35 | 199 |
| Total | 176 | 90 | 144 | 410 |

*Source:* General Social Survey, 2008.

    a. Which is the dependent variable in this table? Which is the independent variable?

    b. Calculate the percentages using church attendance as the independent variable for each cell in the table. Is there a relationship between church attendance and views about homosexual relations? If so, how strong is it?

    c. Suppose that you respond to your classmate by stating that it is not church attendance that explains views about homosexual relations; rather, it is one's opinion about the nature of right and wrong (i.e., morality) that explains attitudes about homosexual relations. Why might there be a potential problem with your argument? Think in terms of assigning variables to the independent and dependent categories.

5. We continue our analysis from Exercise 2, this time examining the relationship between social class (CLASS) and spending on welfare (NATFARE).

    a. What is the number of degrees of freedom for this table?

    b. Calculate the chi-square. Based on an alpha of .01, do you reject the null hypothesis? Explain the reason for your answer.

|  | Social Class | | | | |
| Spending on Welfare | Lower Class | Working Class | Middle Class | Upper Class | Total |
| --- | --- | --- | --- | --- | --- |
| Too little | 23 | 92 | 76 | 8 | 199 |
| About right | 12 | 113 | 133 | 10 | 268 |
| Too much | 12 | 127 | 99 | 7 | 245 |
| Total | 47 | 332 | 308 | 25 | 712 |

6. "Are you afraid to walk in your neighborhood at night?" This question was posed to GSS 2006 respondents. You decide to investigate this variable by sex and race. Do men and women express different fears? Does this change when you consider their race? You obtain the results shown on the following page.

| Fear to Walk Alone at Night | Sex | | |
| --- | --- | --- | --- |
| | *Male* | *Female* | *Total* |
| a. Both whites and blacks | | | |
| Yes | 196 | 523 | 719 |
| No | 667 | 607 | 1,274 |
| Total | 863 | 1,130 | 1,993 |
| b. Whites only | | | |
| Yes | 122 | 360 | 482 |
| No | 504 | 466 | 970 |
| Total | 626 | 826 | 1,452 |
| c. Blacks only | | | |
| Yes | 24 | 96 | 120 |
| No | 75 | 74 | 149 |
| Total | 99 | 170 | 269 |

Use the appropriate PRE measure to summarize the relationship between FEAR and SEX, for all respondents, then for whites and blacks separately. Can you suggest a reason for any differences you find?

7. Do female and male high school seniors have the same college goals? High school seniors were surveyed about their college plans and asked whether they thought that they would graduate from a 4-year program. Data for three time periods are provided in the following table (percentages are displayed).

| | 1980 | | 1990 | | 2001 | |
| --- | --- | --- | --- | --- | --- | --- |
| | *Females (%)* | *Males (%)* | *Females (%)* | *Males (%)* | *Females (%)* | *Males (%)* |
| College plans for seniors | | | | | | |
| Definitely will | 33.6 | 35.6 | 50.8 | 45.8 | 62.4 | 51.1 |
| Graduate college (4-year program) | | | | | | |
| Probably will | 21.3 | 23.5 | 20.5 | 24.0 | 20.2 | 24.8 |
| Definitely/probably won't | 45.0 | 41.0 | 28.8 | 30.2 | 17.5 | 24.1 |

*Source:* U.S. Department of Education, *Trends in Educational Equity of Girls and Women*, NCES 2005–016. Washington, DC: 2005.

   a. Since 1980, how has the pattern of college graduation plans changed for females?
   b. Is the same pattern true for male seniors? Why or why not?

8. What is the relationship between church attendance and views about homosexual relations? Data from the GSS 2006 are presented in the following table, with selected categories from ATTEND (how often do you attend religious services) and HOMOSEX (views about homosexual relations).
   a. Test the null hypothesis that church attendance and views about homosexual relations are independent of each other. Set alpha at .05.
   b. How would you describe the relationship between church attendance and views about homosexual relations?

| Views About Homosexual Relations | Church Attendance | | | |
| --- | --- | --- | --- | --- |
| | Never | Several Times a Year | Every Week | Total |
| Always wrong | 163 | 101 | 266 | 530 |
| Not wrong at all | 215 | 65 | 57 | 337 |
| Total | 378 | 166 | 323 | 867 |

9. In this exercise, we'll explore the relationship between feelings about the Bible and three different types of attitudes. The bivariate tables below present the relationship between feelings about the Bible along with PREMARSX (attitudes toward premarital sex), HOMOSEX (attitudes toward homosexuality), and PORNLAW (should pornography be legalized). As you can see in the following table, response options for the independent and dependent variables were dichotomized (i.e., recoded into two categories). Calculate lambda for each table. What can you conclude? Is belief about the Bible associated with these other variables?

**(a) Feelings About the Bible and PREMARSX**

| Attitudes Toward Premarital Sex | Is the Bible | | |
| --- | --- | --- | --- |
| | Spiritual Word | Book of Fables | Total |
| Wrong | 935 | 82 | 1,017 |
| Not wrong | 590 | 252 | 842 |
| Total | 1,525 | 334 | 1,859 |

**(b) Feelings About the Bible and HOMOSEX**

| Attitudes Toward Homosexuality | Is the Bible | | |
| --- | --- | --- | --- |
| | Spiritual Word | Book of Fables | Total |
| Wrong | 1,116 | 127 | 1,243 |
| Not wrong | 396 | 199 | 595 |
| Total | 1,512 | 326 | 1,838 |

**Exercises**

| (c) Feelings About the Bible and PORNLAW | Is the Bible | | |
|---|---|---|---|
| *Favor Legalization of Pornography* | *Spiritual Word* | *Book of Fables* | *Total* |
| Illegal | 1,529 | 302 | 1,831 |
| Legal | 48 | 18 | 66 |
| Total | 1,577 | 320 | 1,897 |

*Source:* General Social Survey, 2008.

10. In the following table, we present data for MTF 2008 teen girls examining the relationship between race and trying alcohol.

| | Race | | | |
|---|---|---|---|---|
| *Ever Tried Alcohol* | *White* | *Black* | *Hispanic* | *Total* |
| No | 121 | 37 | 22 | 180 |
| Yes | 315 | 57 | 69 | 441 |
| Total | 436 | 94 | 91 | 621 |

*Source:* MTF 2008.

   a. Identify the dependent, independent, and control variables for this table.
   b. What proportion of white, black, and Hispanic female respondents reported trying alcohol?

11. Are teens' attitudes toward trying cocaine dependent on their GPA? Using data from the MTF 2008, test whether attitude toward trying cocaine is independent of high school GPA ($\alpha = .01$). How would you describe the relationship between attitude toward trying cocaine and GPA?

| | GPA | | | |
|---|---|---|---|---|
| *Attitude Toward Trying Cocaine* | *GPA ≤ 2.0* | *2.0 < GPA ≤ 3.0* | *3.0 < GPA ≤ 4.0* | *Total* |
| Don't disapprove | 28 | 63 | 59 | 150 |
| Disapprove | 27 | 89 | 100 | 216 |
| Strongly disapprove | 76 | 388 | 564 | 1,028 |
| Total | 131 | 540 | 723 | 1,394 |

12. Let's examine the relationship between the number of children and attitudes toward abortion (if a woman is poor and cannot afford more children) using 2008 GSS data.

|  | Number of Children | | |
| --- | --- | --- | --- |
| Support Abortion | None or One Child | Two or More Children | Total |
| No | 279 | 467 | 746 |
|  | 49.8% | 63.5% | 57.6% |
| Yes | 281 | 269 | 550 |
|  | 50.2% | 36.5% | 42.4% |
| Total | 560 | 736 | 1,296 |
|  | 100% | 100% | 100% |

Calculate the appropriate measure of association for this table. Does a relationship exist between number of children and support for abortion? Explain using percentages to support your answer.

# Chapter 9

# Regression and Correlation

## Chapter Learning Objectives

❖ Understanding linear relations and prediction rules

❖ Constructing and interpreting straight-line graphs and finding the best-fitting line

❖ Calculating and interpreting $a$ and $b$

❖ Understanding the meaning of prediction errors

❖ Calculating and interpreting the coefficient of determination ($r^2$) and Pearson's correlation coefficient ($r$)

Many research questions require the analysis of relationships between interval-ratio variables. Social scientists, for instance, frequently measure variables such as educational attainment, family size, and household income. Bivariate regression analysis provides us with the tools to express a relationship between two interval-ratio variables in a concise way.[1]

The U.S. Census Bureau collects and reports an array of information about the United States and its residents. Annual household income and educational attainment are two of the many characteristics regularly monitored. Table 9.1 displays the percentage of state residents with a bachelor's degree in 2006 for the 10 most populated states. Also presented are the mean, variance, and range for these data.

In examining Table 9.1 and the descriptive statistics, notice the variability in educational attainment. The percentage of state residents with a bachelor's degree ranges from a low of 23.30% for the state of Ohio to a high of 35.60% for the state of New Jersey.

One possible explanation for the differences in educational attainment is the economic condition of these states. We would expect states with a higher household income to also have a larger percentage of its residents attain a college degree. Table 9.2 displays the median household income in 2006 for 10 of the most populated states. Note that the median household income ranges widely from a low of $44,532 for the state of Ohio to a high of $64,470 for the state of New Jersey.

**Table 9.1**    Percentage of State Residents With a Bachelor's Degree, 2006

| State | Percentage of State Residents With a Bachelor's Degree |
|---|---|
| California | 29.80 |
| Texas | 25.50 |
| New York | 32.20 |
| Florida | 27.20 |
| Illinois | 31.20 |
| Pennsylvania | 26.60 |
| Ohio | 23.30 |
| Michigan | 26.10 |
| Georgia | 28.10 |
| New Jersey | 35.60 |

Mean $\bar{Y} = \dfrac{\Sigma Y}{N} = \dfrac{285.6}{10} = 28.56$

Variance $Y = S_Y^2 = \dfrac{\Sigma(Y - \bar{Y})^2}{N-1} = \dfrac{120.3}{9} = 13.7$

Range $Y = 35.60 - 23.30 = 12.3$

*Source:* U.S. Census Bureau, *Statistical Abstract of the United States*, 2008, Table 221.

## ▣ THE SCATTER DIAGRAM

Let's examine the possible relationship between the interval-ratio variables *percentage of state residents with a bachelor's degree* and *median household income.* One quick visual method used to display such a relationship between two interval-ratio variables is the **scatter diagram** (or **scatterplot**). Often used as a first exploratory step in regression analysis, a scatter diagram can suggest whether two variables are associated.

---

*Scatter diagram (scatterplot)*    A visual method used to display a relationship between two interval-ratio variables.

---

The scatter diagram showing the relationship between educational attainment and household income for the 10 states is shown in Figure 9.1. In a scatter diagram, the scales for the two variables form the vertical and horizontal axes of a graph. Usually, the independent variable, $X$, is arrayed along the horizontal axis and the dependent variable, $Y$, along the vertical axis. Because differences

Table 9.2    Median Household Income ($)

| State | Median Household Income ($) |
|---|---|
| California | 56,645 |
| Texas | 44,922 |
| New York | 51,384 |
| Florida | 45,495 |
| Illinois | 52,006 |
| Pennsylvania | 46,259 |
| Ohio | 44,532 |
| Michigan | 47,182 |
| Georgia | 46,832 |
| New Jersey | 64,470 |

Mean $\bar{Y} = \dfrac{\Sigma Y}{N} = \dfrac{499,727}{10} = \$49,972.70$

Variance $Y = S_Y^2 = \dfrac{\Sigma(Y - \bar{Y})^2}{N-1} = \dfrac{367,421,523}{9} = \$40,824,614$

Range $Y = \$64,470 - \$44,532 = \$19,938$

*Source:* U.S. Census Bureau, American Community Survey, 2006.

in the median household income are hypothesized to account for differences in the percentage of state residents with a bachelor's degree, household income is assumed as the independent variable and is arrayed along the horizontal axis. Educational attainment, the dependent variable, is arrayed along the vertical axis. In Figure 9.1, each dot represents a state; its location lies at the exact intersection of that state's percentage of residents with a bachelor's degree and its median household income.

Note that there is an apparent tendency for states with a lower median household income (e.g., Ohio and Texas) to also have a lower percentage of residents with a bachelor's degree, whereas in states with a higher median household income (e.g., New Jersey and California), there are a higher percentage of residents with a bachelor's degree. In other words, we can say that median household income and educational attainment are positively associated.

Scatter diagrams can also illustrate a negative association between two variables. For example, Figure 9.2 displays the association between median household income and larceny/theft crime rates for the 10 most populated states using data from the *Statistical Abstract of the United States*. Figure 9.2 suggests that a low median household income is associated with a higher rate of larceny/theft. Conversely, high median household income seems to be associated with a lower larceny/theft rate (see Table 9.5 for data).

**Figure 9.1**   Scatter Diagram of Educational Attainment and Median Household Income

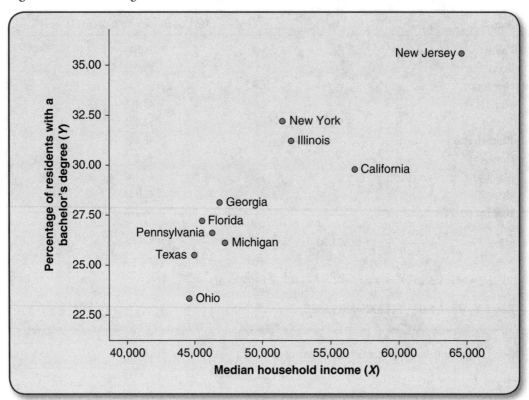

## ▣ LINEAR RELATIONS AND PREDICTION RULES

Scatter diagrams provide a useful but only preliminary step in exploring a relationship between two interval-ratio variables. We need a more systematic way to express this relationship. Let's examine Figures 9.1 and 9.2 again. They allow us to understand how household income is related to educational attainment (Figure 9.1) and criminal behavior (Figure 9.2). The relationships displayed are by no means perfect, but the trends are apparent. In the first case (Figure 9.1), as state income increases, so does the percentage of its residents with a bachelor's degree. In the second case (Figure 9.2), as state income increases, criminal behavior decreases.

One way to evaluate these relationships is by expressing them as *linear relationships*. A **linear relationship** allows us to approximate the observations displayed in a scatter diagram with a straight line. In a perfectly linear relationship, all the observations (the dots) fall along a straight line (a perfect relationship is sometimes called a **deterministic relationship**), and the line itself provides a predicted value of $Y$ (the vertical axis) for any value of $X$ (the horizontal axis). For example, in Figure 9.3, we have superimposed a straight line on the scatterplot originally displayed in Figure 9.1. Using this line, we can obtain a predicted value of the percentage of state residents with a bachelor's degree for any value of

**Figure 9.2**   Median Household Income and Larceny/Theft Crime Rates per 100,000 Population

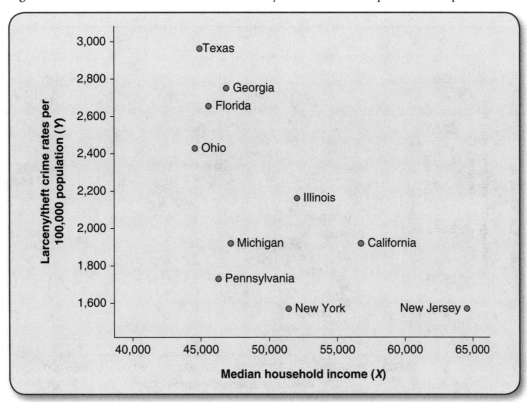

household income, by reading up to the line from the income axis and then over to the percentage with a bachelor's degree axis (indicated by the dotted lines). For example, the predicted value of the percentage of state residents with a bachelor's degree in a state with a $50,000 median household income is just below 30%. Similarly, for a state with a $55,000 median household income, we would predict that 32.50% of its residents would have a bachelor's degree.

---

*Linear relationship*   A relationship between two interval-ratio variables in which the observations displayed in a scatter diagram can be approximated with a straight line.

*Deterministic (perfect) linear relationship*   A relationship between two interval-ratio variables in which all the observations (the dots) fall along a straight line. The line provides a predicted value of *Y* (the vertical axis) for any value of *X* (the horizontal axis).

---

As indicated in Figure 9.3, for the 10 states surveyed, the actual relationship between income and the percentage of state residents with a bachelor's degree is not perfectly linear. Although some of the states

**Figure 9.3**   A Straight-Line Graph for Median Household Income and Percentage of Residents With a Bachelor's Degree

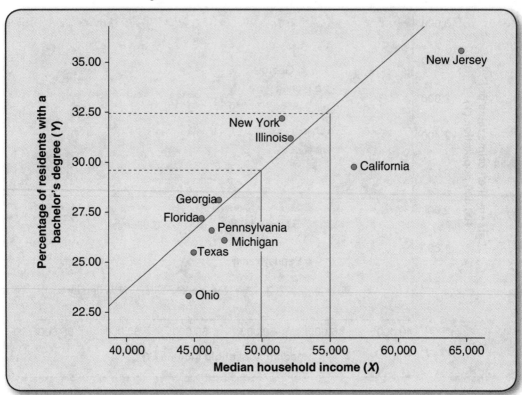

lie very close to the line, none fall exactly on the line and some deviate from it considerably. Are there other lines that provide a better description of the relationship between income and the percentage of state residents with a bachelor's degree?

In Figure 9.4, we have drawn two additional lines that approximate the pattern of relationship shown by the scatter diagram. In each case, notice that even though some of the states lie close to the line, all fall considerably short of perfect linearity. Is there one line that provides the best linear description of the relationship between median household income and the percentage of residents with a bachelor's degree? How do we choose such a line? What are its characteristics? Before we describe a technique for finding the straight line that most accurately describes the relationship between two variables, we first need to review some basic concepts about how straight-line graphs are constructed.

✓ *Learning*
  *Check*

*Use Figure 9.3 to predict the percentage of residents with a bachelor's degree in a state with a median household income of $47,500 and one with a median household income of $50,000.*

**Figure 9.4**  Alternative Straight-Line Graph for Median Household Income and Percentage of Residents With a Bachelor's Degree

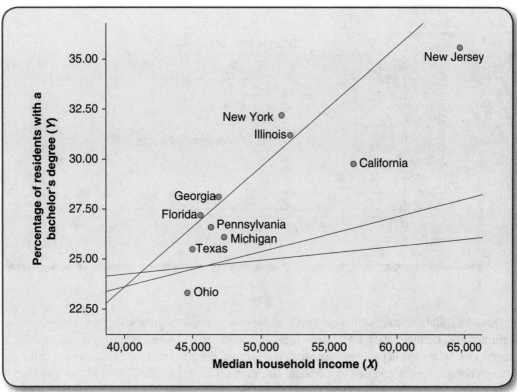

## Constructing Straight-Line Graphs

To illustrate the fundamentals of straight-line graphs, let's take a simple example. Suppose that in a local school system, teachers' salaries are completely determined by seniority. New teachers begin with an annual salary of $12,000, and for each year of seniority, their salary increases by $2,000. The seniority and annual salary of six hypothetical teachers are presented in Table 9.3.

**Table 9.3**  Seniority and Salary of Six Teachers (hypothetical data)

| Seniority (in years) X | Salary (in dollars) Y |
|:---:|:---:|
| 0 | 12,000 |
| 1 | 14,000 |
| 2 | 16,000 |
| 3 | 18,000 |
| 4 | 20,000 |
| 5 | 22,000 |

**Figure 9.5**   A Perfect Linear Relationship Between Seniority (in years) and Annual Salary (in thousand dollars) of Six Teachers (hypothetical)

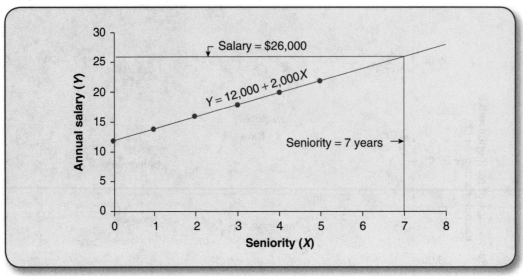

Now, let's plot the values of these two variables on a graph (Figure 9.5). Because seniority is assumed to determine salary, let it be our independent variable ($X$), and let's array it along the horizontal axis. Salary, the dependent variable ($Y$), is arrayed along the vertical axis. Connecting the six observations in Figure 9.5 gives us a straight-line graph. This graph allows us to obtain a predicted salary value for any value of seniority level simply by reading from the specific seniority level up to the line and then over to the salary axis. For instance, we have marked the lines going up from a seniority of 7 years and then over to the salary axis. We can see that a teacher with 7 years of seniority makes $26,000.

The relationship between salary and seniority, as depicted in Table 9.3 and Figure 9.5, can also be described with the following algebraic equation:

$$Y = 12,000 + 2,000(X)$$

where

$X$ = seniority (in years)

$Y$ = salary (in dollars)

This equation allows us to correctly predict salary ($Y$) for any value for seniority ($X$) that we plug into the equation, whether or not it appears in Table 9.3. For example, the salary of a teacher with 5 years of seniority is

$$Y = 12,000 + 2,000(5) = 12,000 + 10,000 = \$22,000$$

The equation describing the relationship between seniority and salary is an equation for a straight line. The equations for all straight-line graphs have the same general form:

$$Y = a + b(X) \qquad\qquad (9.1)$$

where

$Y$ = the predicted score on the dependent variable

$X$ = the score on the independent variable

$a$ = the $Y$-intercept, or the point where the line crosses the $Y$-axis; therefore, $a$ is the value of $Y$ when $X$ is 0. In our example, $a = 12,000$. A teacher makes \$12,000 with 0 years of seniority.

$b$ = the slope of the regression line, or the change in $Y$ with a unit change in $X$. In our example, $a = 12,000$ and $b = 2,000$. A teacher's salary will go up by \$2,000 with each year of seniority.

✓ *Learning Check*

*For each of these four lines, as X goes up by 1 unit, what does Y do? Be sure you can answer this question using both the equation and the line.*

**Four Lines: Illustrating the Slope and the Y-Intercept**

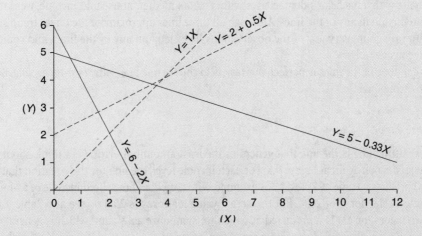

**Slope ($b$)** *The change in variable* X *(the dependent variable) with a unit change in variable* X *(the independent variable).*

**Y-intercept ($a$)** *The point where the line crosses the* Y-*axis, or the value of* Y *when* X *is 0.*

✓ *Learning Check*

*Use the linear equation describing the relationship between seniority and salary of teachers to obtain the predicted salary of a teacher with 12 years of seniority.*

## Finding the Best-Fitting Line

The straight line displayed in Figure 9.5 and the linear equation representing it ($Y = 12{,}000 + 2{,}000(X)$) provide a very simple depiction of the relationship between seniority and salary because salary (the $Y$ variable) is completely determined by seniority (the $X$ variable). When each value of $Y$ is completely determined by $X$, all the points (observations) lie on the line, and the relationship between the two variables is a deterministic, or perfectly linear, relationship.

However, most relationships we study in the social sciences are not deterministic, and we are not able to come up with a linear equation that allows us to predict $Y$ from $X$ with perfect accuracy. We are much more likely to find relationships approximating linearity, but in which numerous cases don't follow this trend perfectly. For instance, in reality, teachers' salaries are not completely determined by seniority, and therefore, knowing years of seniority will not provide us with a perfect prediction of their salary level.

When the dependent variable ($Y$) is not completely determined by the independent variable ($X$), not all (sometimes none) of the observations will lie exactly on the line. Look back at Figure 9.4, our example of the percentage of state residents with a bachelor's degree in relation to the state's median household income. Though each line represents a linear equation showing us how the percentage of state residents with a bachelor's degree rises with a state's median household income, we do not have a perfect prediction in any of the lines. Although all three lines approximate the linear trend suggested by the scatter diagram, very few of the observations lie exactly on any of the lines, and some deviate from them considerably.

Given that none of the lines is perfect, our task is to choose one line—the *best-fitting line*. But which is the best-fitting line?

### Defining Error

The best-fitting line is the one that generates the least amount of error. Let's think about how the error is defined. Look again at Figure 9.3. For each income level, the line (or the equation that this line represents) predicts a value of $Y$. Texas, for example, with a median household income of \$44,922, gives us a predicted value for $Y$ of 26%. But the actual value for Texas is 25.5% (see also Table 9.1). Thus, we have two values for $Y$: (1) a predicted $Y$, which we symbolize as $\hat{Y}$ and which is generated by the prediction equation, also called the *linear regression equation* $\hat{Y} = a + b(X)$ and (2) the observed $Y$, symbolized simply as $Y$. Thus, for Texas, $\hat{Y} = 26\%$, whereas $Y = 25.5\%$.

We can think of the residual as the difference between the observed $Y(Y)$ and the predicted $Y(\hat{Y})$. If we symbolize the error as $e$, then

$$e = Y - \hat{Y}$$

The error for Texas is 25.5% − 26% = −0.5 percentage points.

### The Residual Sum of Squares ($\Sigma e^2$)

We want a line or a prediction equation that minimizes $e$ for each individual observation. However, any line we choose will minimize the error for some observations but may maximize it for others. We want to find a prediction equation that minimizes the errors over all observations.

There are many mathematical ways of defining the errors. Statisticians prefer to square and sum the errors over all observations. The result is the *residual sum of squares*, or $\Sigma e^2$. Symbolically, $\Sigma e^2$ is expressed as

$$\Sigma e^2 = \Sigma(Y - \hat{Y})^2$$

### The Least Squares Line

The best-fitting regression line is that line where the sum of the squared errors, or $\Sigma e^2$, is at a minimum. Such a line is called the **least squares line** (or **best-fitting line**), and the technique that produces this line is called the **least squares method**. The technique involves choosing $a$ and $b$ for the equation $\hat{Y} = a + bx$ such that $\Sigma e^2$ will have the smallest possible value.

---

***Least squares line (best-fitting line)***   A line where the residual sum of squares, or $\Sigma e^2$, is at a minimum.

***Least squares method***   The technique that produces the least squares line.

---

## Computing *a* and *b* for the Prediction Equation

To figure out the values of $a$ and $b$ in a way that minimizes $\Sigma e^2$, we need to apply the following formulas:

$$b = \frac{S_{YX}}{S_X^2} \tag{9.2}$$

$$a = \bar{Y} - b(\bar{X}) \tag{9.3}$$

where

$S_{YX}$ = the covariance of $X$ and $Y$

$S_X^2$ = the variance of $X$

$\bar{Y}$ = the mean of $Y$

$\bar{X}$ = the mean of $X$

$a$ = the $Y$-intercept

$b$ = the slope of the line

These formulas assume that $X$ is the independent variable and $Y$ is the dependent variable.

Before we compute $a$ and $b$, let's examine these formulas. The denominator for $b$ is the variance of the variable $X$. It is defined as follows:

$$\text{Variance } (X) = S_X^2 = \frac{\Sigma(X - \bar{X})^2}{N - 1}$$

This formula should be familiar to you from Chapter 4. The numerator $(S_{YX})$, however, is a new term. It is the covariance of $X$ and $Y$ and is defined as

$$\text{Covariance } (X, Y) = S_{YX} = \frac{\Sigma(X - \bar{X})(Y - \bar{Y})}{N - 1} \tag{9.4}$$

The covariance is a measure of how $X$ and $Y$ vary together. Basically, the covariance tells us to what extent higher values of one variable "go together" with higher values on the second variable (in which case we have a positive covariation) or with lower values on the second variable (which is a negative covariation). Take a look at this formula. It tells us to subtract the mean of $X$ from each $X$ score and the mean of $Y$ from each $Y$ score, and then take the product of the two deviations. The results are then summed for all the cases and divided by $N - 1$.

In Table 9.4, we show the computations necessary to calculate the values of $a$ and $b$ for our 10 states. To calculate the covariance, we first subtract $\bar{X}$ from each $X$ score (Column 3) and $\bar{Y}$ from each $Y$ score (Column 5). We then multiply these deviations for every observation. The products of the mean deviations are shown in Column 7. For example, for the first observation, California, the mean deviation for median household income is 6,672.3 (56,645 − 49,972.7 = 6,672.3); for the percentage of residents with a bachelor's degree, it is 1.24 (29.8 − 28.56 = 1.24). The product of these deviations, 8,273.65 (6,672.3 × 1.24 = 8,273.65), is shown in Column 7. The sum of these products, shown at the bottom of Column 7, is 186,591.26. Dividing it by 9 ($N - 1$), we get the covariance of 20,732.36.

The covariance is a measure of the linear relationship between two variables, and its value reflects both the strength and the direction of the relationship. The covariance will be close to zero when $X$ and $Y$ are unrelated; it will be larger than zero when the relationship is positive and smaller than zero when the relationship is negative.

Now, let's substitute the values for the covariance and the variance from Table 9.4 to calculate $b$:

$$b = \frac{S_{YX}}{S_X^2} = \frac{20{,}732.36}{40{,}824{,}614} = 0.0005$$

Once $b$ has been calculated, finding $a$, the intercept, is simple:

$$a = \bar{Y} - b(\bar{X}) = 28.56 - 0.0005(49{,}972.7) = 3.57$$

The prediction equation is therefore

$$\hat{Y} = 3.57 + 0.0005(X)$$

This equation can be used to obtain a predicted value for the percentage of state residents who have a bachelor's degree given a state's median household income. For example, for a state with a median household income of $48,000, the predicted percentage is

$$\hat{Y} = 3.57 + 0.0005(48,000) = 27.57$$

**Table 9.4** Worksheet for Calculating $a$ and $b$ for the Regression Equation

| | (1) | (2) | (3) | (4) | (5) | (6) | (7) |
|---|---|---|---|---|---|---|---|
| **State** | **Median Household Income** X | **Percentage With a Bachelor's Degree** Y | $(X - \bar{X})$ | $(X - \bar{X})^2$ | $(Y - \bar{Y})$ | $(Y - \bar{Y})^2$ | $(X - \bar{X})(Y - \bar{Y})$ |
| California | 56,645 | 29.8 | 6,672.3 | 44,519,587 | 1.24 | 1.54 | 8,273.65 |
| Texas | 44,922 | 25.5 | −5,050.7 | 25,509,570 | −3.06 | 9.36 | 15,455.14 |
| New York | 51,384 | 32.2 | 1,411.3 | 1,991,767 | 3.64 | 13.25 | 5,137.13 |
| Florida | 45,495 | 27.2 | −4,477.7 | 20,049,797 | −1.36 | 1.85 | 6,089.67 |
| Illinois | 52,006 | 31.2 | 2,033.3 | 4,134,309 | 2.64 | 6.97 | 5,367.91 |
| Pennsylvania | 46,259 | 26.6 | −3,713.7 | 13,791,568 | −1.96 | 3.84 | 7,278.85 |
| Ohio | 44,532 | 23.3 | −5,440.7 | 29,601,216 | −5.26 | 27.67 | 28,618.08 |
| Michigan | 47,182 | 26.1 | −2,790.7 | 7,788,006 | −2.46 | 6.05 | 6,865.12 |
| Georgia | 46,832 | 28.1 | −3,140.7 | 9,863,996 | −0.46 | 0.21 | 1,444.72 |
| New Jersey | 64,470 | 35.6 | 14,497.3 | 210,171,707 | 7.04 | 49.56 | 102,060.99 |
| | $\sum X = 499,727$ | $\sum Y = 285.6$ | 0.00[a] | 367,421,526 | 0.00[a] | 120.3 | 186,591.26 |

Mean $X = \bar{X} = \dfrac{\sum X}{N} = \dfrac{499,727}{10} = 49,972.7$  $\qquad$ Mean $Y = \bar{Y} = \dfrac{\sum Y}{N} = \dfrac{285.6}{10} = 28.56$

Variance $(Y) = S_Y^2 = \dfrac{\sum(Y - \bar{Y})^2}{N - 1} = \dfrac{120.3}{9} = 13.37$

Standard deviation $(Y) = S_Y = \sqrt{13.37} = 3.66$

Variance $(X) = S_X^2 = \dfrac{\sum(X - \bar{X})^2}{N - 1} = \dfrac{367,421,523}{9} = 40,824,614$

Standard deviation $(X) = S_X = \sqrt{40,824,614} = 6,389.41$

Covariance $(X, Y) = S_{YX} = \dfrac{\sum(X - \bar{X})(Y - \bar{Y})}{N - 1} = \dfrac{186,591.26}{9} = 20,732.36$

a. Answers may differ due to rounding; however, the exact value of these column totals, properly calculated, will always be equal to zero.

Similarly, for a state with a median household income of $56,000, the predicted value is

$$\hat{Y} = 3.57 + 0.0005(56,000) = 31.57$$

Now, we can plot the straight-line graph corresponding to the regression equation. To plot a straight line, we need only two points, where each point corresponds to an $X$, $Y$ value predicted by the equation. We can use the two points we just obtained: (1) $X = \$48,000$, $\hat{Y} = 27.57\%$ and (2) $X = \$56,000$, $\hat{Y} = 31.57\%$. In Figure 9.6, the regression line is plotted over the scatter diagram we first displayed in Figure 9.1.

✓ *Learning Check*

> Use the prediction equation to calculate the predicted values of Y for New York, Georgia, and Ohio. Verify that the regression line in Figure 9.6 passes through these points.

## Interpreting *a* and *b*

Now, let's interpret the coefficients $a$ and $b$ in our equation. The $b$ coefficient is equal to 0.0005%. This tells us that the percentage of state residents with a bachelor's degree will increase by 0.0005% for every $1 increment in their state's median household income. Similarly, an increase of $10,000 in a state's median household income corresponds to a 5% increase in the percentage of state residents with a bachelor's degree.

Note that because the relationships between variables in the social sciences are inexact, we don't expect our regression equation to make perfect predictions for every individual case. However, even though the pattern suggested by the regression equation may not hold for every individual state, it gives us a tool by which to make the best possible guess about how a state's median household income is associated, *on average*, with the percentage of state residents with a bachelor's degree. We can say that the slope of 0.0005% is the estimate of this underlying relationship.

The $Y$-intercept ($a$) is the predicted value of $Y$, when $X = 0$. Thus, it is the point at which the regression line and the $Y$-axis intersect. With $a = 3.57$, we would predict that very few residents (3.57%) of a state with a median household income equal to zero would have obtained a bachelor's degree. Note, however, that no state has an income as low as zero. As a general rule, be cautious when making predictions for $Y$ based on values of $X$ that are outside the range of the data. Thus, when the lowest value for $X$ is far above zero, the intercept may not have a clear substantive interpretation.

---

*Y-intercept* (a)   The point where the regression line crosses the $Y$-axis and where $X = 0$.

---

## 🔳 STATISTICS IN PRACTICE: MEDIAN HOUSEHOLD INCOME AND CRIMINAL BEHAVIOR

In our ongoing example, we have looked at the association between median household income and educational attainment measured by the percentage of state residents with a bachelor's degree. The regression equation we have estimated from the data collected by the U.S. Census Bureau in 10 states show that as a state's median household income rises, so does its percentage of residents with a bachelor's degree. This finding confirms that household income is related to educational attainment.

**Figure 9.6** The Best-Fitting Line for Median Household Income and Percentage of State Residents With a Bachelor's Degree

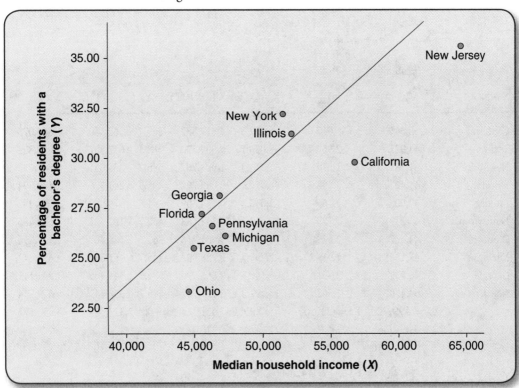

Now, let's examine the relationship between median household income and criminal behavior. The U.S. Census Bureau regularly collects a variety of information from U.S. residents. In the most recent *Statistical Abstract of the United States*, crime rates were tabulated and reported for all 50 states and the District of Columbia. Let's focus on the larceny/theft crime rate per 100,000 population. The first two columns of Table 9.5 display the larceny/theft crime rate and median household income for the 10 most populated states. The scatter diagram for these data was displayed earlier, in Figure 9.2.

Let's examine Figure 9.2 once again. The scatter diagram seems to indicate that the two variables— larceny/theft crime rate and median household income—are linearly related. It also illustrates that these variables are negatively associated; that is, as median household income rises, the larceny/theft crime rate declines.

For a more systematic analysis of the association, we need to estimate the least squares regression equation for these data. Since we want to predict the larceny/theft crime rate, we treat this variable as our dependent variable ($Y$).

Table 9.5 also shows the calculations necessary to find $a$ and $b$ for our data on median household income in relation to the larceny/theft crime rate.

Now, let's substitute the values for the covariance and the variance from Table 9.5 to calculate $b$:

$$b = \frac{S_{YX}}{S_X^2} = \frac{-2,038,953}{40,824,614} = -0.05$$

**Table 9.5** Median Household Income and the Larceny/Theft Crime Rate for 10 States

| State | (1) Income $X$ | (2) Larceny/ Theft Crime Rate $Y$ | (3) $(X - \bar{X})$ | (4) $(X - \bar{X})^2$ | (5) $(Y - \bar{Y})$ | (6) $(Y - \bar{Y})^2$ | (7) $(X - \bar{X})(Y - \bar{Y})$ |
|---|---|---|---|---|---|---|---|
| California | 56,645 | 1,917 | 6,672.3 | 44,519,587 | −249.7 | 62,350 | −1,666,073 |
| Texas | 44,922 | 2,962 | −5,050.7 | 25,509,570 | 795.3 | 632,502 | −4,016,822 |
| New York | 51,384 | 1,570 | 1,411.3 | 1,991,767 | −596.7 | 356,051 | −842,123 |
| Florida | 45,495 | 2,658 | −4,477.7 | 20,049,797 | 491.3 | 241,376 | −2,199,894 |
| Illinois | 52,006 | 2,165 | 2,033.3 | 4,134,309 | −1.7 | 2.89 | −3,457 |
| Pennsylvania | 46,259 | 1,729 | −3,713.7 | 13,791,568 | −437.7 | −191,581 | 1,625,486 |
| Ohio | 44,532 | 2,429 | −5,440.7 | 29,601,216 | 262.3 | 68,801 | −1,427,096 |
| Michigan | 47,182 | 1,918 | −2,790.7 | 7,788,006 | −248.7 | −61,852 | 694,047 |
| Georgia | 46,832 | 2,751 | −3,140.7 | 9,863,996 | 584.3 | 341,406 | −1,835,111 |
| New Jersey | 64,470 | 1,568 | 14,497.3 | 210,171,707 | −598.7 | 358,442 | −8,679,534 |
| | $\Sigma X = 499,727$ | $\Sigma Y = 21,667$ | | 0.00[a] 367,421,523 | | 0.00[a] 2,314,364 | −18,350,577 |

Mean $X = \bar{X} = \dfrac{\Sigma X}{N} = \dfrac{499,727}{10} = 49,972.7$  $\quad$ Mean $Y = \bar{Y} = \dfrac{\Sigma Y}{N} = \dfrac{21,667}{10} = 2,166.7$

Variance $(Y) = S_Y^2 = \dfrac{\Sigma(Y - \bar{Y})^2}{N - 1} = \dfrac{2,314,364}{9} = 257,152$

Standard deviation $(Y) = S_Y = \sqrt{257,152} = 507.10$

Variance $(X) = S_X^2 = \dfrac{\Sigma(X - \bar{X})^2}{N - 1} = \dfrac{367,421,523}{9} = 40,824,614$

Standard deviation $(X) = S_X = \sqrt{40,824,614} = 6,389.41$

Covariance $(X, Y)$: $S_{YX} = \dfrac{\Sigma(X - \bar{X})(Y - \bar{Y})}{N - 1} = \dfrac{-18,350,577}{9} = -2,038,953$

a. Answers may differ due to rounding; however, the exact value of these column totals, properly calculated, will always be equal to zero.

Once $b$ has been calculated, finding $a$, the intercept, is simple:

$$a = \bar{Y} - b(\bar{X}) = 2,166.7 - (-0.05)(49,972.7) = 4,665$$

The prediction equation is therefore

$$\hat{Y} = 4{,}665 + (-0.05)X$$

This equation can be used to obtain a predicted value for a state's larceny/theft crime rate given a state's median household income.

Now, let's interpret the coefficients $a$ and $b$ in our equation. The $b$ coefficient is equal to −0.05. This tells us that the larceny/theft crime rate will decrease by 0.05 for every $1 increase in a state's median household income. Similarly, an increase of $10,000 in a state's median household income corresponds to a 500-unit decrease ($-0.05 \times 10{,}000$) in the state's predicted larceny/theft crime rate.

The intercept $a$ is the predicted value of $Y$, when $X = 0$. Thus, it is the point at which the regression line and the $Y$-axis intersect. With $a = 4{,}665$, a state with a median household income equal to zero is predicted to have a 4,665 larceny/theft crime rate.

## ◙ METHODS FOR ASSESSING THE ACCURACY OF PREDICTIONS

So far, we have developed two regression equations that are helping us make state-level predictions about educational attainment and criminal behavior. But in both cases, our predictions are far from perfect. If we examine Figures 9.6 and 9.7, we can see that we fail to make accurate predictions in every case. Though some of the states lie pretty close to the regression line, hardly any lie directly on the line—an indication that some error of prediction was made.

We saw earlier that one way to judge the accuracy of the predictions is to "eyeball" the scatterplot. The closer the observations are to the regression line, the better the "fit" between the predictions and the actual observations. Still, we want a more systematic method for making such a judgment. We need a measure that tells us how accurate a prediction the regression model provides. The *coefficient of determination*, or $r^2$, is such a measure. It tells us how well the bivariate regression model fits the data. Both $r^2$ and $r$ measure the strength of the association between two interval-ratio variables.

The **coefficient of determination** ($r^2$) reflects the proportion of the total variation in the dependent variable, $Y$, explained by the independent variable, $X$. An $r^2$ of 0.79 means that by using median household income and the linear prediction rule to predict $Y$—the percentage of state residents with a bachelor's degree—we have reduced the error of prediction by 79% ($0.79 \times 100$). We can also say that the independent variable (median household income) explains about 79% of the variation in the dependent variable (the percentage of state residents with a bachelor's degree), as illustrated in Figure 9.8.

---

*Coefficient of determination ($r^2$)*   A PRE measure reflecting the proportional reduction of error that results from using the linear regression model. It reflects the proportion of the total variation in the dependent variable, $Y$, explained by the independent variable, $X$.

---

**Figure 9.7**  Regression Line for Median Household Income and Larceny/Theft Crime Rate per 100,000 Population

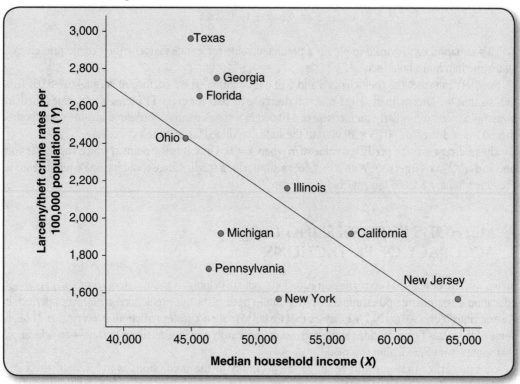

The coefficient of determination ranges from 0.0 to 1.0. An $r^2$ of 1.0 means that by using the linear regression model, we have reduced uncertainty by 100%. It also means that the independent variable accounts for 100% of the variation in the dependent variable. With an $r^2$ of 1.0, all the observations fall along the regression line. An $r^2$ of 0.0 means that using the regression equation to predict $Y$ does not improve the prediction of $Y$. Figure 9.9 shows $r^2$ values near 0.0 and near 1.0. In Figure 9.9a, where $r^2$ is approximately 1.0, the regression model provides a good fit. In contrast, a very poor fit is evident in Figure 9.9b, where $r^2$ is near zero. An $r^2$ near zero indicates either poor fit or a well-fitting line with a $b$ of zero.

## Calculating $r^2$

An easy method for calculating $r^2$ uses the following equation:

$$r^2 = \frac{[\text{Covariance } (X, Y)]^2}{[\text{Variance } (X)][\text{Variance } (Y)]} = \frac{S^2_{YX}}{S^2_X S^2_Y} \qquad (9.5)$$

**Figure 9.8**    A Pie Graph Approach to $r^2$

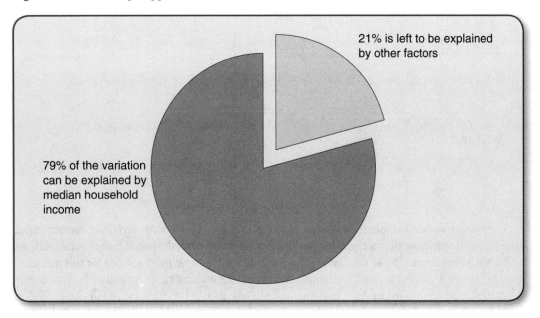

21% is left to be explained by other factors

79% of the variation can be explained by median household income

**Figure 9.9**    Examples Showing $r^2$ (a) Near 1.0 and (b) Near 0

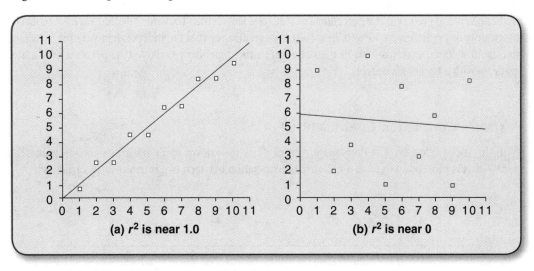

(a) $r^2$ is near 1.0

(b) $r^2$ is near 0

This formula tells us to divide the square of the covariance of $X$ and $Y$ by the product of the variance of $X$ and the variance of $Y$.

To calculate $r^2$ for our example, we can go back to Table 9.4, where the covariance and the variances for the two variables have already been calculated:

$$S_{YX} = 20{,}732.36$$

$$S_X^2 = 40{,}824{,}614$$

$$S_Y^2 = 13.37$$

Therefore,

$$r^2 = \frac{(20{,}732.36)^2}{(40{,}824{,}614)(13.37)} = \frac{429{,}830{,}751}{545{,}825{,}089} = 0.79$$

Since we are working with actual values for median household income, its metric, or measurement, values are different from the metric values for the dependent variable, the percentage of state residents with a bachelor's degree. While this hasn't been an issue until now, we must account for this measurement difference if we elect to use the variances and covariance to calculate $r^2$ (Formula 9.5). The remedy is actually quite simple. All we have to do is multiply our obtained $r^2$, 0.79, by 100 to obtain 79. You might be wondering, "Why multiply the obtained $r^2$ by 100?" The answer is we multiply by 100 because the dependent variable, percentage of residents with a bachelor's degree, is measured as a percentage ranging from 1 to 100.

We can multiply $r^2$ by 100 to obtain the percentage of variation in the dependent variable explained by the independent variable. An $r^2$ of 0.79 means that by using median household income and the linear prediction rule to predict $Y$, the percentage of state residents with a bachelor's degree, we have reduced uncertainty of prediction by 79% ($0.79 \times 100$). We can also say that the independent variable (median household income) explains 79% of the variation in the dependent variable (the percentage of state residents with a bachelor's degree).

## Pearson's Correlation Coefficient (r)

In the social sciences, it is the square root of $r^2$, or $r$—known as **Pearson's correlation coefficient**—that is most often used as a measure of association between two interval-ratio variables:

$$r = \sqrt{r^2}$$

---

*Pearson's correlation coefficient* (r)   The square root of $r^2$; it is a measure of association for interval-ratio variables, reflecting the strength of the linear association between two interval-ratio variables. It can be positive or negative in sign.

---

Pearson's *r* is usually computed directly by using the following definitional formula:

$$r = \frac{[\text{Covariance } (X, Y)]}{[\text{Standard deviation } (X)][\text{Standard deviation } (Y)]} = \frac{S_{YX}}{S_X S_Y} \quad (9.6)$$

Thus, *r* is defined as the ratio of the covariance of *X* and *Y* to the product of the standard deviations of *X* and *Y*.

### Characteristics of Pearson's r

Pearson's *r* is a measure of relationship or association for interval-ratio variables. Like gamma (introduced in Chapter 8), it ranges from 0.0 to ±1.0, with 0.0 indicating no association between the two variables. An *r* of +1.0 means that the two variables have a perfect positive association; −1.0 indicates that it is a perfect negative association. The absolute value of *r* indicates the strength of the linear association between two variables. (Refer to the guide to interpreting the strength of the association on page 214.) Thus, a correlation of −0.75 demonstrates a stronger association than a correlation of 0.50. Figure 9.10 illustrates a strong positive relationship, a strong negative relationship, a moderate positive relationship, and a weak negative relationship.

Unlike the *b* coefficient, *r* is a symmetrical measure. That is, the correlation between *X* and *Y* is identical to the correlation between *Y* and *X*. In contrast, *b* may be different when the variables are switched—for example, when we use *Y* as the independent variable rather than as the dependent variable.

To calculate *r* for our example of the relationship between median household income and the percentage of state residents with a bachelor's degree, let's return to Table 9.4, where the covariance and the standard deviations for *X* and *Y* have already been calculated:

$$r = \frac{S_{YX}}{S_X S_Y} = \frac{20,732.36}{(6,389.41)(3.66)} = \frac{20,732.36}{23,385.24} = 0.89$$

A correlation coefficient of 0.89 indicates that there is a strong positive linear relationship between median household income and the percentage of state residents with a bachelor's degree.

Note that we could have just taken the square root of $r^2$ to calculate *r*, because $r = \sqrt{r^2}$ or $\sqrt{0.79} = 0.89$. Similarly, if we first calculate *r*, we can obtain $r^2$ simply by squaring *r* (be careful not to lose the sign of *r*).[2]

## ▣ STATISTICS IN PRACTICE: TEEN PREGNANCY AND SOCIAL INEQUALITY

The United States has by far the highest rate of teenage pregnancy of any industrialized nation. The pregnancy rate for U.S. teens aged between 15 and 19 years was 95.9 pregnancies per 1,000 women in 1990. Although, in 2005, the teen pregnancy rate reached its lowest point in more than 30 years

**Figure 9.10**   Scatter Diagrams Illustrating Weak, Moderate, and Strong Relationships as Indicated by the Absolute Value of $r$

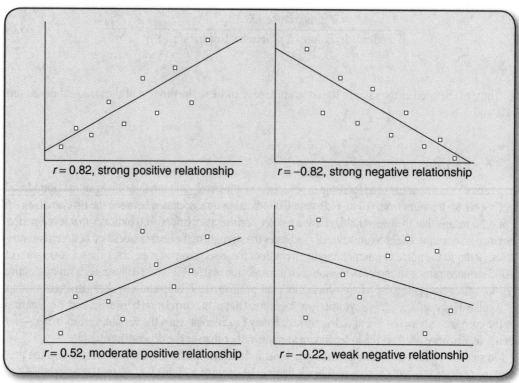

$r = 0.82$, strong positive relationship

$r = -0.82$, strong negative relationship

$r = 0.52$, moderate positive relationship

$r = -0.22$, weak negative relationship

(69.5 per 1,000 women aged 15–19),[3] this rate is still almost twice as high as in other industrialized nations. These high rates have been attributed, among other factors, to the high rate of poverty and inequality in the United States.

The association between teen pregnancy and poverty and social inequality has been well documented both nationally and internationally. Teen pregnancy rates are higher among people living in poverty, and industrial societies that have done the most to reduce social inequality tend to have the lowest rates of teen pregnancy.[4] The noted sociologist William Wilson has claimed that the disappearance of hundreds of low-skilled jobs in the past 25 years and the resulting increase in unemployment, especially in the inner cities, has led to the increase in teenage pregnancy rates and to welfare dependency.[5] Teenagers living in areas of high unemployment, poverty, and lack of opportunities are six to seven times more likely to become unwed parents.[6]

To examine the degree to which economic factors influence teenage pregnancy rates, we analyze state-by-state data on unemployment rates in 2004 and teenage pregnancy rates in 2005. Using unemployment rate and pregnancy rate, both interval-ratio variables, we can examine the hypothesis that states with higher unemployment rates will tend to have higher teenage pregnancy rates.

Figure 9.11 shows the scatter diagram for unemployment rate and teenage pregnancy rate. Because we are assuming that the unemployment rate in 2004 can predict the teen pregnancy rate in 2005, we are going to treat unemployment rate as our independent variable, *X*. Teen pregnancy rate, then, is our dependent variable, *Y*. The scatter diagram seems to suggest that the two variables are linearly related. It also illustrates that these variables are positively associated; that is, as the state's unemployment rate rises, the teen pregnancy rate rises as well.

Our bivariate regression equation for 50 states[7] is

$$\hat{Y} = 23.925 + 8.127(X)$$

In this prediction equation, the **slope** (*b*) is 8.127 and the intercept (*a*) is 23.925. The positive slope, 8.127, confirms our earlier impression, based on the scatter diagram, that the relationship between the unemployment rate and teenage pregnancy rate is positive. In other words, the higher the unemployment rate, the higher the pregnancy rate. A *b* equal to 8.127 means that for every 1 percentage point increase in the unemployment rate, the pregnancy rate for teens aged 15 to 19 years will increase by 8.127 pregnancies per 1,000 women. The intercept, *a*, of 23.925 indicates that with full employment (an unemployment rate of 0), the teen pregnancy rate will be 23.925 pregnancies per 1,000 women. The regression line corresponding to this linear regression equation is shown in Figure 9.12.

**Figure 9.11** Scatter Diagram for Unemployment Rate and Teenage Pregnancy Rate

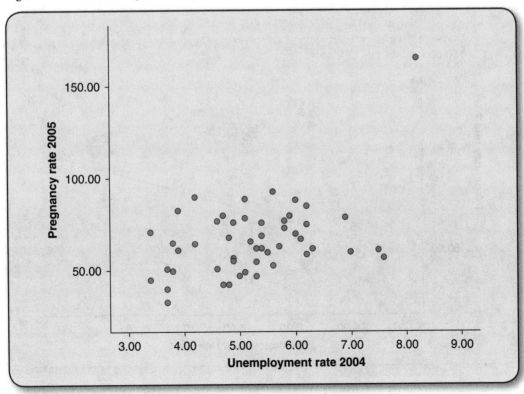

*Source:* Guttmacher Institute, *U.S. Teenage Pregnancies, Births and Abortions: National and State Trends and Trends by Race and Ethnicity,* January 2010, and *Statistical Abstract of the United States,* 2007.

*Slope* (b)   The amount of change in a dependent variable per unit change in an independent variable.

Based on the linear regression equation, we could predict the teenage pregnancy rate for any state based on its unemployment rate in 2004. For example, with a 2004 unemployment rate of 5.8%, Alabama's predicted 2005 teen pregnancy rate is

$$\hat{Y} = 23.925 + 8.127(5.8) = 71.06$$

With a higher unemployment rate of 7.5%, the predicted 2005 teen pregnancy rate for Alaska is

$$\hat{Y} = 23.925 + 8.127(7.5) = 84.88$$

**Figure 9.12**   Scatter Diagram Showing Regression Line for Unemployment Rate and Teenage Pregnancy Rate

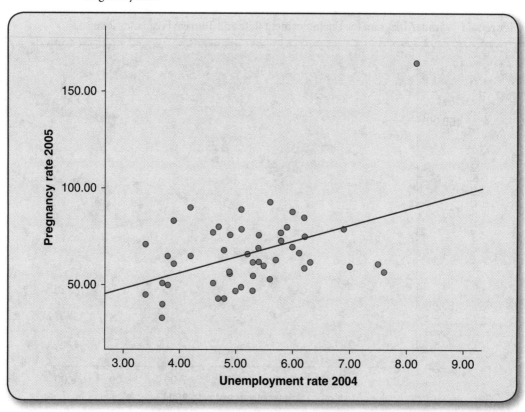

*Source:* Guttmacher Institute, *U.S. Teenage Pregnancies, Births and Abortions: National and State Trends and Trends by Race and Ethnicity,* January 2010, and *Statistical Abstract of the United States,* 2007.

We also calculated $r$ and $r^2$ for these data. We obtained an $r$ of 0.44 and an $r^2$ of $0.44^2 = 0.19$. An $r$ of 0.44 indicates that there is a moderate positive relationship between the unemployment rate in 2004 and the teen pregnancy rate in 2005.

The coefficient of determination, $r^2$, measures the proportional reduction of error that results from using the linear regression model to predict teen pregnancy rates. An $r^2$ of 0.19 means that by using the regression equation, our prediction of teen pregnancy rates is improved by 19% ($0.19 \times 100$) over the prediction we would make using the mean pregnancy rate alone. We can also say that the independent variable (unemployment rate) explains about 19% of the variation in the dependent variable (teen pregnancy rate).

This analysis deals with only one factor affecting teen pregnancy rates. Other important socioeconomic indicators likely to affect teen pregnancy rates—such as poverty rates, welfare policies, and expenditures on education—would also need to be considered for a complete analysis of the determinants of teenage pregnancy.

## ▣ STATISTICS IN PRACTICE: THE MARRIAGE PENALTY IN EARNING

Among factors commonly associated with earnings are human capital variables (e.g., age, education, work experience, and health) and labor market variables (such as the unemployment rate and the structure of occupations). Individual characteristics, such as gender, race, and ethnicity, also explain disparities in earnings. In addition, marital status has been linked to differences in earnings. Although marriage is associated with higher earnings for men, for women it carries a penalty; married women tend to earn less at every educational level than single women.

The lower earnings of married women have been related to differences in labor force experience. Getting married and being the mother of young children tend to limit women's choice of jobs to those that may offer flexible working hours but are generally low paying and offer fewer opportunities for promotion. Moreover, married women tend to be out of the labor market longer and have fewer years on the job than single women. When they reenter the job market or begin their career after their children are grown, they compete with coworkers with considerably more work experience and on-the-job training. (Women who need to become financially independent after divorce or widowhood may share some of the same liabilities as married women.)

All these suggest that the returns for formal education will be generally lower for married women. Thus, we would expect single women to earn more for each year of formal education than married women. We explore this issue by analyzing the bivariate relationship between level of education and personal income among single and married females (working full-time) that were included in a 2008 GSS sample of 673 respondents.[8] We are assuming that level of education (measured in years) can predict personal income, and therefore, we treat education as our independent variable, $X$. Personal income (measured in dollars), then, is the dependent variable, $Y$. Since both are interval-ratio variables, we can use bivariate regression analysis to examine the difference in returns for education.

Our bivariate regression equation for single females working full-time is

$$\hat{Y}(\text{single}) = -47{,}254.46 + 6{,}541.19(X)$$

The regression equation tells us that for every unit increase in education—the unit is 1 year—we can predict an increase of $6,541.19 in the annual income of single women in our sample who work full-time.

The bivariate regression equation for married females working full-time is

$$\hat{Y}(\text{married}) = -9{,}536.24 + 4{,}014.63(X)$$

The regression equation tells us that for every unit increase in education, we can predict an increase of $4,014.63 in the annual income of married women in our sample who work full-time.

This analysis indicates that, as we suggested, the returns for education are lower for married women. For every year of education, the earnings of single women increase by $2,526.56 ($6,541.19 − $4,014.63) more than those of married women.

Let's use these regression equations to predict the difference in annual income between a single woman and a married woman, both with 16 years of education and working full-time:

$$\hat{Y}(\text{single}) = -47{,}254.46 + 6{,}541.19(16) = 57{,}404.58$$

$$\hat{Y}(\text{married}) = -9{,}536.24 + 4{,}014.63(16) = 54{,}697.84$$

The predicted difference in annual income between a single and a married woman with a college education, both working full-time, is $2,706.74 ($57,404.58 − $54,697.84).

We also calculated the $r$ and $r^2$ for these data. For single women, $r = 0.48$; for married women, $r = 0.23$. These coefficients indicate that for both groups there is a weak-to-moderate (the relationship is substantially stronger for single women) relationship between education and earnings.

To determine how much of the variation in income can be explained by education, we need to calculate $r^2$. For single women,

$$r^2 (\text{single}) = 0.48^2 = 0.23$$

and for married women,

$$r^2 (\text{married}) = 0.23^2 = 0.05$$

Using the regression equation, our prediction of income for single women is improved by 23% ($0.23 \times 100$) over the prediction we would make using the mean alone. For married women, there is less of an improvement in prediction, 5% ($0.05 \times 100$).

This analysis deals with only one factor affecting earnings—the level of education. Other important factors associated with earnings—including occupation, seniority, race/ethnicity, and age—would need to be considered for a complete analysis of the differences in earnings between single and married women.

## MAIN POINTS

- A scatter diagram (also called scatterplot) is a quick visual method used to display relationships between two interval-ratio variables. It is used as a first exploratory step in regression analysis and can suggest to us whether two variables are associated.

- Equations for all straight lines have the same general form:

$$\hat{Y} = a + b(X)$$

where

$\hat{Y}$ = the predicted score on the dependent variable

$X$ = the score on the independent variable

$a$ = the $Y$-intercept, or the point where the line crosses the $Y$-axis; therefore, $a$ is the value of $Y$ when $X$ is zero

$b$ = the slope of the line, or the change in $Y$ with a unit change in $X$

- The best-fitting regression line is that line where the residual sum of squares, or $\Sigma e^2$, is at a minimum. Such a line is called the least squares line, and the technique that produces this line is called the least squares method.

- The coefficient of determination ($r^2$) and Pearson's correlation coefficient ($r$) measure how well the regression model fits the data. Pearson's $r$ also measures the strength of the association between the two variables. The coefficient of determination, $r^2$, can be interpreted as a PRE measure. It reflects the proportional reduction of error resulting from use of the linear regression model.

## KEY TERMS

coefficient of
    determination ($r^2$)
deterministic
    (perfect) linear
    relationship

least squares line (best-
    fitting line)
least squares
    method
linear relationship

Pearson's correlation
    coefficient ($r$)
scatter diagram (scatterplot)
slope ($b$)
$Y$-intercept ($a$)

## ON YOUR OWN

Log on to the web-based student study site at **www.sagepub.com/ssdsessentials** for additional study questions, web quizzes, web resources, flashcards, codebooks and datasets, web exercises, appendices, and links to social science journal articles reflecting the statistics used in this chapter.

## CHAPTER EXERCISES

1. Based on the following eight countries, examine the data to determine the extent of the relationship between simply being concerned about the environment and actually giving money to environmental groups.

Exercises

| Country | Percentage Concerned | Percentage Donating Money |
|---|---|---|
| Austria | 35.5 | 27.8 |
| Denmark | 27.2 | 22.3 |
| Netherlands | 30.1 | 44.8 |
| Philippines | 50.1 | 6.8 |
| Russia | 29.0 | 1.6 |
| Slovenia | 50.3 | 10.7 |
| Spain | 35.9 | 7.4 |
| United States | 33.8 | 22.8 |

*Source:* International Social Survey Programme, 2000.

a. Construct a scatterplot of the two variables, placing percentage concerned about the environment on the horizontal or X-axis and the percentage donating money to environmental groups on the vertical or Y-axis.

b. Does the relationship between the two variables seem linear? Describe the relationship.

c. Find the value of the Pearson correlation coefficient that measures the association between the two variables and offer an interpretation.

2. There is often thought to be a relationship between a person's educational attainment and the number of children he or she has. The hypothesis is that as one's educational level increases, he or she has fewer children. Investigate this conjecture with 25 cases drawn from the 2006 GSS file. The following table displays educational attainment, in years, and the number of children for each respondent.

| Education | Children | Education | Children |
|---|---|---|---|
| 16 | 0 | 12 | 2 |
| 12 | 1 | 12 | 3 |
| 12 | 3 | 11 | 1 |
| 6 | 6 | 12 | 2 |
| 14 | 2 | 11 | 2 |
| 14 | 2 | 12 | 0 |
| 16 | 2 | 12 | 2 |
| 12 | 2 | 12 | 3 |
| 17 | 2 | 12 | 4 |
| 12 | 3 | 12 | 1 |
| 14 | 4 | 14 | 0 |
| 13 | 0 | 12 | 3 |
| 12 | 1 | | |

a. Calculate the Pearson correlation coefficient for these two variables. Does its value support the hypothesized relationship?

b. Calculate the least squares regression equation using education as a predictor variable. What is the value of the slope, $b$? What is the value of the intercept, $a$?

c. What is the predicted number of children for a person with a college degree (16 years of education)?

d. Does any respondent actually have this number of children? If so, what is his or her level of education? If not, is this a problem or an indication that the regression equation you calculated is incorrect? Why or why not?

3. The condition and health of our environment is a growing concern. Let's examine the relationship between a country's gross national product (GNP) and the percentage of respondents willing to pay higher prices for goods to protect the environment. The following table displays information for five countries selected at random.

   a. Calculate the correlation coefficient between a country's GNP and the percentage of its residents willing to pay higher prices to protect the environment. What is its value?

   b. Provide an interpretation for the coefficient.

| Country | GNP per Capita | Percentage Willing to Pay |
|---|---|---|
| United States | 29.24 | 44.9 |
| Ireland | 18.71 | 53.3 |
| Netherlands | 24.78 | 61.2 |
| Norway | 34.31 | 40.7 |
| Sweden | 25.58 | 32.6 |

*Source:* International Social Survey Programme, 2000.

4. In Chapter 4, Exercise 5, we studied the variability of crime rates and police expenditures in the eastern and midwestern United States. We've now been asked to investigate the hypothesis that the number of crimes is related to police expenditures because states with higher crime rates are likely to increase their police force, thereby spending more on the number of officers on the street.

   a. Construct a scatter diagram of the number of crimes and police expenditures, with number of crimes as the predictor variable. What can you say about the relationship between these two variables based on the scatterplot?

   b. Find the least squares regression equation that predicts police expenditures from the number of crimes. What is the slope? What is the intercept?

   c. Calculate the coefficient of determination ($r^2$), and provide an interpretation.

   d. If the number of crimes increased by 2,000 for a state, by how much would you predict police expenditures to increase?

   e. Does it make sense to predict police expenditures when the number of crimes is equal to zero? Why or why not?

| State | Number of Crimes (per 100,000 population) | Police Protection Expenditures (in millions of dollars) |
|---|---|---|
| Maine | 2,635 | 221 |
| New Hampshire | 2,013 | 274 |
| Vermont | 2,442 | 136 |
| Massachusetts | 2,838 | 1,673 |
| Rhode Island | 2,815 | 286 |

*Source:* U.S. Census Bureau, *The 2010 Statistical Abstract*, Tables 297 and 431.

5. Before calculating a correlation coefficient or a regression equation, it is always important to examine a scatter diagram between two variables to see how well a straight line fits the data. If a straight line does not appear to fit, other curves can be used to describe the relationship (this subject is not discussed in our text).

Exercises

The SPSS scatterplot in Figure 9.13 and output shown in Figure 9.14 display the relationship between education (measured in years) and television viewing (measured in hours) based on 2008 GSS data. We can hypothesize that as educational attainment increases, hours of television viewing will decrease, indicating a negative relationship between the two variables.

**Figure 9.13** Scatterplot of Hours of Television Viewing per Day by Highest Year of School Completed

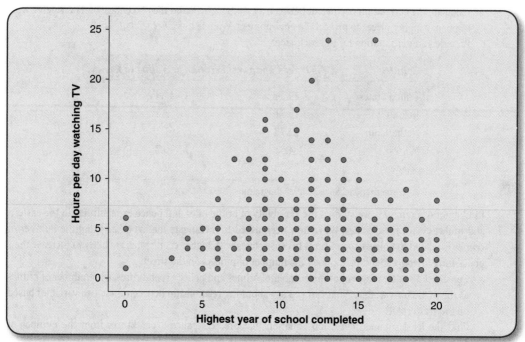

**Figure 9.14** Linear Regression Output Specifying the Relationship Between Education and Hours Spent per Day Watching Television

**Coefficients[a]**

| Model | | Unstandardized Coefficients | | Standardized Coefficients | t | Sig. |
|---|---|---|---|---|---|---|
| | | B | Std. Error | Beta | | |
| 1 | (Constant) | 5.575 | .367 | | 15.175 | .000 |
| | HIGHEST YEAR OF SCHOOL COMPLETED | -.196 | .027 | -.229 | -7.335 | .000 |

a. Dependent Variable: HOURS PER DAY WATCHING TV

**Model Summary**

| Model | R | R Square | Adjusted R Square | Std. Error of the Estimate |
|---|---|---|---|---|
| 1 | .229[a] | .052 | .051 | 2.516 |

a. Predictors: (Constant), HIGHEST YEAR OF SCHOOL COMPLETED

a. Assess the relationship between the two variables based on the scatterplot and output for the *b* coefficient. Is there a relationship between these two variables as hypothesized? Is it a negative or a positive relationship?

b. Describe the relationship between these two variables using representative values of years of education and hours of television viewing. For example, if an individual has 16 years of education, what are the predicted hours of television viewing? How can you determine this?

c. Does a straight line adequately represent (visually) the relationship between these two variables? Why or why not?

6. Based on the statistical data obtained from the countries in South America, let's analyze the relationship between gross domestic product (GDP) per capita and infant mortality rate (IMR).

| Country | GDP per Capita in 2008 (in U.S. dollars) | Infant Mortality in 2010 (estimated) (per 1,000 births) |
|---|---|---|
| Argentina | 8,236 | 11.1 |
| Bolivia | 1,720 | 43.4 |
| Brazil | 8,205 | 21.9 |
| Chile | 10,084 | 7.5 |
| Colombia | 5,416 | 16.9 |
| Ecuador | 4,056 | 20.3 |
| Paraguay | 2,561 | 23.8 |
| Peru | 4,477 | 27.7 |
| Uruguay | 9,654 | 11.0 |
| Venezuela | 11,246 | 21.1 |

*Sources:* World Bank, *World Development Indicators*, 2010, and *The World Factbook*, 2010.

a. Construct a scatterplot from the following data, predicting IMR from GDP. What is the relationship between GDP and IMR for these 10 countries in South America?

b. Does it appear (visually) that a straight line fits these data? Why or why not?

c. Calculate the correlation coefficient and coefficient of determination. Do these values offer further support for your answer to (b)? How?

7. Social scientists have long been interested in the aspirations and achievements of people in the United States. Research on social mobility, status, and educational attainment has provided convincing evidence on the relationship between parents' and children's socioeconomic achievement. The GSS 2006 data set has information on the educational level of respondents and their mothers. Use this information for the following selected nonrandom subsample of respondents to see whether those whose mothers had more education are more likely to have more education themselves.

a. Construct a scatterplot, predicting the highest year of respondent's schooling with the highest year of the mother's schooling.

b. Calculate the regression equation with mother's education as the predictor variable, and draw the regression line on the scatterplot. What is the slope? What is the intercept? Describe how the straight line "fits" the data.

c. What is the error of prediction for the second case (the person with 13 years of education and mother's education 15 years)? What is the error of prediction for the person with 14 years of education and mother's education 18 years?

d. What is the predicted years of education for someone whose mother received 4 years of education? How about for someone whose mother received 12 years of education?

e. Calculate the mean number of years of education for respondents and for respondents' mothers. Plot this point on the scatterplot. Where does it fall? Can you think of a reason why this should be true?

| Mother's Highest School Year Completed | Respondent's Highest School Year Completed |
|:---:|:---:|
| 0 | 12 |
| 15 | 13 |
| 6 | 9 |
| 9 | 12 |
| 16 | 16 |
| 12 | 12 |
| 6 | 16 |
| 18 | 14 |
| 12 | 13 |
| 14 | 12 |
| 14 | 18 |
| 7 | 12 |

8. In Exercise 6, we investigated the relationship between infant mortality rate and GDP in South America. The estimated birthrates (number of live births per 1,000 inhabitants) in these same countries in 2010 are shown in the following table:

| Country | Birth Rate in 2010 (estimated) |
|:---|:---:|
| Argentina | 18 |
| Bolivia | 25 |
| Brazil | 18 |
| Chile | 15 |
| Colombia | 18 |
| Ecuador | 20 |
| Paraguay | 18 |
| Peru | 19 |
| Uruguay | 14 |
| Venezuela | 13 |

Source: The World Factbook, 2010.

a. Construct a scatterplot for GDP and birthrate and one for infant mortality rate and birthrate. Do you think each can be characterized by a linear relationship?

b. Calculate the coefficient of determination and correlation coefficient for each relationship.

c. Use this information to describe the relationship between the variables.

9. In 2010, a U.S. Census Bureau report revealed that approximately 13% of all Americans were living below the poverty line in 2007. This figure is higher than in 2000, when the poverty rate was 12.2%. This translates to an increase of approximately 4.75 million Americans living below the poverty line. Individuals and families living below the poverty line face many obstacles, the least of which is access to health care. In many cases, those living below the poverty line are without any form of health insurance. Using data from the U.S. Census Bureau, analyze the relationship between living below the poverty line and access to health care.

| State | Percentage Below Poverty Line (2007) | Percentage Without Health Insurance (2007) |
|---|---|---|
| Alabama | 16.9 | 12.0 |
| California | 12.4 | 18.2 |
| Idaho | 12.1 | 13.9 |
| Louisiana | 18.6 | 18.5 |
| New Jersey | 8.6 | 15.8 |
| New York | 13.7 | 13.2 |
| Pennsylvania | 11.6 | 9.5 |
| Rhode Island | 12.0 | 10.8 |
| South Carolina | 15.0 | 16.4 |
| Texas | 16.3 | 25.2 |
| Washington | 11.4 | 11.3 |
| Wisconsin | 10.8 | 8.2 |

*Source:* U.S. Census Bureau, *The 2010 Statistical Abstract,* 2010, Tables 693 and 150.

a. Construct a scatterplot, predicting the percentage without health insurance with the percentage living below the poverty level. Does it appear that a straight-line relationship will fit the data?

b. Calculate the regression equation with percentage of the population without health insurance as the dependent variable, and draw the regression line on the scatterplot. What is its slope? What is the intercept? Has your opinion changed about whether a straight line seems to fit the data? Are there any states that fall far from the regression line? Which one(s)?

c. What percentage of the population must be living below the poverty line to obtain a predicted value of 5% without health insurance?

d. Predicting a value that falls beyond the observed range of the two variables in a regression is problematic at best, so your answer in (c) isn't necessarily statistically believable. However, what is a nonstatistical, or substantive, reason? Why might making such a prediction be important?

10. Using the table below, we will examine the relationship between GNP per capita and the percentage of respondents willing to pay more in taxes (using Table 9.4 as a model for your calculation).

a. Calculate *a* and *b* and write out the regression equation (i.e., prediction equation).

b. About what percentage of citizens are willing to pay higher taxes for a country with a GNP per capita of 3.0 (i.e., $3,000)? For a GNP per capita of 30.0 (i.e., $30,000)?

| Country | GNP per Capita | Percentage Willing to Pay Higher Taxes |
|---------|----------------|----------------------------------------|
| Canada | 19.71 | 24.0 |
| Chile | 4.99 | 29.1 |
| Finland | 24.28 | 12.0 |
| Ireland | 18.71 | 34.3 |
| Japan | 32.35 | 37.2 |
| Latvia | 2.42 | 17.3 |
| Mexico | 3.84 | 34.7 |
| Netherlands | 24.78 | 51.9 |
| New Zealand | 14.60 | 31.1 |
| Norway | 34.31 | 22.8 |
| Portugal | 10.67 | 17.1 |
| Russia | 2.66 | 29.9 |
| Spain | 14.10 | 22.2 |
| Sweden | 25.58 | 19.5 |
| Switzerland | 39.98 | 33.5 |
| United States | 29.24 | 31.6 |

*Sources:* The World Bank Group, *Development Education Program Learning Module: Economics, GNP per Capita,* 2004. International Social Survey Programme, 2000.

11. On completing this chapter, you should be able to correctly answer the following questions.
   a. True or false: It is possible, in fact it often is the case, that your slope, $b$, will be a positive value and your correlation coefficient, $r$, will be a negative value.
   b. Both $a$ and $b$ refer to changes in which variable, the independent or dependent?
   c. The coefficient of determination, $r^2$, is a PRE measure. What does this mean?
   d. True or false: All regression equations reflect *causal* relationships expressed as linear functions.

# Chapter 10

# Analysis of Variance

## Chapter Learning Objectives

❖ Understanding the application of an analysis of variance (ANOVA) model
❖ Assessing the significance and interpretation of the *F* statistics

Many research questions require us to look at multiple samples or groups, at least more than two at a time. We may be interested in studying the influence of ethnic identity (white, African American, Asian American, Latino/a) on church attendance, the influence of social class (lower, working, middle, and upper) on President Barack Obama's job approval ratings, or the effect of educational attainment (less than high school, high school graduate, some college, and college graduate) on household income. Note that each of these examples requires a comparison between multiple demographic or ethnic groups, more than the two-group comparisons that we reviewed in Chapter 7. While it would be easy to confine our analyses between two groups, our social world is much more complex and diverse.

Let's say that we're interested in examining educational attainment—on average, how many years of education do Americans achieve? The 2007 Census reported that 84.5% of adults (25 years and older) completed at least a high school degree, and more than 27% of all adults attained at least a bachelor's degree (Crissey, 2009).[1] Early in his term, President Obama pledged that the United States will have the world's highest proportion of college graduates by 2020. Special attention has been paid to the educational achievement of Latino students. Data from the U.S. Census, as well as from the U.S. Department of Education, confirm that Latino students continue to have lower levels of educational achievement than other racial or ethnic groups.

In Chapter 7, Testing Hypotheses, we introduced statistical techniques to assess the difference between two sample means or proportions. For our example in Table 7.3, we compared the difference in educational attainment for blacks and Hispanics. But what if we wanted to examine separate groups of men and women by their race or ethnicity? Is there a significant variation in educational attainment among black women, Hispanic women, black men, and Hispanic men?

Table 10.1  Educational Attainment (measured in years) for Four GSS 2006 Groups

| Black Males $n_1 = 6$ | Hispanic Males $n_2 = 4$ | Black Females $n_3 = 6$ | Hispanic Females $n_4 = 5$ |
|---|---|---|---|
| 16 | 14 | 16 | 14 |
| 12 | 12 | 18 | 12 |
| 14 | 11 | 16 | 12 |
| 12 | 11 | 14 | 13 |
| 12 |  | 16 | 14 |
| 12 |  | 12 |  |

We've taken a random sample of 21 men and women from the GSS, grouped them into four demographic categories, and included their educational attainment in Table 10.1. With the *t*-test statistic we covered in Chapter 7, we would have to analyze the mean educational attainment of black women versus Hispanic women, black women versus black men, and black women versus Hispanic men, and so on. In the end, we would have a tedious series of *t*-test statistic calculations, and we still wouldn't be able to answer our original question: Is there a difference in educational attainment among all *four* demographic groups?

There is a statistical technique that will allow us to examine all the four groups or samples simultaneously. This technique is called **analysis of variance (ANOVA)**. ANOVA follows the same five-step model of hypothesis testing that we used with *t* test and *Z* test for proportions (in Chapter 7) and chi-square (in Chapter 8). In this chapter, we review the calculations for ANOVA, discuss how we can test the significance of $r^2$ (the coefficient of determination) and $R^2$ (multiple coefficient of determination) using ANOVA, and discuss two applications of ANOVA from the research literature.

---

*Analysis of variance (ANOVA)*   An inferential statistics technique designed to test for a significant relationship between two variables in two or more groups or samples.

---

## ▣ UNDERSTANDING ANALYSIS OF VARIANCE

Recall that the *t* test examines the difference between two means $\bar{Y}_1 - \bar{Y}_2$, while the null hypothesis assumed that there was no difference between them: $\mu_1 = \mu_2$. Rejecting the null hypothesis meant that there was a significant difference between the two mean scores (or the populations from which the samples were drawn). In our Chapter 7 example, we analyzed the difference between mean years of education for black men and Hispanic men. Based on our *t*-test statistic, we rejected the null hypothesis, concluding that black men, on average, have significantly more years of education than Hispanic men do.

The logic of ANOVA is the same but extending to two or more groups. For the data presented in Table 10.1, ANOVA will allow us to examine the variation among four means ($\bar{Y}_1$, $\bar{Y}_2$, $\bar{Y}_3$, $\bar{Y}_4$) and the null hypothesis can be stated as follows: $\mu_1 = \mu_2 = \mu_3 = \mu_4$. Rejecting the null hypothesis for ANOVA indicates that there is a significant variation among the four samples (or the four populations from which the samples were drawn) and that at least one of the sample means is significantly different from the others. In our example, it suggests that years of education (dependent variable) do vary by group membership (independent variable). When ANOVA procedures are applied to data with one dependent and one independent variable, it is called a **one-way ANOVA**.

---

*One-way ANOVA*  Analysis of variance application with one dependent and one independent variable.

---

The means, standard deviations, and variances for the samples have been calculated and are shown in Table 10.2. Note that the four mean educational years are not identical, with black women having the highest educational attainment. Also, based on the standard deviations, we can tell that the samples are relatively homogeneous with deviations within 1.00 to 2.07 years of the mean. We already know that there is a difference between the samples, but the question remains: Is this difference significant? Do the samples reflect a relationship between demographic group membership and educational attainment in the general population?

To determine whether the differences are significant, ANOVA examines the differences *between* our four samples, as well as the differences *within* a single sample. The differences can also be referred to as variance or variation, which is why ANOVA is the analysis of *variance*. What is the difference between one sample's mean score and the overall mean? What is the variation of individual scores within one

**Table 10.2**  Means, Variances, and Standard Deviations for Four GSS 2006 Groups

| *Black Males* $n_1 = 6$ | *Hispanic Males* $n_2 = 4$ | *Black Females* $n_3 = 6$ | *Hispanic Females* $n_4 = 5$ |
|:---:|:---:|:---:|:---:|
| 16 | 14 | 16 | 14 |
| 12 | 12 | 18 | 12 |
| 14 | 11 | 16 | 12 |
| 12 | 11 | 14 | 13 |
| 12 | | 16 | 14 |
| 12 | | 12 | |
| $\bar{Y}_1 = 13.00$ | $\bar{Y}_2 = 12.00$ | $\bar{Y}_3 = 15.33$ | $\bar{Y}_4 = 13.00$ |
| $S_1 = 1.67$ | $S_2 = 1.41$ | $S_3 = 2.07$ | $S_4 = 1.00$ |
| $S_1^2 = 2.79$ | $S_2^2 = 1.99$ | $S_3^2 = 4.28$ | $S_4^2 = 1.00$ |

$$\bar{Y} = 13.48$$

sample? Are all the scores alike (no variation), or is there a broad variation in scores? ANOVA allows us to determine whether the variance between samples is larger than the variance within the samples. If the variance is larger between samples than the variance within samples, we know that educational attainment varies significantly across the samples. It would support the notion that group membership explains the variation in educational attainment.

# ▣ THE STRUCTURE OF HYPOTHESIS TESTING WITH ANOVA

## The Assumptions

ANOVA requires several assumptions regarding the method of sampling, the level of measurement, the shape of the population distribution, and the homogeneity of variance.

1. Independent random samples are used. Our choice of sample members from one population has no effect on the choice of sample members from the other populations.
2. The dependent variable, years of education, is an interval-ratio level of measurement. Some researchers also apply ANOVA to ordinal-level measurements.
3. The population is normally distributed. Although we cannot confirm whether the populations are normal, given that our $N$ is so small, we must assume so to proceed with our analysis.
4. The population variances are equal. Based on our calculations in Table 10.2, we see that the sample variances, although not identical, are relatively homogeneous.[2]

## Stating the Research and the Null Hypotheses and Setting Alpha

The research hypothesis $(H_1)$ proposes that at least one of the means is different. We do not identify which one(s) will be different, or larger or smaller, we only predict that a difference does exist.

$H_1$: At least one mean is different from the others.

ANOVA is a test of the null hypothesis of no difference between any of the means. Since we're working with four samples, we include four $\mu$s in our null hypothesis.

$H_0$: $\mu_1 = \mu_2 = \mu_3 = \mu_4$

As we did in other models of hypothesis testing, we'll have to set our alpha. Alpha is the level of probability at which we'll reject our null hypothesis. For this example, we'll set alpha at .05.

## The Concepts of Between and Within Total Variance

A word of caution before we proceed: Since we're working with four different samples and a total of 21 respondents, we'll have a lot of calculations. It's important to be consistent with your notations (don't mix up numbers for the different samples) and be careful with your calculations.

Our primary set of calculations has to do with the two types of variance: between-group variance and within-group variance. The estimate of each variance has two parts, the sum of squares and degrees of freedom (*df*).

The **between-group sum of squares or *SSB*** measures the difference in average years of education between our four groups. Sum of squares is the short form for "sum of squared deviations." For *SSB*, what we're measuring is the sum of squared deviations between each sample mean to the overall mean score. The formula for the *SSB* can be presented as follows:

$$SSB = \Sigma n_k (\bar{Y}_k - \bar{Y})^2 \qquad (10.1)$$

where

$n_k$ = the number of cases in a sample (*k* represents the number of different samples)

$\bar{Y}_k$ = the mean of a sample

$\bar{Y}$ = the overall mean

*SSB* can also be understood as the amount of variation in the dependent variable (years of education) that can be attributed to or explained by the independent variable (the four demographic groups).

---

**Between-group sum of squares or SSB**   The sum of squared deviations between each sample mean to the overall mean score.

---

**Within-group sum of squares or *SSW*** measures the variation of scores within a single sample or, as in our example, the variation in years of education within one group. *SSW* is also referred to as the amount of unexplained variance, since this is what remains after we consider the effect of the specified independent variable. The formula for *SSW* measures the sum of squared deviations within each group, between each individual score with its sample mean.

$$SSW = \Sigma(Y_i - \bar{Y}_k)^2 \qquad (10.2)$$

where

$Y_i$ = each individual score in a sample

$\bar{Y}_k$ = the mean of a sample

Even with our small sample size, if we were to use Formula 10.2, we'd have a tedious and cumbersome set of calculations. Instead, we suggest using the following computational formula for within-group variation or *SSW*:

$$SSW = \Sigma Y_i^2 - \Sigma \frac{(\Sigma Y_k)^2}{n_k} \qquad (10.3)$$

where

$Y_i^2$ = the squared scores from each sample

$\Sigma Y_k$ = the sum of the scores of each sample

$n_k$ = the number of cases in a sample

---

**Within-group sum of squares or** SSW    Sum of squared deviations within each group, calculated between each individual score and the sample mean.

---

Together, the explained (*SSB*) and unexplained (*SSW*) variances compose the amount of total variation in scores. The **total sum of squares or** *SST* can be represented by

$$SST = \sum (Y_i - \bar{Y})^2 = SSB + SSW \qquad (10.4)$$

where

$Y_i$ = each individual score

$\bar{Y}$ = the overall mean

---

*Total sum of squares or* SST    The total variation in scores, calculated by adding *SSB* and *SSW*.

---

The second part of estimating the between-group and within-group variances is calculating the degrees of freedom. Degrees of freedom are also discussed in Chapters 7 and 8. For ANOVA, we have to calculate two degrees of freedom. For *SSB*, the degrees of freedom are determined by

$$df_b = k-1 \qquad (10.5)$$

where $k$ is the number of samples.

For *SSW*, the degrees of freedom are determined by

$$df_w = N - k \qquad (10.6)$$

where

$N$ = total number of cases

$k$ = number of samples

Finally, we can estimate the between-group variance by calculating **mean square between**. Simply stated, mean squares are averages computed by dividing each sum of squares by its corresponding degrees of freedom. Mean square between can be represented by

$$\text{Mean square between} = SSB/df_b \qquad (10.7)$$

and the within-group variance or **mean square within** can be represented by

$$\text{Mean square within} = SSW/df_w \qquad (10.8)$$

---

*Mean square between*   Sum of squares between divided by its corresponding degrees of freedom.

*Mean square within*   Sum of squares within divided by its corresponding degrees of freedom.

---

## ◼ DECOMPOSITION OF *SST*

According to Formula 10.4, sum of squares total (*SST*) is equal to

$$SST = \Sigma(Y_i - \bar{Y})^2 = SSB + SSW$$

You can see that the between sum of squares (explained variance) and within sum of squares (unexplained variance) account for the total variance (*SST*) in a particular dependent variable. How does that apply to a single case in our educational attainment example? Let's take the first black female in Table 10.1 with 16 years of education.

Her total deviation (corresponding to *SST*) is based on the difference between her score from the overall mean (Formula 10.4). Her score is quite a bit higher than the overall mean education of 13.48 years. The difference of her score from the overall mean is 2.52 years (16 − 13.48). Between-group deviation (corresponding to *SSB*) can be determined by measuring the difference between her group average from the overall mean (Formula 10.1). We've already commented on the higher educational attainment for black females (average of 15.33 years) when compared with the other three demographic groups. The deviation between the group average and overall average for black females is 1.85 years (15.33 − 13.48). Finally, the within-group deviation (corresponding to *SSW*, Formula 10.2) is based on the difference between the first black female's years of education and the group average for black females: 0.67 years (16 − 15.33). So for the first black female in our sample, *SSB* + *SSW* = *SST* or 1.85 + 0.67 = 2.52. In a complete ANOVA problem, we're computing these two sources of deviation (*SSB* and *SSW*) to obtain *SST* (Formula 10.4) for everyone in the sample.

## The *F* Statistic

Together the mean square between (Formula 10.7) and mean square within (Formula 10.8) compose the obtained *F* ratio or *F* statistic. Developed by R. A. Fisher, the *F* statistic is the ratio of between-group variance to within-group variance and is determined by Formula 10.9:

$$F = \frac{\text{Mean square between}}{\text{Mean square within}} = \frac{SSB/df_b}{SSW/df_w} \tag{10.9}$$

We know that a larger obtained *F* statistic means that there is more between-group variance than within-group variance, increasing the chances of rejecting our null hypothesis. In Table 10.3, we present additional calculations to compute *F*.

Let's calculate between-group sum of squares and degrees of freedom based on Formulas 10.1 and 10.5. The calculation for *SSB* is

$$\Sigma n_k (\bar{Y}_k - \bar{Y})^2 = 6(13.00 - 13.48)^2 + 4(12.00 - 13.48)^2$$
$$+ 6(15.33 - 13.48)^2 + 5(13.00 - 13.48)^2$$
$$= 31.83$$

**Table 10.3** Computational Worksheet for ANOVA

| Black Males $n_1 = 6$ | Hispanic Males $n_2 = 4$ | Black Females $n_3 = 6$ | Hispanic Females $n_4 = 5$ |
|:---:|:---:|:---:|:---:|
| 16 | 14 | 16 | 14 |
| 12 | 12 | 18 | 12 |
| 14 | 11 | 16 | 12 |
| 12 | 11 | 14 | 13 |
| 12 | | 16 | 14 |
| 12 | | 12 | |
| $\bar{Y}_1 = 13.00$ | $\bar{Y}_2 = 12.00$ | $\bar{Y}_3 = 15.33$ | $\bar{Y}_4 = 13.00$ |
| $S_1 = 1.67$ | $S_2 = 1.41$ | $S_3 = 2.07$ | $S_4 = 1.00$ |
| $S_1^2 = 2.79$ | $S_2^2 = 1.99$ | $S_3^2 = 4.28$ | $S_4^2 = 1.00$ |
| $\Sigma Y_1 = 78$ | $\Sigma Y_2 = 48$ | $\Sigma Y_3 = 92$ | $\Sigma Y_4 = 65$ |
| $\Sigma Y_1^2 = 1{,}028$ | $\Sigma Y_2^2 = 582$ | $\Sigma Y_3^2 = 1{,}432$ | $\Sigma Y_4^2 = 849$ |

$$\bar{Y} = 13.48$$

The degrees of freedom for *SSB* is $k - 1$ or $4 - 1 = 3$. Based on Formula 10.7, the mean square between is

$$\text{Mean square between} = \frac{31.83}{3} = 10.61$$

The within-group sum of squares and degrees of freedom are based on Formulas 10.3 and 10.6. The calculation for *SSW* is

$$SSW = \Sigma Y_i^2 - \Sigma \frac{(\Sigma Y_k)^2}{n_k} = (1{,}028 + 582 + 1{,}432 + 849)$$

$$- \left( \frac{78^2}{6} + \frac{48^2}{4} + \frac{92^2}{6} + \frac{65^2}{5} \right)$$

$$= 3{,}891 - 3{,}845.67$$

$$= 45.33$$

The degrees of freedom for *SSW* is $N - k = 21 - 4 = 17$. Based on Formula 10.8, the mean square within is

$$\text{Mean square within} = \frac{45.33}{17} = 2.67$$

Finally, our calculation of *F* is based on Formula 10.9:

$$F = \frac{10.61}{2.67} = 3.97$$

---

**F *ratio or* F *statistic***   The test statistic for ANOVA, calculated by the ratio of mean square between to mean square within.

---

## Making a Decision

To determine the probability of calculating an *F* statistic of 3.97, we rely on Appendix D, the distribution of the *F* statistic. Appendix D lists the corresponding values of the *F* distribution for various degrees of freedom and two levels of significance, .05 and .01. Table 10.4 displays the distribution of *F* for .05 level of significance.

Since we set alpha at .05, we'll refer to the table marked "$\alpha = .05$." Note that Appendix D includes two *dfs*. These refer to our degrees of freedom, $df_1 = df_b$ and $df_2 = df_w$.

Because of the two degrees of freedom, we'll have to determine the probability of our obtained *F* differently than we did with *t* test or chi-square. For this ANOVA example, we'll have to determine the corresponding *F*, also called the critical *F*, when $df_b = 3$ and $df_w = 17$, and $\alpha = .05$.

Based on Appendix D, the critical *F* is 3.20, while our obtained *F* (the one that we calculated) is 3.97. Since our obtained *F* is greater than the critical *F* ($3.97 > 3.20$), we know that its probability is $<.05$,

**Table 10.4** Distribution of *F* for .05 Level of Significance

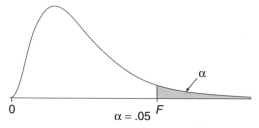

$\alpha$

0     $\alpha = .05$   *F*

| df₂ | df₁ | | | | | | | | | |
|---|---|---|---|---|---|---|---|---|---|---|
| | *1* | *2* | *3* | *4* | *5* | *6* | *8* | *12* | *24* | ∞ |
| 1 | 161.40 | 199.50 | 215.70 | 224.60 | 230.20 | 234.00 | 238.90 | 243.90 | 249.00 | 254.30 |
| 2 | 18.51 | 19.00 | 19.16 | 19.25 | 19.30 | 19.33 | 19.37 | 19.41 | 19.45 | 19.50 |
| 3 | 10.13 | 9.55 | 9.28 | 9.12 | 9.01 | 8.94 | 8.84 | 8.74 | 8.64 | 8.53 |
| 4 | 7.71 | 6.94 | 6.59 | 6.39 | 6.26 | 6.16 | 6.04 | 5.91 | 5.77 | 5.63 |
| 5 | 6.61 | 5.79 | 5.41 | 5.19 | 5.05 | 4.95 | 4.82 | 4.68 | 4.53 | 4.36 |
| 6 | 5.99 | 5.14 | 4.76 | 4.53 | 4.39 | 4.28 | 4.15 | 4.00 | 3.84 | 3.67 |
| 7 | 5.59 | 4.74 | 4.35 | 4.12 | 3.97 | 3.87 | 3.73 | 3.57 | 3.41 | 3.23 |
| 8 | 5.32 | 4.46 | 4.07 | 3.84 | 3.69 | 3.58 | 3.44 | 3.28 | 3.12 | 2.93 |
| 9 | 5.12 | 4.26 | 3.86 | 3.63 | 3.48 | 3.37 | 3.23 | 3.07 | 2.90 | 2.71 |
| 10 | 4.96 | 4.10 | 3.71 | 3.48 | 3.33 | 3.22 | 3.07 | 2.91 | 2.74 | 2.54 |
| 11 | 4.84 | 3.98 | 3.59 | 3.36 | 3.20 | 3.09 | 2.95 | 2.79 | 2.61 | 2.40 |
| 12 | 4.75 | 3.88 | 3.49 | 3.26 | 3.11 | 3.00 | 2.85 | 2.69 | 2.50 | 2.30 |
| 13 | 4.67 | 3.80 | 3.41 | 3.18 | 3.02 | 2.92 | 2.77 | 2.60 | 2.42 | 2.21 |
| 14 | 4.60 | 3.74 | 3.34 | 3.11 | 2.96 | 2.85 | 2.70 | 2.53 | 2.35 | 2.13 |
| 15 | 4.54 | 3.68 | 3.29 | 3.06 | 2.90 | 2.79 | 2.64 | 2.48 | 2.29 | 2.07 |
| 16 | 4.49 | 3.63 | 3.24 | 3.01 | 2.85 | 2.74 | 2.59 | 2.42 | 2.24 | 2.01 |
| 17 | 4.45 | 3.59 | 3.20 | 2.96 | 2.81 | 2.70 | 2.55 | 2.38 | 2.19 | 1.96 |
| 18 | 4.41 | 3.55 | 3.16 | 2.93 | 2.77 | 2.66 | 2.51 | 2.34 | 2.15 | 1.92 |
| 19 | 4.38 | 3.52 | 3.13 | 2.90 | 2.74 | 2.63 | 2.48 | 2.31 | 2.11 | 1.88 |
| 20 | 4.35 | 3.49 | 3.10 | 2.87 | 2.71 | 2.60 | 2.45 | 2.28 | 2.08 | 1.84 |
| 21 | 4.32 | 3.47 | 3.07 | 2.84 | 2.68 | 2.57 | 2.42 | 2.25 | 2.05 | 1.81 |
| 22 | 4.30 | 3.44 | 3.05 | 2.82 | 2.66 | 2.55 | 2.40 | 2.23 | 2.03 | 1.78 |
| 23 | 4.28 | 3.42 | 3.03 | 2.80 | 2.64 | 2.53 | 2.38 | 2.20 | 2.00 | 1.76 |
| 24 | 4.26 | 3.40 | 3.01 | 2.78 | 2.62 | 2.51 | 2.36 | 2.18 | 1.98 | 1.73 |
| 25 | 4.24 | 3.38 | 2.99 | 2.76 | 2.60 | 2.49 | 2.34 | 2.16 | 1.96 | 1.71 |
| 26 | 4.22 | 3.37 | 2.98 | 2.74 | 2.59 | 2.47 | 2.32 | 2.15 | 1.95 | 1.69 |
| 27 | 4.21 | 3.35 | 2.96 | 2.73 | 2.57 | 2.46 | 2.30 | 2.13 | 1.93 | 1.67 |
| 28 | 4.20 | 3.34 | 2.95 | 2.71 | 2.56 | 2.44 | 2.29 | 2.12 | 1.91 | 1.65 |
| 29 | 4.18 | 3.33 | 2.93 | 2.70 | 2.54 | 2.43 | 2.28 | 2.10 | 1.90 | 1.64 |
| 30 | 4.17 | 3.32 | 2.92 | 2.69 | 2.53 | 2.42 | 2.27 | 2.09 | 1.89 | 1.62 |
| 40 | 4.08 | 3.23 | 2.84 | 2.61 | 2.45 | 2.34 | 2.18 | 2.00 | 1.79 | 1.51 |
| 60 | 4.00 | 3.15 | 2.76 | 2.52 | 2.37 | 2.25 | 2.10 | 1.92 | 1.70 | 1.39 |
| 120 | 3.92 | 3.07 | 2.68 | 2.45 | 2.29 | 2.17 | 2.02 | 1.83 | 1.61 | 1.25 |
| ∞ | 3.84 | 2.99 | 2.60 | 2.37 | 2.21 | 2.09 | 1.94 | 1.75 | 1.52 | 1.00 |

*Source:* R. A. Fisher and F. Yates, *Statistical Tables for Biological, Agricultural and Medical Research*, 6th ed. Copyright © R. A. Fisher and F. Yates, 1963. Pearson Education Limited.

**Figure 10.1** Comparing Obtained Versus Critical *F*

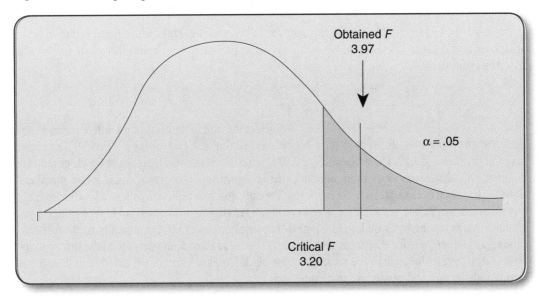

extending into the shaded area. (If our obtained *F* was <3.20, we could determine that its probability was greater than our alpha [$\alpha$] of .05, in the unshaded area of the *F* distribution curve. Refer to Figure 10.1.) We can reject the null hypothesis of no difference and conclude that there is a significant difference in educational attainment between the four groups.

## ◉ THE FIVE STEPS IN HYPOTHESIS TESTING: A SUMMARY

To summarize, we've calculated an analysis of variance test examining the difference between four demographic groups and their average years of education.

*Making Assumptions.*

1. Independent random samples are used.
2. The dependent variable, years of education, is an interval-ratio level of measurement.
3. The population is normally distributed.
4. The population variances are equal.

*Stating the Research and Null Hypothesis and Selecting Alpha.*

$H_1$: At least one mean is different from the others.

$H_0$: $\mu_1 = \mu_2 = \mu_3 = \mu_4$

$\alpha = .05$

*Selecting the Sampling Distribution and Specifying the Test Statistic.* The *F* distribution and *F* statistic are used to test the significance of the difference between the four sample means.

*Computing the Test Statistic.* We need to calculate the between-group and within-group variation (sum of squares and degrees of freedom). We estimate $SSB = 10.61$ ($df_b = 3$) and $SSW = 2.67$ ($df_w = 17$). Based on Formula 10.9,

$$F = \frac{10.61}{2.67} = 3.97$$

*Making a Decision and Interpreting the Results.* We reject the null hypothesis of no difference and conclude that the groups are different in their educational attainment. Our **obtained** *F* of 3.97 is greater than the **critical** *F* of 3.20. The probability of 3.97 is <.05. *F* doesn't advise us about which groups are different, only that educational attainment does differ significantly by demographic group members. However, based on the sample data, we know that the only group to achieve a college education average was black females (15.33 years). If we were to rank the remaining means, second highest educational attainment was among black males (13.00) and Hispanic females (13.00), and finally, the lowest educational attainment was for Hispanic males (12.00). The mean education years for all the three groups were at least 2 years lower than the mean score for black females. Educational attainment does differ significantly by race and gender group membership.

---

**F *critical*** The *F*-test statistic that corresponds to the alpha level, $df_w$, and $df_b$ (as in Appendix E).

**F *obtained*** The *F*-test statistic that is calculated.

---

## 🔲 ASSESSING THE RELATIONSHIP BETWEEN VARIABLES

Based on our five-step model of *F*, we've determined that there is a significant difference between the four demographic groups in their educational attainment. We rejected the null hypothesis and concluded that the years of education (our dependent variable) do vary by group membership (our independent variable). But can we say anything about how strong the relationship is between the variables?

The correlation ratio or eta square ($\eta^2$) allows us to make a statement about the strength of the relationship. Eta square is determined by the following:

$$\eta^2 = \frac{SSB}{SST} \tag{10.10}$$

The ratio of *SSB* to *SST* (*SSB* + *SSW*) represents the proportion of variance that is explained by the group (or independent) variable. Eta square indicates the strength of the relationship between the independent and dependent variables, ranging in value from 0 to 1.0. As eta square approaches 0, the relationship between the variables is weaker, and as eta square approaches 1, the relationship between the variables is stronger.

Based on our ANOVA example,

$$\eta^2 = \frac{10.61}{10.61 + 2.67} = \frac{10.61}{13.28} = 0.80$$

We can state that 80% of the variation in educational attainment can be attributed to demographic group membership. Or, phrased another way, 80% of the variation in the dependent variable (educational attainment) can be explained by the independent variable (group membership). So how strong is this relationship? We can base our determination of the strength on the same scale that we used to assess gamma. We can conclude that there is a very strong relationship between group membership and educational attainment.

## ▣ READING THE RESEARCH LITERATURE: STRESSES AND STRAINS AMONG GRANDMOTHER CAREGIVERS

Musil and colleagues (2009) examined the family life stresses and strains affecting grandmothers involved in caregiving to grandchildren.[3] Previous studies suggested that grandmother caregivers have more depressive symptoms than their noncaregiving peers. The sample comprises grandmothers, divided into three caregiving groups: primary, multigenerational, or noncaregiver. The groups were defined as

> Primary caregiver grandmothers had responsibility for raising their grandchildren without parents living in the home. Multigenerational grandmothers lived in a home with one or more grandchildren and the grandchild(ren)'s parent(s). Noncaregiver grandmothers did not live with or provide regular babysitting for grandchildren but lived within 1 hour or 50 miles of grandchildren and had an ongoing relationship with them. (p. 395)[4]

The researchers measured family life stresses and strains based on several existing scales.

- The level of strain and stresses (conflict, difficulty) was measured for general intrafamily strain (conflict among children, difficulty in managing children), family life stresses, financial (increasing financial debts), transitions (a member lost or quit a job, moved into a new home), family legal (incidents of physical abuse or aggression), family loss (child died), family care (child became seriously ill or injured), and pregnancy (teenager became pregnant).
- Social support was assessed based on the Duke Social Support Index, measuring both subjective and instrumental dimensions of support. Instrumental support items measured the extent to which friends and family offered assistance or help in specific situations. A higher score indicates high instrumental support. Subjective support was measured by items about feelings of support and involvement with friends and family. A higher score indicates a high level of subjective support.
- Resourcefulness was measured by the Self-Control Schedule. A higher score indicates greater resourcefulness.
- Depressive symptoms were evaluated based on a 20-item Center for Epidemiological Studies—Depression Scale. Higher scores indicate an increased clinical depression.

**Table 10.5**　Means, Standard Deviations, and ANOVA Results by Caregiver Group

| Variables | Primary (n = 183) | | Multigenerational (n = 136) | | Noncaregivers (n = 167) | | F Test |
|---|---|---|---|---|---|---|---|
| | M | SD | M | SD | M | SD | |
| Intrafamily strain | 4.4 | 2.8 | 3.9 | 2.9 | 2.7 | 2.3 | 18.4*** |
| Family life stresses (aggregate) | 5.2 | 3.2 | 5.4 | 3.1 | 4.6 | 2.9 | 4.1 |
| Financial | 1.2 | 0.8 | 1.1 | 0.8 | 0.9 | 0.8 | 5.8* |
| Transitions | 1.7 | 1.5 | 2.3 | 1.7 | 1.9 | 1.5 | 4.7** |
| Family legal | 0.7 | 0.9 | 0.4 | 0.7 | 0.4 | 0.7 | 9.9*** |
| Family loss | 0.7 | 0.7 | 0.7 | 0.7 | 0.6 | 0.7 | 0.2 |
| Family care | 0.8 | 1.1 | 0.8 | 1.0 | 0.7 | 1.0 | 0.2 |
| Pregnancy | 0.1 | 0.2 | 0.1 | 0.2 | 0.0 | 0.2 | 2.0 |
| Support—Instrumental | 7.7 | 3.4 | 9.7 | 2.2 | 8.6 | 2.8 | 19.2*** |
| Support—Subjective | 11.1 | 3.1 | 11.8 | 2.4 | 12.2 | 2.6 | 7.9*** |
| Resourcefulness | 3.2 | 0.6 | 3.2 | 0.6 | 3.3 | 0.6 | 2.2 |
| Depressive symptoms | 15.8 | 11.3 | 12.4 | 10.4 | 11.5 | 10.6 | 8.0*** |

Note: $*P <= .05, **P < .01, ***P < .001$.

A portion of Musil and her colleagues' findings are presented in Table 10.5. In the table, they report the mean score and standard deviation for each stress/strain area, social support, resourcefulness, and depressive symptoms, using analysis of variance to compare the results for each grandmother group. They highlight the significant differences in the following paragraph.

There were significant differences between groups in intrafamily strain: Primary caregivers reported the most strain (Table 1, Table 10.5 this chapter); there were no differences in the family life stresses summary score. There were significant differences between grandmother caregiver groups on specific family life stresses.... Post hoc tests showed that noncaregivers reported fewer financial strains than primary and multigenerational grandmothers, and primary caregivers reported significantly more family legal problems. Multigenerational grandmothers reported more transitions than primary caregivers. There were no significant between-group differences on family-care strains, loss, or pregnancy strain. There were significant between-group differences in support, but not resourcefulness. Noncaregivers reported more subjective support than primary caregivers. Grandmothers in multigenerational homes reported the most instrumental support and primary caregivers reported the least. Primary caregivers reported higher depressive symptoms than grandmothers in the other two groups. (p. 399)[5]

Notice how *F*-test results are not reported in their summary. However, we know from Table 10.5 which model was significant. For example, the ANOVA model for intrafamily strain produced an obtained *F* test of 18.4 (significant at the .001 level). As we review the mean scores for each group, the highest level of intrafamily strain was reported by primary caregivers (a mean score of 4.4), followed by multigenerational caregivers (3.9). Musil and colleagues conclude that these results reflect the "more complex family situations in these homes." Apart from the need to coordinate the schedules of grandchildren and the adults in the household, they identify the additional relationship strains of lack of privacy, less discretionary time, and conflict with birth parents as sources of intrafamily strain.

## MAIN POINTS

- Analysis of variance (ANOVA) procedures allow us to examine the variation in means in more than two samples. To determine whether the difference in mean scores is significant, ANOVA examines the differences between multiple samples, as well as the differences within a single sample.

- One-way ANOVA is a procedure using one dependent variable and one independent variable. The five-step hypothesis testing model is applied to one-way ANOVA.

- The test statistic for ANOVA is $F$. The $F$ statistic is the ratio of between-group variance to within-group variance.

## KEY TERMS

analysis of variance
between-group sum of squares (*SSB*)
$F$ critical

$F$ obtained
$F$ ratio or $F$ statistic
mean square between
mean square within

one-way ANOVA
total sum of squares (*SST*)
within-group sum of squares (*SSW*)

## ON YOUR OWN

Log on to the web-based student study site at**www.sagepub.com/ssdsessentials** for additional study questions, web quizzes, web resources, flashcards, codebooks and datasets, web exercises, appendices, and links to social science journal articles reflecting the statistics used in this chapter.

## CHAPTER EXERCISES

1. In this exercise, we examine the relationship between respondent's sex and respondent's age when first child was born using data from the 2006 GSS. SPSS ANOVA output is presented in Figure 10.2.

**Figure 10.2**  ANOVA Output for Sex and Age When First Child Was Born

| Descriptives | | | | | | | | |
|---|---|---|---|---|---|---|---|---|
| RS AGE WHEN 1ST CHILD BORN | | | | | | | | |
| | | | | | 95% Confidence Interval for Mean | | | |
| | N | Mean | Std. Deviation | Std. Error | Lower Bound | Upper Bound | Minimum | Maximum |
| Male | 306 | 25.59 | 5.654 | .323 | 24.95 | 26.22 | 13 | 50 |
| Female | 428 | 22.57 | 5.194 | .251 | 22.08 | 23.06 | 14 | 51 |
| Total | 734 | 23.83 | 5.589 | .206 | 23.42 | 24.23 | 13 | 51 |

**ANOVA**

RS AGE WHEN 1ST CHILD BORN

| | Sum of Squares | df | Mean Square | F | Sig. |
|---|---|---|---|---|---|
| Between Groups | 1625.356 | 1 | 1625.356 | 55.933 | .000 |
| Within Groups | 21271.015 | 732 | 29.059 | | |
| Total | 22896.371 | 733 | | | |

a. On average, are men or women older at the birth of their first child?
b. Based on the SPSS output, what can you conclude about the relationship between sex and age when first child was born? Assume $\alpha = .05$.

2. In several bivariate tables in Chapter 8, we examined the relationship between support for abortion and preferred family size. We found that there was a relationship between larger preferred family size and no support for abortion. We extend this analysis to an ANOVA model based on GSS 2008 data, using support for abortion under any circumstance ($1 = yes$, $2 = no$) as our independent grouping variable. For our dependent variable, we use CHILDS (number of actual children). SPSS output tables (both Descriptives and ANOVA) are presented in Figure 10.3.

**Figure 10.3   CHILDS Number of Children**

**Descriptives**

NUMBER OF CHILDREN

| | N | Mean | Std. Deviation | Std. Error | 95% Confidence Interval for Mean Lower Bound | 95% Confidence Interval for Mean Upper Bound | Minimum | Maximum |
|---|---|---|---|---|---|---|---|---|
| YES | 417 | 1.52 | 1.550 | .076 | 1.37 | 1.67 | 0 | 8 |
| NO | 553 | 2.17 | 1.828 | .078 | 2.02 | 2.33 | 0 | 8 |
| Total | 970 | 1.89 | 1.744 | .056 | 1.78 | 2.00 | 0 | 8 |

**ANOVA**

NUMBER OF CHILDREN

| | Sum of Squares | df | Mean Square | F | Sig. |
|---|---|---|---|---|---|
| Between Groups | 101.438 | 1 | 101.438 | 34.509 | .000 |
| Within Groups | 2845.411 | 968 | 2.939 | | |
| Total | 2946.849 | 969 | | | |

a. Which group had the highest average number of children?
b. Is there a significant difference in the number of children between the two groups? (Assume that $\alpha = .05$.) Provide evidence to support your answer.
c. Calculate $\eta^2$.

3. In Chapter 8, we analyzed the relationship between social class and health assessment. We repeat the analysis here for a random sample of 32 cases from the GSS 2006. Health is measured according to a four-point scale: 1 = *excellent*, 2 = *good*, 3 = *fair*, and 4 = *poor*. Four social classes are reported here: lower, working, middle, and upper. Complete the five-step model for these data, using α = .05.

| Lower Class | Working Class | Middle Class | Upper Class |
|:---:|:---:|:---:|:---:|
| 3 | 2 | 2 | 2 |
| 2 | 1 | 3 | 1 |
| 2 | 3 | 1 | 1 |
| 2 | 2 | 1 | 2 |
| 3 | 2 | 2 | 1 |
| 3 | 2 | 3 | 1 |
| 4 | 3 | 3 | 1 |
| 4 | 3 | 1 | 2 |

4. We extend our analysis in Exercise 2 to include CHLDIDEL as our dependent variable. The grouping variable remains the same. SPSS output tables (both Descriptives and ANOVA) are presented in Figure 10.4.
   a. Which group had the highest average of ideal number of children?
   b. Is there a significant difference in the ideal number of children between the two groups? (Assume α = .05.) Explain the reason for your answer.
   c. Would your answer change if alpha were set at .01?

**Figure 10.4** CHLDIDEL Ideal Number of Children

**Descriptives**

IDEAL NUMBER OF CHILDREN

| | N | Mean | Std. Deviation | Std. Error | 95% Confidence Interval for Mean | | Minimum | Maximum |
|---|---|---|---|---|---|---|---|---|
| | | | | | Lower Bound | Upper Bound | | |
| YES | 185 | 2.84 | 1.755 | .129 | 2.59 | 3.10 | 0 | 8 |
| NO | 273 | 3.11 | 1.866 | .113 | 2.89 | 3.34 | 0 | 8 |
| Total | 458 | 3.00 | 1.825 | .085 | 2.84 | 3.17 | 0 | 8 |

**ANOVA**

IDEAL NUMBER OF CHILDREN

| | Sum of Squares | df | Mean Square | F | Sig. |
|---|---|---|---|---|---|
| Between Groups | 8.057 | 1 | 8.057 | 2.427 | .120 |
| Within Groups | 1513.934 | 456 | 3.320 | | |
| Total | 1521.991 | 457 | | | |

5. In this exercise, let's examine the relationship between educational degree and church attendance. We selected a sample of 30 International Social Science Programme respondents, noting their educational status (no degree, secondary degree, and university degree) and their level of church attendance (0 = *never*, 1 = *infrequently*, or 2 = *two to three times per month or more*).

   Complete the five-step model for these data, using α = .01.

| No Degree | Secondary Degree | University Degree |
|:---:|:---:|:---:|
| 2 | 2 | 0 |
| 1 | 2 | 0 |
| 1 | 2 | 0 |
| 2 | 1 | 0 |
| 2 | 1 | 1 |
| 2 | 0 | 1 |
| 0 | 2 | 0 |
| 2 | 1 | 1 |
| 2 | 1 | 2 |
| 2 | 2 | 1 |

6. Based on a sample of 21 Monitoring the Future 2006 respondents, we present their racial/ethnic background and the numbers of school days missed in the past 4 weeks.
   a. Complete the five-step model for these data; set $\alpha$ at .05.
   b. If alpha were set at .01, would your decision change? Explain.

| White | Black | Hispanic |
|:---:|:---:|:---:|
| 4 | 1 | 4 |
| 5 | 2 | 3 |
| 3 | 2 | 5 |
| 4 | 1 | 1 |
| 4 | 3 | 5 |
| 4 | 4 | 2 |
| 6 | 3 | 2 |

7. We selected a sample of 14 respondents from the MTF 2006. We present their number of moving (traffic) violations in the last 12 months along with their residential area (residential area is the independent variable). Complete the five-step model for these data, using $\alpha = .05$.

| Small Town | Medium-Sized City | Large City |
|:---:|:---:|:---:|
| 0 | 2 | 3 |
| 0 | 3 | 4 |
| 1 | 1 | 4 |
| 2 | 1 | 3 |
| 1 |  | 2 |

# APPENDIX A
# THE STANDARD NORMAL TABLE

The values in Column A are Z scores. Column B lists the proportion of area between the mean and a given Z. Column C lists the proportion of area beyond a given Z. Only positive Z scores are listed. Because the normal curve is symmetrical, the areas for negative Z scores will be exactly the same as the areas for positive Z scores.

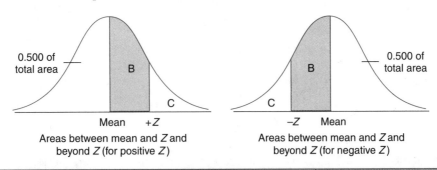

Areas between mean and Z and beyond Z (for positive Z)

Areas between mean and Z and beyond Z (for negative Z)

| A | B | C | A | B | C | A | B | C |
|---|---|---|---|---|---|---|---|---|
| Z | Area Between Mean and Z | Area Beyond Z | Z | Area Between Mean and Z | Area Beyond Z | Z | Area Between Mean and Z | Area Beyond Z |
| 0.00 | 0.0000 | 0.5000 | 0.11 | 0.0438 | 0.4562 | 0.21 | 0.0832 | 0.4168 |
| 0.01 | 0.0040 | 0.4960 | 0.12 | 0.0478 | 0.4522 | 0.22 | 0.0871 | 0.4129 |
| 0.02 | 0.0080 | 0.4920 | 0.13 | 0.0517 | 0.4483 | 0.23 | 0.0910 | 0.4090 |
| 0.03 | 0.0120 | 0.4880 | 0.14 | 0.0557 | 0.4443 | 0.24 | 0.0948 | 0.4052 |
| 0.04 | 0.0160 | 0.4840 | 0.15 | 0.0596 | 0.4404 | 0.25 | 0.0987 | 0.4013 |
| 0.05 | 0.0199 | 0.4801 | 0.16 | 0.0636 | 0.4364 | 0.26 | 0.1026 | 0.3974 |
| 0.06 | 0.0239 | 0.4761 | 0.17 | 0.0675 | 0.4325 | 0.27 | 0.1064 | 0.3936 |
| 0.07 | 0.0279 | 0.4721 | 0.18 | 0.0714 | 0.4286 | 0.28 | 0.1103 | 0.3897 |
| 0.08 | 0.0319 | 0.4681 | 0.19 | 0.0753 | 0.4247 | 0.29 | 0.1141 | 0.3859 |
| 0.09 | 0.0359 | 0.4641 | 0.20 | 0.0793 | 0.4207 | 0.30 | 0.1179 | 0.3821 |
| 0.10 | 0.0398 | 0.4602 | | | | | | |

*(Continued)*

(Continued)

| A | B | C | A | B | C | A | B | C |
|---|---|---|---|---|---|---|---|---|
| Z | Area Between Mean and Z | Area Beyond Z | Z | Area Between Mean and Z | Area Beyond Z | Z | Area Between Mean and Z | Area Beyond Z |
| 0.31 | 0.1217 | 0.3783 | 0.71 | 0.2611 | 0.2389 | 1.11 | 0.3665 | 0.1335 |
| 0.32 | 0.1255 | 0.3745 | 0.72 | 0.2642 | 0.2358 | 1.12 | 0.3686 | 0.1314 |
| 0.33 | 0.1293 | 0.3707 | 0.73 | 0.2673 | 0.2327 | 1.13 | 0.3708 | 0.1292 |
| 0.34 | 0.1331 | 0.3669 | 0.74 | 0.2703 | 0.2297 | 1.14 | 0.3729 | 0.1271 |
| 0.35 | 0.1368 | 0.3632 | 0.75 | 0.2734 | 0.2266 | 1.15 | 0.3749 | 0.1251 |
| 0.36 | 0.1406 | 0.3594 | 0.76 | 0.2764 | 0.2236 | 1.16 | 0.3770 | 0.1230 |
| 0.37 | 0.1443 | 0.3557 | 0.77 | 0.2794 | 0.2206 | 1.17 | 0.3790 | 0.1210 |
| 0.38 | 0.1480 | 0.3520 | 0.78 | 0.2823 | 0.2177 | 1.18 | 0.3810 | 0.1190 |
| 0.39 | 0.1517 | 0.3483 | 0.79 | 0.2852 | 0.2148 | 1.19 | 0.3830 | 0.1170 |
| 0.40 | 0.1554 | 0.3446 | 0.80 | 0.2881 | 0.2119 | 1.20 | 0.3849 | 0.1151 |
| 0.41 | 0.1591 | 0.3409 | 0.81 | 0.2910 | 0.2090 | 1.21 | 0.3869 | 0.1131 |
| 0.42 | 0.1628 | 0.3372 | 0.82 | 0.2939 | 0.2061 | 1.22 | 0.3888 | 0.1112 |
| 0.43 | 0.1664 | 0.3336 | 0.83 | 0.2967 | 0.2033 | 1.23 | 0.3907 | 0.1093 |
| 0.44 | 0.1700 | 0.3300 | 0.84 | 0.2995 | 0.2005 | 1.24 | 0.3925 | 0.1075 |
| 0.45 | 0.1736 | 0.3264 | 0.85 | 0.3023 | 0.1977 | 1.25 | 0.3944 | 0.1056 |
| 0.46 | 0.1772 | 0.3228 | 0.86 | 0.3051 | 0.1949 | 1.26 | 0.3962 | 0.1038 |
| 0.47 | 0.1808 | 0.3192 | 0.87 | 0.3078 | 0.1992 | 1.27 | 0.3980 | 0.1020 |
| 0.48 | 0.1844 | 0.3156 | 0.88 | 0.3106 | 0.1894 | 1.28 | 0.3997 | 0.1003 |
| 0.49 | 0.1879 | 0.3121 | 0.89 | 0.3133 | 0.1867 | 1.29 | 0.4015 | 0.0985 |
| 0.50 | 0.1915 | 0.3085 | 0.90 | 0.3159 | 0.1841 | 1.30 | 0.4032 | 0.0968 |
| 0.51 | 0.1950 | 0.3050 | 0.91 | 0.3186 | 0.1814 | 1.31 | 0.4049 | 0.0951 |
| 0.52 | 0.1985 | 0.3015 | 0.92 | 0.3212 | 0.1788 | 1.32 | 0.4066 | 0.0934 |
| 0.53 | 0.2019 | 0.2981 | 0.93 | 0.3238 | 0.1762 | 1.33 | 0.4082 | 0.0918 |
| 0.54 | 0.2054 | 0.2946 | 0.94 | 0.3264 | 0.1736 | 1.34 | 0.4099 | 0.0901 |
| 0.55 | 0.2088 | 0.2912 | 0.95 | 0.3289 | 0.1711 | 1.35 | 0.4115 | 0.0885 |
| 0.56 | 0.2123 | 0.2877 | 0.96 | 0.3315 | 0.1685 | 1.36 | 0.4131 | 0.0869 |
| 0.57 | 0.2157 | 0.2843 | 0.97 | 0.3340 | 0.1660 | 1.37 | 0.4147 | 0.0853 |
| 0.58 | 0.2190 | 0.2810 | 0.98 | 0.3365 | 0.1635 | 1.38 | 0.4612 | 0.0838 |
| 0.59 | 0.2224 | 0.2776 | 0.99 | 0.3389 | 0.1611 | 1.39 | 0.4177 | 0.0823 |
| 0.60 | 0.2257 | 0.2743 | 1.00 | 0.3413 | 0.1587 | 1.40 | 0.4192 | 0.0808 |
| 0.61 | 0.2291 | 0.2709 | 1.01 | 0.3438 | 0.1562 | 1.41 | 0.4207 | 0.0793 |
| 0.62 | 0.2324 | 0.2676 | 1.02 | 0.3461 | 0.1539 | 1.42 | 0.4222 | 0.0778 |
| 0.63 | 0.2357 | 0.2643 | 1.03 | 0.3485 | 0.1515 | 1.43 | 0.4236 | 0.0764 |
| 0.64 | 0.2389 | 0.2611 | 1.04 | 0.3508 | 0.1492 | 1.44 | 0.4251 | 0.0749 |
| 0.65 | 0.2422 | 0.2578 | 1.05 | 0.3531 | 0.1469 | 1.45 | 0.4265 | 0.0735 |
| 0.66 | 0.2454 | 0.2546 | 1.06 | 0.3554 | 0.1446 | 1.46 | 0.4279 | 0.0721 |
| 0.67 | 0.2486 | 0.2514 | 1.07 | 0.3577 | 0.1423 | 1.47 | 0.4292 | 0.0708 |
| 0.68 | 0.2517 | 0.2483 | 1.08 | 0.3599 | 0.1401 | 1.48 | 0.4306 | 0.0694 |
| 0.69 | 0.2549 | 0.2451 | 1.09 | 0.3621 | 0.1379 | 1.49 | 0.4319 | 0.0681 |
| 0.70 | 0.2580 | 0.2420 | 1.10 | 0.3643 | 0.1357 | 1.50 | 0.4332 | 0.0668 |

| A | B | C | A | B | C | A | B | C |
|---|---|---|---|---|---|---|---|---|
| Z | Area Between Mean and Z | Area Beyond Z | Z | Area Between Mean and Z | Area Beyond Z | Z | Area Between Mean and Z | Area Beyond Z |
| 1.51 | 0.4345 | 0.0655 | 1.91 | 0.4719 | 0.0281 | 2.31 | 0.4896 | 0.0104 |
| 1.52 | 0.4357 | 0.0643 | 1.92 | 0.4726 | 0.0274 | 2.32 | 0.4898 | 0.0102 |
| 1.53 | 0.4370 | 0.0630 | 1.93 | 0.4732 | 0.0268 | 2.33 | 0.4901 | 0.0099 |
| 1.54 | 0.4382 | 0.0618 | 1.94 | 0.4738 | 0.0262 | 2.34 | 0.4904 | 0.0096 |
| 1.55 | 0.4394 | 0.0606 | 1.95 | 0.4744 | 0.0256 | 2.35 | 0.4906 | 0.0094 |
| 1.56 | 0.4406 | 0.0594 | 1.96 | 0.4750 | 0.0250 | 2.36 | 0.4909 | 0.0091 |
| 1.57 | 0.4418 | 0.0582 | 1.97 | 0.4756 | 0.0244 | 2.37 | 0.4911 | 0.0089 |
| 1.58 | 0.4429 | 0.0571 | 1.98 | 0.4761 | 0.0239 | 2.38 | 0.4913 | 0.0087 |
| 1.59 | 0.4441 | 0.0559 | 1.99 | 0.4767 | 0.0233 | 2.39 | 0.4916 | 0.0084 |
| 1.60 | 0.4452 | 0.0548 | 2.00 | 0.4772 | 0.0228 | 2.40 | 0.4918 | 0.0082 |
| 1.61 | 0.4463 | 0.0537 | 2.01 | 0.4778 | 0.0222 | 2.41 | 0.4920 | 0.0080 |
| 1.62 | 0.4474 | 0.0526 | 2.02 | 0.4783 | 0.0217 | 2.42 | 0.4922 | 0.0078 |
| 1.63 | 0.4484 | 0.0516 | 2.03 | 0.4788 | 0.0212 | 2.43 | 0.4925 | 0.0075 |
| 1.64 | 0.4495 | 0.0505 | 2.04 | 0.4793 | 0.0207 | 2.44 | 0.4927 | 0.0073 |
| 1.65 | 0.4505 | 0.0495 | 2.05 | 0.4798 | 0.0202 | 2.45 | 0.4929 | 0.0071 |
| 1.66 | 0.4515 | 0.0485 | 2.06 | 0.4803 | 0.0197 | 2.46 | 0.4931 | 0.0069 |
| 1.67 | 0.4525 | 0.0475 | 2.07 | 0.4808 | 0.0192 | 2.47 | 0.4932 | 0.0068 |
| 1.68 | 0.4535 | 0.0465 | 2.08 | 0.4812 | 0.0188 | 2.48 | 0.4934 | 0.0066 |
| 1.69 | 0.4545 | 0.0455 | 2.09 | 0.4817 | 0.0183 | 2.49 | 0.4936 | 0.0064 |
| 1.70 | 0.4554 | 0.0466 | 2.10 | 0.4821 | 0.0179 | 2.50 | 0.4938 | 0.0062 |
| 1.71 | 0.4564 | 0.0436 | 2.11 | 0.4826 | 0.0174 | 2.51 | 0.4940 | 0.0060 |
| 1.72 | 0.4573 | 0.0427 | 2.12 | 0.4830 | 0.0170 | 2.52 | 0.4941 | 0.0059 |
| 1.73 | 0.4582 | 0.0418 | 2.13 | 0.4834 | 0.0166 | 2.53 | 0.4943 | 0.0057 |
| 1.74 | 0.4591 | 0.0409 | 2.14 | 0.4838 | 0.0162 | 2.54 | 0.4945 | 0.0055 |
| 1.75 | 0.4599 | 0.0401 | 2.15 | 0.4842 | 0.0158 | 2.55 | 0.4946 | 0.0054 |
| 1.76 | 0.4608 | 0.0392 | 2.16 | 0.4846 | 0.0154 | 2.56 | 0.4948 | 0.0052 |
| 1.77 | 0.4616 | 0.0384 | 2.17 | 0.4850 | 0.0150 | 2.57 | 0.4949 | 0.0051 |
| 1.78 | 0.4625 | 0.0375 | 2.18 | 0.4854 | 0.0146 | 2.58 | 0.4951 | 0.0049 |
| 1.79 | 0.4633 | 0.0367 | 2.19 | 0.4857 | 0.0143 | 2.59 | 0.4952 | 0.0048 |
| 1.80 | 0.4641 | 0.0359 | 2.20 | 0.4861 | 0.0139 | 2.60 | 0.4953 | 0.0047 |
| 1.81 | 0.4649 | 0.0351 | 2.21 | 0.4864 | 0.0136 | 2.61 | 0.4955 | 0.0045 |
| 1.82 | 0.4656 | 0.0344 | 2.22 | 0.4868 | 0.0132 | 2.62 | 0.4956 | 0.0044 |
| 1.83 | 0.4664 | 0.0336 | 2.23 | 0.4871 | 0.0129 | 2.63 | 0.4957 | 0.0043 |
| 1.84 | 0.4671 | 0.0329 | 2.24 | 0.4875 | 0.0125 | 2.64 | 0.4959 | 0.0041 |
| 1.85 | 0.4678 | 0.0322 | 2.25 | 0.4878 | 0.0122 | 2.65 | 0.4960 | 0.0040 |
| 1.86 | 0.4686 | 0.0314 | 2.26 | 0.4881 | 0.0119 | 2.66 | 0.4961 | 0.0039 |
| 1.87 | 0.4693 | 0.0307 | 2.27 | 0.4884 | 0.0116 | 2.67 | 0.4962 | 0.0038 |
| 1.88 | 0.4699 | 0.0301 | 2.28 | 0.4887 | 0.0113 | 2.68 | 0.4963 | 0.0037 |
| 1.89 | 0.4706 | 0.0294 | 2.29 | 0.4890 | 0.0110 | 2.69 | 0.4964 | 0.0036 |
| 1.90 | 0.4713 | 0.0287 | 2.30 | 0.4893 | 0.0107 | 2.70 | 0.4965 | 0.0035 |

*(Continued)*

(Continued)

| A | B | C | A | B | C | A | B | C |
|---|---|---|---|---|---|---|---|---|
| Z | Area Between Mean and Z | Area Beyond Z | Z | Area Between Mean and Z | Area Beyond Z | Z | Area Between Mean and Z | Area Beyond Z |
| 2.71 | 0.4966 | 0.0034 | 3.01 | 0.4987 | 0.0013 | 3.31 | 0.4995 | 0.0005 |
| 2.72 | 0.4967 | 0.0033 | 3.02 | 0.4987 | 0.0013 | 3.32 | 0.4995 | 0.0005 |
| 2.73 | 0.4968 | 0.0032 | 3.03 | 0.4988 | 0.0012 | 3.33 | 0.4996 | 0.0004 |
| 2.74 | 0.4969 | 0.0031 | 3.04 | 0.4988 | 0.0012 | 3.34 | 0.4996 | 0.0004 |
| 2.75 | 0.4970 | 0.0030 | 3.05 | 0.4989 | 0.0011 | 3.35 | 0.4996 | 0.0004 |
| 2.76 | 0.4971 | 0.0029 | 3.06 | 0.4989 | 0.0011 | 3.36 | 0.4996 | 0.0004 |
| 2.77 | 0.4972 | 0.0028 | 3.07 | 0.4989 | 0.0011 | 3.37 | 0.4996 | 0.0004 |
| 2.78 | 0.4973 | 0.0027 | 3.08 | 0.4990 | 0.0010 | 3.38 | 0.4996 | 0.0004 |
| 2.79 | 0.4974 | 0.0026 | 3.09 | 0.4990 | 0.0010 | 3.39 | 0.4997 | 0.0003 |
| 2.80 | 0.4974 | 0.0026 | 3.10 | 0.4990 | 0.0010 | 3.40 | 0.4997 | 0.0003 |
| 2.81 | 0.4975 | 0.0025 | 3.11 | 0.4991 | 0.0009 | 3.41 | 0.4997 | 0.0003 |
| 2.82 | 0.4976 | 0.0024 | 3.12 | 0.4991 | 0.0009 | 3.42 | 0.4997 | 0.0003 |
| 2.83 | 0.4977 | 0.0023 | 3.13 | 0.4991 | 0.0009 | 3.43 | 0.4997 | 0.0003 |
| 2.84 | 0.4977 | 0.0023 | 3.14 | 0.4992 | 0.0008 | 3.44 | 0.4997 | 0.0003 |
| 2.85 | 0.4978 | 0.0022 | 3.15 | 0.4992 | 0.0008 | 3.45 | 0.4997 | 0.0003 |
| 2.86 | 0.4979 | 0.0021 | 3.16 | 0.4992 | 0.0008 | 3.46 | 0.4997 | 0.0003 |
| 2.87 | 0.4979 | 0.0021 | 3.17 | 0.4992 | 0.0008 | 3.47 | 0.4997 | 0.0003 |
| 2.88 | 0.4980 | 0.0020 | 3.18 | 0.4993 | 0.0007 | 3.48 | 0.4997 | 0.0003 |
| 2.89 | 0.4981 | 0.0019 | 3.19 | 0.4993 | 0.0007 | 3.49 | 0.4998 | 0.0002 |
| 2.90 | 0.4981 | 0.0019 | 3.20 | 0.4993 | 0.0007 | 3.50 | 0.4998 | 0.0002 |
| 2.91 | 0.4982 | 0.0018 | 3.21 | 0.4993 | 0.0007 | 3.60 | 0.4998 | 0.0002 |
| 2.92 | 0.4982 | 0.0018 | 3.22 | 0.4994 | 0.0006 | 3.70 | 0.4999 | 0.0001 |
| 2.93 | 0.4983 | 0.0017 | 3.23 | 0.4994 | 0.0006 | 3.80 | 0.4999 | 0.0001 |
| 2.94 | 0.4984 | 0.0016 | 3.24 | 0.4994 | 0.0006 | 3.90 | 0.4999 | <0.0001 |
| 2.95 | 0.4984 | 0.0016 | 3.25 | 0.4994 | 0.0006 | 4.00 | 0.4999 | <0.0001 |
| 2.96 | 0.4985 | 0.0015 | 3.26 | 0.4994 | 0.0006 | | | |
| 2.97 | 0.4985 | 0.0015 | 3.27 | 0.4995 | 0.0005 | | | |
| 2.98 | 0.4986 | 0.0014 | 3.28 | 0.4995 | 0.0005 | | | |
| 2.99 | 0.4986 | 0.0014 | 3.29 | 0.4995 | 0.0005 | | | |
| 3.00 | 0.4986 | 0.0014 | 3.30 | 0.4995 | 0.0005 | | | |

# Appendix B
# Distribution of *t*

| | | Level of Significance for One-Tailed Test | | | | |
|---|---|---|---|---|---|---|
| | .10 | .05 | .025 | .01 | .005 | .0005 |
| | | Level of Significance for Two-Tailed Test | | | | |
| df | .20 | .10 | .05 | .02 | .01 | .001 |
| 1 | 3.078 | 6.314 | 12.706 | 31.821 | 63.657 | 636.619 |
| 2 | 1.886 | 2.920 | 4.303 | 6.965 | 9.925 | 31.598 |
| 3 | 1.638 | 2.353 | 3.182 | 4.541 | 5.841 | 12.941 |
| 4 | 1.533 | 2.132 | 2.776 | 3.747 | 4.604 | 8.610 |
| 5 | 1.476 | 2.015 | 2.571 | 3.365 | 4.032 | 6.859 |
| 6 | 1.440 | 1.943 | 2.447 | 3.143 | 3.707 | 5.959 |
| 7 | 1.415 | 1.895 | 2.365 | 2.998 | 3.499 | 5.405 |
| 8 | 1.397 | 1.860 | 2.306 | 2.896 | 3.355 | 5.041 |
| 9 | 1.383 | 1.833 | 2.262 | 2.821 | 3.250 | 4.781 |
| 10 | 1.372 | 1.812 | 2.228 | 2.764 | 3.169 | 4.587 |
| 11 | 1.363 | 1.796 | 2.201 | 2.718 | 3.106 | 4.437 |
| 12 | 1.356 | 1.782 | 2.179 | 2.681 | 3.055 | 4.318 |
| 13 | 1.350 | 1.771 | 2.160 | 2.650 | 3.012 | 4.221 |
| 14 | 1.345 | 1.761 | 2.145 | 2.624 | 2.977 | 4.140 |
| 15 | 1.341 | 1.753 | 2.131 | 2.602 | 2.947 | 4.073 |
| 16 | 1.337 | 1.746 | 2.120 | 2.583 | 2.921 | 4.015 |
| 17 | 1.333 | 1.740 | 2.110 | 2.567 | 2.898 | 3.965 |
| 18 | 1.330 | 1.734 | 2.101 | 2.552 | 2.878 | 3.922 |
| 19 | 1.328 | 1.729 | 2.093 | 2.539 | 2.861 | 3.883 |
| 20 | 1.325 | 1.725 | 2.086 | 2.528 | 2.845 | 3.850 |

*(Continued)*

(Continued)

| df | Level of Significance for One-Tailed Test | | | | | |
|---|---|---|---|---|---|---|
| | .10 | .05 | .025 | .01 | .005 | .0005 |
| | Level of Significance for Two-Tailed Test | | | | | |
| | .20 | .10 | .05 | .02 | .01 | .001 |
| 21 | 1.323 | 1.721 | 2.080 | 2.518 | 2.831 | 3.819 |
| 22 | 1.321 | 1.717 | 2.074 | 2.508 | 2.819 | 3.792 |
| 23 | 1.319 | 1.714 | 2.069 | 2.500 | 2.807 | 3.767 |
| 24 | 1.318 | 1.711 | 2.064 | 2.492 | 2.797 | 3.745 |
| 25 | 1.316 | 1.708 | 2.060 | 2.485 | 2.787 | 3.725 |
| 26 | 1.315 | 1.706 | 2.056 | 2.479 | 2.779 | 3.707 |
| 27 | 1.314 | 1.703 | 2.052 | 2.473 | 2.771 | 3.690 |
| 28 | 1.313 | 1.701 | 2.048 | 2.467 | 2.763 | 3.674 |
| 29 | 1.311 | 1.699 | 2.045 | 2.462 | 2.756 | 3.659 |
| 30 | 1.310 | 1.697 | 2.042 | 2.457 | 2.750 | 3.646 |
| 40 | 1.303 | 1.684 | 2.021 | 2.423 | 2.704 | 3.551 |
| 60 | 1.296 | 1.671 | 2.000 | 2.390 | 2.660 | 3.460 |
| 120 | 1.289 | 1.658 | 1.980 | 2.358 | 2.617 | 3.373 |
| ∞ | 1.282 | 1.645 | 1.960 | 2.326 | 2.576 | 3.291 |

*Source:* Abridged from R. A. Fisher and F. Yates, *Statistical Tables for Biological, Agricultural and Medical Research*, 6th ed. Copyright © R. A. Fisher and F. Yates, 1963. Reprinted by permission of Pearson Education Limited.

# Appendix C
# Distribution of Chi-Square

| df | .99 | .98 | .95 | .90 | .80 | .70 | .50 | .30 | .20 | .10 | .05 | .02 | .01 | .001 |
|----|-----|-----|-----|-----|-----|-----|-----|-----|-----|-----|-----|-----|-----|------|
| 1 | .03157 | .03628 | .00393 | .0158 | .0642 | .148 | .455 | 1.074 | 1.642 | 2.706 | 3.841 | 5.412 | 6.635 | 10.827 |
| 2 | .0201 | .0404 | .103 | .211 | .446 | .713 | 1.386 | 2.408 | 3.219 | 4.605 | 5.991 | 7.824 | 9.210 | 13.815 |
| 3 | .115 | .185 | .352 | .584 | 1.005 | 1.424 | 2.366 | 3.665 | 4.642 | 6.251 | 7.815 | 9.837 | 11.341 | 16.268 |
| 4 | .297 | .429 | .711 | 1.064 | 1.649 | 2.195 | 3.357 | 4.878 | 5.989 | 7.779 | 9.488 | 11.668 | 13.277 | 18.465 |
| 5 | .554 | .752 | 1.145 | 1.610 | 2.343 | 3.000 | 4.351 | 6.064 | 7.289 | 9.236 | 11.070 | 13.388 | 15.086 | 20.517 |
| 6 | .872 | 1.134 | 1.635 | 2.204 | 3.070 | 3.828 | 5.348 | 7.231 | 8.558 | 10.645 | 12.592 | 15.033 | 16.812 | 22.457 |
| 7 | 1.239 | 1.564 | 2.167 | 2.833 | 3.822 | 4.671 | 6.346 | 8.383 | 9.803 | 12.017 | 14.067 | 16.622 | 18.475 | 24.322 |
| 8 | 1.646 | 2.032 | 2.733 | 3.490 | 4.594 | 5.527 | 7.344 | 9.524 | 11.030 | 13.362 | 15.507 | 18.168 | 20.090 | 26.125 |
| 9 | 2.088 | 2.532 | 3.325 | 4.168 | 5.380 | 6.393 | 8.343 | 10.656 | 12.242 | 14.684 | 16.919 | 19.679 | 21.666 | 27.877 |
| 10 | 2.558 | 3.059 | 3.940 | 4.865 | 6.179 | 7.267 | 9.342 | 11.781 | 13.442 | 15.987 | 18.307 | 21.161 | 23.209 | 29.588 |
| 11 | 3.053 | 3.609 | 4.575 | 5.578 | 6.989 | 8.148 | 10.341 | 12.899 | 14.631 | 17.275 | 19.675 | 22.618 | 24.725 | 31.264 |
| 12 | 3.571 | 4.178 | 5.226 | 6.304 | 7.807 | 9.034 | 11.340 | 14.011 | 15.812 | 18.549 | 21.026 | 24.054 | 26.217 | 32.909 |
| 13 | 4.107 | 4.765 | 5.892 | 7.042 | 8.634 | 9.926 | 12.340 | 15.119 | 16.985 | 19.812 | 22.362 | 25.472 | 27.688 | 34.528 |
| 14 | 4.660 | 5.368 | 6.571 | 7.790 | 9.467 | 10.821 | 13.339 | 16.222 | 18.151 | 21.064 | 23.685 | 26.873 | 29.141 | 36.123 |
| 15 | 5.229 | 5.985 | 7.261 | 8.547 | 10.307 | 11.721 | 14.339 | 17.322 | 19.311 | 22.307 | 24.996 | 28.259 | 30.578 | 37.697 |
| 16 | 5.812 | 6.614 | 7.962 | 9.312 | 11.152 | 12.624 | 15.338 | 18.418 | 20.465 | 23.542 | 26.296 | 29.633 | 32.000 | 39.252 |
| 17 | 6.408 | 7.255 | 8.672 | 10.085 | 12.002 | 13.531 | 16.338 | 19.511 | 21.615 | 24.769 | 27.587 | 30.995 | 33.409 | 40.790 |
| 18 | 7.015 | 7.906 | 9.390 | 10.865 | 12.857 | 14.440 | 17.338 | 20.601 | 22.760 | 25.989 | 28.869 | 32.346 | 34.805 | 42.312 |
| 19 | 7.633 | 8.567 | 10.117 | 11.651 | 13.716 | 15.352 | 18.338 | 21.689 | 23.900 | 27.204 | 30.144 | 33.687 | 36.191 | 43.820 |
| 20 | 8.260 | 9.237 | 10.851 | 12.443 | 14.578 | 16.266 | 19.337 | 22.775 | 25.038 | 28.412 | 31.410 | 35.020 | 37.566 | 45.315 |
| 21 | 8.897 | 9.915 | 11.591 | 13.240 | 15.445 | 17.182 | 20.337 | 23.858 | 26.171 | 29.615 | 32.671 | 36.343 | 38.932 | 46.797 |
| 22 | 9.542 | 10.600 | 12.338 | 14.041 | 16.314 | 18.101 | 21.337 | 24.939 | 27.301 | 30.813 | 33.924 | 37.659 | 40.289 | 48.268 |
| 23 | 10.196 | 11.293 | 13.091 | 14.848 | 17.187 | 19.021 | 22.337 | 26.018 | 28.429 | 32.007 | 35.172 | 38.968 | 41.638 | 49.728 |
| 24 | 10.856 | 11.992 | 13.848 | 15.659 | 18.062 | 19.943 | 23.337 | 27.096 | 29.553 | 33.196 | 36.415 | 40.270 | 42.980 | 51.179 |
| 25 | 11.524 | 12.697 | 14.611 | 16.473 | 18.940 | 20.867 | 24.337 | 28.172 | 30.675 | 34.382 | 37.652 | 41.566 | 44.314 | 52.620 |
| 26 | 12.198 | 13.409 | 15.379 | 17.292 | 19.820 | 21.792 | 25.336 | 29.246 | 31.795 | 35.563 | 38.885 | 42.856 | 45.642 | 54.052 |
| 27 | 12.879 | 14.125 | 16.151 | 18.114 | 20.703 | 22.719 | 26.336 | 30.319 | 32.912 | 36.741 | 40.113 | 44.140 | 46.963 | 55.476 |
| 28 | 13.565 | 14.847 | 16.928 | 18.939 | 21.588 | 23.647 | 27.336 | 31.391 | 34.027 | 37.916 | 41.337 | 45.419 | 48.278 | 56.893 |
| 29 | 14.256 | 15.574 | 17.708 | 19.768 | 22.475 | 24.577 | 28.336 | 32.461 | 35.139 | 39.087 | 42.557 | 46.693 | 49.588 | 58.302 |
| 30 | 14.953 | 16.306 | 18.493 | 20.599 | 23.364 | 25.508 | 29.336 | 33.530 | 36.250 | 40.256 | 43.773 | 47.962 | 50.892 | 59.703 |

*Source:* R. A. Fisher & F. Yates, *Statistical Tables for Biological, Agricultural and Medical Research*, 6th ed. Copyright © R. A. Fisher and F. Yates, 1963. Reprinted by permission of Pearson Education Limited.

# APPENDIX D
# DISTRIBUTION OF F

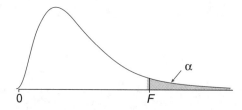

| | | | | | $\alpha = .05$ | | | | | |
|---|---|---|---|---|---|---|---|---|---|---|
| | | | | | $df_1$ | | | | | |
| $df_2$ | 1 | 2 | 3 | 4 | 5 | 6 | 8 | 12 | 24 | ∞ |
| 1 | 161.4 | 199.5 | 215.7 | 224.6 | 230.2 | 234.0 | 238.9 | 243.9 | 249.0 | 254.3 |
| 2 | 18.51 | 19.00 | 19.16 | 19.25 | 19.30 | 19.33 | 19.37 | 19.41 | 19.45 | 19.50 |
| 3 | 10.13 | 9.55 | 9.28 | 9.12 | 9.01 | 8.94 | 8.84 | 8.74 | 8.64 | 8.53 |
| 4 | 7.71 | 6.94 | 6.59 | 6.39 | 6.26 | 6.16 | 6.04 | 5.91 | 5.77 | 5.63 |
| 5 | 6.61 | 5.79 | 5.41 | 5.19 | 5.05 | 4.95 | 4.82 | 4.68 | 4.53 | 4.36 |
| 6 | 5.99 | 5.14 | 4.76 | 4.53 | 4.39 | 4.28 | 4.15 | 4.00 | 3.84 | 3.67 |
| 7 | 5.59 | 4.74 | 4.35 | 4.12 | 3.97 | 3.87 | 3.73 | 3.57 | 3.41 | 3.23 |
| 8 | 5.32 | 4.46 | 4.07 | 3.84 | 3.69 | 3.58 | 3.44 | 3.28 | 3.12 | 2.93 |
| 9 | 5.12 | 4.26 | 3.86 | 3.63 | 3.48 | 3.37 | 3.23 | 3.07 | 2.90 | 2.71 |
| 10 | 4.96 | 4.10 | 3.71 | 3.48 | 3.33 | 3.22 | 3.07 | 2.91 | 2.74 | 2.54 |
| 11 | 4.84 | 3.98 | 3.59 | 3.36 | 3.20 | 3.09 | 2.95 | 2.79 | 2.61 | 2.40 |
| 12 | 4.75 | 3.88 | 3.49 | 3.26 | 3.11 | 3.00 | 2.85 | 2.69 | 2.50 | 2.30 |
| 13 | 4.67 | 3.80 | 3.41 | 3.18 | 3.02 | 2.92 | 2.77 | 2.60 | 2.42 | 2.21 |
| 14 | 4.60 | 3.74 | 3.34 | 3.11 | 2.96 | 2.85 | 2.70 | 2.53 | 2.35 | 2.13 |
| 15 | 4.54 | 3.68 | 3.29 | 3.06 | 2.90 | 2.79 | 2.64 | 2.48 | 2.29 | 2.07 |
| 16 | 4.49 | 3.63 | 3.24 | 3.01 | 2.85 | 2.74 | 2.59 | 2.42 | 2.24 | 2.01 |
| 17 | 4.45 | 3.59 | 3.20 | 2.96 | 2.81 | 2.70 | 2.55 | 2.38 | 2.19 | 1.96 |
| 18 | 4.41 | 3.55 | 3.16 | 2.93 | 2.77 | 2.66 | 2.51 | 2.34 | 2.15 | 1.92 |
| 19 | 4.38 | 3.52 | 3.13 | 2.90 | 2.74 | 2.63 | 2.48 | 2.31 | 2.11 | 1.88 |
| 20 | 4.35 | 3.49 | 3.10 | 2.87 | 2.71 | 2.60 | 2.45 | 2.28 | 2.08 | 1.84 |
| 21 | 4.32 | 3.47 | 3.07 | 2.84 | 2.68 | 2.57 | 2.42 | 2.25 | 2.05 | 1.81 |
| 22 | 4.30 | 3.44 | 3.05 | 2.82 | 2.66 | 2.55 | 2.40 | 2.23 | 2.03 | 1.78 |
| 23 | 4.28 | 3.42 | 3.03 | 2.80 | 2.64 | 2.53 | 2.38 | 2.20 | 2.00 | 1.76 |
| 24 | 4.26 | 3.40 | 3.01 | 2.78 | 2.62 | 2.51 | 2.36 | 2.18 | 1.98 | 1.73 |
| 25 | 4.24 | 3.38 | 2.99 | 2.76 | 2.60 | 2.49 | 2.34 | 2.16 | 1.96 | 1.71 |
| 26 | 4.22 | 3.37 | 2.98 | 2.74 | 2.59 | 2.47 | 2.32 | 2.15 | 1.95 | 1.69 |
| 27 | 4.21 | 3.35 | 2.96 | 2.73 | 2.57 | 2.46 | 2.30 | 2.13 | 1.93 | 1.67 |
| 28 | 4.20 | 3.34 | 2.95 | 2.71 | 2.56 | 2.44 | 2.29 | 2.12 | 1.91 | 1.65 |
| 29 | 4.18 | 3.33 | 2.93 | 2.70 | 2.54 | 2.43 | 2.28 | 2.10 | 1.90 | 1.64 |
| 30 | 4.17 | 3.32 | 2.92 | 2.69 | 2.53 | 2.42 | 2.27 | 2.09 | 1.89 | 1.62 |
| 40 | 4.08 | 3.23 | 2.84 | 2.61 | 2.45 | 2.34 | 2.18 | 2.00 | 1.79 | 1.51 |
| 60 | 4.00 | 3.15 | 2.76 | 2.52 | 2.37 | 2.25 | 2.10 | 1.92 | 1.70 | 1.39 |
| 120 | 3.92 | 3.07 | 2.68 | 2.45 | 2.29 | 2.17 | 2.02 | 1.83 | 1.61 | 1.25 |
| ∞ | 3.84 | 2.99 | 2.60 | 2.37 | 2.21 | 2.09 | 1.94 | 1.75 | 1.52 | 1.00 |

*Source:* R. A. Fisher and F. Yates, *Statistical Tables for Biological, Agricultural and Medical Research,* 6th ed. Copyright © R. A. Fisher and F. Yates, 1963. Reprinted by permission of Pearson Education Limited.

| | | | | | α = .01 | | | | | |
|---|---|---|---|---|---|---|---|---|---|---|
| | | | | | df₁ | | | | | |
| **df₂** | *1* | *2* | *3* | *4* | *5* | *6* | *8* | *12* | *24* | *∞* |
| 1 | 4052 | 4999 | 5403 | 5625 | 5764 | 5859 | 5981 | 6106 | 6234 | 6366 |
| 2 | 98.49 | 99.01 | 99.17 | 99.25 | 99.30 | 99.33 | 99.36 | 99.42 | 99.46 | 99.50 |
| 3 | 34.12 | 30.81 | 29.46 | 28.71 | 28.24 | 27.91 | 27.49 | 27.05 | 26.60 | 26.12 |
| 4 | 21.20 | 18.00 | 16.69 | 15.98 | 15.52 | 15.21 | 14.80 | 14.37 | 13.93 | 13.46 |
| 5 | 16.26 | 13.27 | 12.06 | 11.39 | 10.97 | 10.67 | 10.27 | 9.89 | 9.47 | 9.02 |
| 6 | 13.74 | 10.92 | 9.78 | 9.15 | 8.75 | 8.47 | 8.10 | 7.72 | 7.31 | 6.88 |
| 7 | 12.25 | 9.55 | 8.45 | 7.85 | 7.46 | 7.19 | 6.84 | 6.47 | 6.07 | 5.65 |
| 8 | 11.26 | 8.65 | 7.59 | 7.01 | 6.63 | 6.37 | 6.03 | 5.67 | 5.28 | 4.86 |
| 9 | 10.56 | 8.02 | 6.99 | 6.42 | 6.06 | 5.80 | 5.47 | 5.11 | 4.73 | 4.31 |
| 10 | 10.04 | 7.56 | 6.55 | 5.99 | 5.64 | 5.39 | 5.06 | 4.71 | 4.33 | 3.91 |
| 11 | 9.65 | 7.20 | 6.22 | 5.67 | 5.32 | 5.07 | 4.74 | 4.40 | 4.02 | 3.60 |
| 12 | 9.33 | 6.93 | 5.95 | 5.41 | 5.06 | 4.82 | 4.50 | 4.16 | 3.78 | 3.36 |
| 13 | 9.07 | 6.70 | 5.74 | 5.20 | 4.86 | 4.62 | 4.30 | 3.96 | 3.59 | 3.16 |
| 14 | 8.86 | 6.51 | 5.56 | 5.03 | 4.69 | 4.46 | 4.14 | 3.80 | 3.43 | 3.00 |
| 15 | 8.68 | 6.36 | 5.42 | 4.89 | 4.56 | 4.32 | 4.00 | 3.67 | 3.29 | 2.87 |
| 16 | 8.53 | 6.23 | 5.29 | 4.77 | 4.44 | 4.20 | 3.89 | 3.55 | 3.18 | 2.75 |
| 17 | 8.40 | 6.11 | 5.18 | 4.67 | 4.34 | 4.10 | 3.79 | 3.45 | 3.08 | 2.65 |
| 18 | 8.28 | 6.01 | 5.09 | 4.58 | 4.25 | 4.01 | 3.71 | 3.37 | 3.00 | 2.57 |
| 19 | 8.18 | 5.93 | 5.01 | 4.50 | 4.17 | 3.94 | 3.63 | 3.30 | 2.92 | 2.49 |
| 20 | 8.10 | 5.85 | 4.94 | 4.43 | 4.10 | 3.87 | 3.56 | 3.23 | 2.86 | 2.42 |
| 21 | 8.02 | 5.78 | 4.87 | 4.37 | 4.04 | 3.81 | 3.51 | 3.17 | 2.80 | 2.36 |
| 22 | 7.94 | 5.72 | 4.82 | 4.31 | 3.99 | 3.76 | 3.45 | 3.12 | 2.75 | 2.31 |
| 23 | 7.88 | 5.66 | 4.76 | 4.23 | 3.94 | 3.71 | 3.41 | 3.07 | 2.70 | 2.26 |
| 24 | 7.82 | 5.61 | 4.72 | 4.22 | 3.90 | 3.67 | 3.36 | 3.03 | 2.66 | 2.21 |
| 25 | 7.77 | 5.57 | 4.68 | 4.18 | 3.86 | 3.63 | 3.32 | 2.99 | 2.62 | 2.17 |
| 26 | 7.72 | 5.53 | 4.64 | 4.14 | 3.82 | 3.59 | 3.29 | 2.96 | 2.58 | 2.13 |
| 27 | 7.68 | 5.49 | 4.60 | 4.11 | 3.78 | 3.56 | 3.26 | 2.93 | 2.55 | 2.10 |
| 28 | 7.64 | 5.45 | 4.57 | 4.07 | 3.75 | 3.53 | 3.23 | 2.90 | 2.52 | 2.06 |
| 29 | 7.60 | 5.42 | 4.54 | 4.04 | 3.73 | 3.50 | 3.20 | 2.87 | 2.49 | 2.03 |
| 30 | 7.56 | 5.39 | 4.51 | 4.02 | 3.70 | 3.47 | 3.17 | 2.84 | 2.47 | 2.01 |
| 40 | 7.31 | 5.18 | 4.31 | 3.83 | 3.51 | 3.29 | 2.99 | 2.66 | 2.29 | 1.80 |
| 60 | 7.08 | 4.98 | 4.13 | 3.65 | 3.34 | 3.12 | 2.82 | 2.50 | 2.12 | 1.60 |
| 120 | 6.85 | 4.79 | 3.95 | 3.48 | 3.17 | 2.96 | 2.66 | 2.34 | 1.95 | 1.38 |
| ∞ | 6.64 | 4.60 | 3.78 | 3.32 | 3.02 | 2.80 | 2.51 | 2.18 | 1.79 | 1.00 |

# Appendix E
# A Basic Math Review

*by James Harris*

You have probably already heard that there is a lot of math in statistics and for this reason you are somewhat anxious about taking a statistics course. Although it is true that courses in statistics can involve a great deal of mathematics, you should be relieved to hear that this course will stress interpretation rather than the ability to solve complex mathematical problems. With that said, however, you will still need to know how to perform some basic mathematical operations as well as understand the meanings of certain symbols used in statistics. Following is a review of the symbols and math you will need to know to successfully complete this course.

## ▣ SYMBOLS AND EXPRESSIONS USED IN STATISTICS

Statistics provides us with a set of tools for describing and analyzing *variables*. A variable is an attribute that can vary in some way. For example, a person's age is a variable because it can range from just born to over one hundred years old. "Race" and "gender" are also variables, though with fewer categories than the variable "age." In statistics, variables you are interested in measuring are often given a symbol. For example, if we wanted to know something about the age of students in our statistics class, we would use the symbol $Y$ to represent the variable "age." Now let's say for simplicity we asked only the students sitting in the first row their ages—19, 21, 23, and 32. These four ages would be scores of the $Y$ variable.

Another symbol that you will frequently encounter in statistics is $\Sigma$, or uppercase sigma. Sigma is a Greek letter that stands for summation in statistics. In other words, when you see the symbol $\Sigma$, it means you should sum all of the scores. An example will make this clear. Using our sample of students' ages represented by $Y$, the use of sigma as in the expression $\Sigma Y$ (read as: the sum of $Y$) tells us to sum all the scores of the variable $Y$. Using our example, we would find the sum of the set of scores from the variable "age" by adding the scores together:

$$19 + 21 + 23 + 32 = 95$$

So, for the variable "age," $\Sigma Y = 95$.

Sigma is also often used in expressions with an exponent, as in the expression $\sum Y^2$ (read as: the sum of squared scores). This means that we should first square all the scores of the $Y$ variable and then sum the squared products. So using the same set of scores, we would solve the expression by squaring each score first and then adding them together:

$$19^2 + 21^2 + 23^2 + 32^2 = 361 + 441 + 529 + 1{,}024 = 2{,}355$$

So, for the variable "age," $\sum Y^2 = 2{,}355$.

A similar, but slightly different, expression, which illustrates the function of parentheses, is $(\sum Y)^2$ (read as: the sum of scores, squared). In this expression, the parentheses tell us to first sum all the scores and then square this summed total. Parentheses are often used in expressions in statistics, and they always tell us to perform the expression within the parentheses first and then the part of the problem that is outside of the parentheses. To solve this expression, we need to sum all the scores first. However, we already found that $\sum Y = 95$, so to solve the expression $(\sum Y)^2$, we simply square this summed total,

$$95^2 = 9{,}025$$

So, for the variable "age," $(\sum Y)^2 = 9{,}025$.

You should also be familiar with the different symbols that denote multiplication and division. Most students are familiar with the times sign ($\times$); however, there are several other ways to express multiplication. For example,

$$3(4) \quad \textit{eg. } (5)6 \quad \textit{eg. } (4)(2) \quad \textit{eg. } 7 \cdot 8 \quad \textit{eg. } 9 * 6$$

all symbolize the operation of multiplication. In this text, the first three are most often used to denote multiplication. There are also several ways division can be expressed. You are probably familiar with the conventional division sign ($\div$), but division can also be expressed in these other ways:

$$4/6 \quad \frac{6}{3}$$

This text uses the latter two forms to express division.

In statistics you are likely to encounter greater than and less than signs ($>, <$), greater than or equal to and less than or equal to signs ($\geq, \leq$), and not equal to signs ($\neq$). It is important you understand what each sign means, though admittedly it is easy to confuse them. Use the following expressions for review. Notice that numerals and symbols are often used together:

$4 > 2$ means 4 is greater than 2

$H_1 > 10$ means $H_1$ is greater than 10

$7 < 9$ means 7 is less than 9

$a < b$ means $a$ is less than $b$

$Y \geq 10$ means that the value for $Y$ is a value greater than or equal to 10

$a \leq b$ means that the value for $a$ is less than or equal to the value for $b$

$8 \neq 10$ means 8 does not equal 10

$H_1 \neq H_2$ means $H_1$ does not equal $H_2$

# ▣ PROPORTIONS AND PERCENTAGES

Proportions and percentages are commonly used in statistics and provide a quick way to express information about the relative frequency of some value. You should know how to find proportions and percentages.

Proportions are identified by $P$; to find a proportion, apply this formula:

$$P = \frac{f}{N}$$

where $f$ stands for the frequency of cases in a category and $N$ the total number of cases in all categories. So, in our sample of four students, if we wanted to know the proportion of males in the front row, there would be a total of two categories, female and male. Because there are 3 females and 1 male in our sample, our $N$ is 4; and the number of cases in our category "male" is 1. To get the proportion, divide 1 by 4:

$$P = \frac{f}{N} \quad P = \frac{1}{4} = .25$$

So, the proportion of males in the front row is .25. To convert this to a percentage, simply multiply the proportion by 100 or use the formula for percentaging:

$$\% = \frac{f}{N} \times 100 \quad \% = \frac{1}{4} \times 100 = 25\%$$

# ▣ WORKING WITH NEGATIVES

Addition, subtraction, multiplication, division, and squared numbers are not difficult for most people; however, there are some important rules to know when working with negatives that you may need to review.

1.  When adding a number that is negative, it is the same as subtracting:
$$5 + (-2) = 5 - 2 = 3$$

2.  When subtracting a negative number, the sign changes:
$$8 - (-4) = 8 + 4 = 12$$

3.  When multiplying or dividing a negative number, the product or quotient is always negative:
$$6 \times -4 = -24, -10 \div 5 = -2$$

4.  When multiplying or dividing two negative numbers, the product or quotient is always positive:
$$-3 \times -7 = 21, -12 \div -4 = 3$$

5.  Squaring a number that is negative always gives a positive product because it is the same as multiplying two negative numbers:

$$-5^2 = 25 \text{ is the same as } -5 \times -5 = 25$$

## ▣ ORDER OF OPERATIONS AND COMPLEX EXPRESSIONS

In statistics, you are likely to encounter some fairly lengthy equations that require several steps to solve. To know what part of the equation to work out first, follow two basic rules. The first is called the rules of precedence. They state that you should solve all squares and square roots first, then multiplication and division, and finally, all addition and subtraction from left to right. The second rule is to solve expressions in parentheses first. If there are brackets in the equation, solve the expression within parentheses first and then the expression within the brackets. This means that parentheses and brackets can override the rules of precedence. In statistics, it is common for parentheses to control the order of calculations. These rules may seem somewhat abstract here, but a brief review of their application should make them more clear.

To solve this problem,

$$4 + 6 \cdot 8 = 4 + 48 = 52$$

do the multiplication first and then the addition. Not following the rules of precedence will lead to a substantially different answer:

$$4 + 6 \cdot 8 = 10 \cdot 8 = 80$$

which is incorrect.

To solve this problem,

$$6 - 4(6)/3^2$$

first, find the square of 3,

$$6 - 4(6)/9$$

then do the multiplication and division from left to right,

$$6 - \frac{24}{9} = 6 - 2.67$$

and finally, work out the subtraction,

$$6 - 2.67 = 3.33$$

To work out the following equation, do the expressions within parentheses first:

$$(4 + 3) - 6(2)/(3 - 1)^2$$

First, solve the addition and subtraction in the parentheses,

$$(7) - 6(2)/(2)^2$$

Now that you have solved the expressions within parentheses, work out the rest of the equation based on the rules of precedence, first squaring the 2,

$$(7) - 6(2)/4$$

Then do the multiplication and division next:

$$(7) - \frac{12}{4} = (7) - 3$$

Finally, work out the subtraction to solve the equation:

$$7 - 3 = 4$$

The following equation may seem intimidating at first, but by solving it in steps and following the rules, even these complex equations should become manageable:

$$\sqrt{(8(4 - 2)^2)/(12/4)^2}$$

For this equation, work out the expressions within parentheses first; note that there are parentheses within parentheses. In this case, work out the inner parentheses first,

$$\sqrt{(8(2)^2)/3^2}$$

Now do the outer parentheses, making sure to follow the rules of precedence within the parentheses—square first and then multiply:

$$\sqrt{\frac{32}{3^2}}$$

Now, work out the square of 3 first and then divide:

$$\sqrt{\frac{32}{9}} = \sqrt{3.55}$$

Last, take the square root:

$$1.88$$

# LEARNING CHECK SOLUTIONS

## Chapter 3

(p. 65)

*Learning Check.* Listed below are the political party affiliations of 15 individuals. Find the mode.

| | | | | |
|---|---|---|---|---|
| Democrat | Republican | Democrat | Republican | Republican |
| Independent | Democrat | Democrat | Democrat | Republican |
| Independent | Democrat | Independent | Republican | Democrat |

*Answer:*

The mode is "Democrat," because this category has the highest frequency, which is 7.

(p. 66)

*Learning Check.* Find the median of the following distribution of an interval-ratio variable: 22, 15, 18, 33, 17, 5, 11, 28, 40, 19, 8, 20.

*Answer:*

First, we need to arrange the numbers: 5, 8, 11, 15, 17, 18, 19, 20, 22, 28, 33, 40.

$(N + 1)/2 = (12 + 1)/2 = 6.5$. So the median is the average of the sixth and the seventh numbers, which are 18 and 19.

$$\text{Median} = \frac{18 + 19}{2} = 18.5.$$

So the median is 18.5.

(p. 72)

*Learning Check.* The following distribution is the same as the one you used to calculate the median in an earlier Learning Check: 22, 15, 18, 33, 17, 5, 11, 28, 40, 19, 8, 20. Can you calculate the mean? Is it the same as the median, or is it different?

*Answer:*

$$\text{Mean} = \frac{22 + 15 + 18 + 33 + 17 + 5 + 11 + 28 + 40 + 19 + 8 + 20}{12} = 19.67$$

So the mean, 19.67, is larger than the median, 18.5.

# Chapter 4

(p. 92)

*Learning Check . Why can't we use the range to describe diversity in nominal variables? The range can be used to describe diversity in ordinal variables (e.g., we can say that responses to a question ranged from "somewhat satisfied" to "very dissatisfied"), but it has no quantitative meaning. Why not?*

*Answer:*

*In nominal variables, the numbers are used only to represent the different categories of a variable without implying anything about the magnitude or quantitative difference between these categories. Therefore, the range, being a measure of variability that gives the quantitative difference between two values that a variable takes, is not an appropriate measure for nominal variables. Similarly, in ordinal variables, numbers corresponding with the categories of a variable are used only to rank order these categories without having any meaning in terms of the quantitative difference between these categories. Therefore, the range does not convey any quantitative meaning when used to describe the diversity in ordinal variables.*

(p. 94)

*Learning Check. Why is the IQR better than the range as a measure of variability, especially when there are extreme scores in the distribution? To answer this question, you may want to examine Figure 4.2.*

*Answer:*

*Extreme scores directly impact the range, which is by definition the difference between the highest and the lowest scores. Therefore, if a distribution has extreme (very high and/or very low) scores, the range does not provide an accurate description of the distribution. IQR, on the other hand, is not affected by extreme scores. Thus, it is a better measure of variability than the range when there are extreme scores in the distribution.*

(p. 98)

*Learning Check. Examine Table 4.4 again and note the disproportionate contribution of the western region to the sum of the squared deviations from the mean (it actually accounts for about 45% of the sum of squares). Can you explain why? (Hint: It has something to do with the sensitivity of the mean to extreme values.)*

*Answer:*

*The western region has the highest projected percentage change in the elderly population between 2008 and 2015, which is 27%. Therefore, it deviates more from the mean than the other regions. The more a category of a variable deviates from the mean, the larger the square of the deviation gets, and hence the more this category contributes to the sum of the squared deviations from the mean.*

# Chapter 5

(p. 115)

*Learning Check. Transform the Z scores in Table 5.2 back into raw scores. Your answers should agree with the raw scores listed in the table.*

*Answer:*

| Z Score | Raw Score |
|---------|-----------|
| −2.93 | $Y = 70.07 - 2.93(10.27) = 40$ |
| −0.98 | $Y = 70.07 - 0.98(10.27) = 60$ |
| 0.97 | $Y = 70.07 + 0.97(10.27) = 80$ |
| 2.91 | $Y = 70.07 + 2.91(10.27) = 100$ |

(p. 122)

*Learning Check: Can you find the number of students who got a score of at least 90 in the statistics course? How many students got a score below 60?*

Answer:

$Z = (90 - 70.07)/10.27 = 1.94$

$C = 0.0262$

$(0.0262)1200 =$ about 34 students

$Z = (60 - 70.07)/10.27 = -0.98$

$C = 0.1635$

$(0.1635)1200 =$ about 196 students

# Chapter 6

(p. 139)

*Learning Check. Suppose a population distribution has a mean $\mu_Y = 150$ and a standard deviation $\sigma_Y = 30$ and you draw a simple random sample of N = 100 cases. What is the probability that the mean is between 147 and 153? What is the probability that the sample mean exceeds 153? Would you be surprised to find a mean score of 159? Why? (Hint: To answer these questions, you need to apply what you learned in Chapter 5 about Z scores and areas under the normal curve [Appendix A].) Remember, to translate a raw score into a Z score we used this formula:*

$$Z = \frac{Y - \bar{Y}}{S_Y}$$

*However, because here we are dealing with a sampling distribution, replace Y with the sample mean $\bar{Y}$, $\bar{Y}$ with the sampling distribution's mean $\mu_{\bar{Y}}$, and $S_Y$ with the standard error of the mean $\sigma_Y/\sqrt{N}$.*

$$Z = \frac{\bar{Y} - \mu_{\bar{Y}}}{\sigma_Y/\sqrt{N}}$$

Answer:

*Z score equivalent of 147 is*

$$Z = \frac{\bar{Y} - \mu_{\bar{Y}}}{\sigma_Y/\sqrt{N}} = \frac{147 - 150}{30/\sqrt{100}} = \frac{-3}{3} = -1$$

*Z score equivalent of 153 is*

$$Z = \frac{\bar{Y} - \mu_{\bar{Y}}}{\sigma_Y/\sqrt{N}} = \frac{153 - 150}{30/\sqrt{100}} = \frac{3}{3} = 1$$

*Using the standard normal table (Appendix A), we can see that the probability of the area between the mean and a score 1 standard deviation above or below the mean is 0.3413. So the probability that the mean is between 147 and 153, both of which deviate from the mean by 1 standard deviation, is 0.6826 (0.3413 + 0.3413), or 68.26%.*

*The probability of the area beyond 1 standard deviation from the mean is 0.1587. So the probability that the mean exceeds 153 is 0.1587, or 15.87%.*

*Z score equivalent of 159 is*

$$Z = \frac{\bar{Y} - \mu_{\bar{Y}}}{\sigma_Y/\sqrt{N}} = \frac{159 - 150}{30/\sqrt{100}} = \frac{9}{3} = 3$$

*The probability of the area beyond 3 standard deviations from the mean, according to the standard normal table, is 0.0014. Therefore, it would be surprising to find a mean score of 159, as the probability is very low (0.14%).*

(p. 141)

*Learning Check. What is the difference between a point estimate and a confidence interval?*

*Answer:*

*When the estimate of a population parameter is a single number, it is called a point estimate. When the estimate is a range of scores, it is called an interval estimate. Confidence intervals are used for interval estimates.*

(p. 143)

*Learning Check. To understand the relationship between the confidence level and Z, review the material in Chapter 5. What would be the appropriate Z value for a 98% confidence interval?*

*Answer:*

*The appropriate Z value for a 98% confidence interval is 2.33.*

(p. 144)

*Learning Check. What is the 90% confidence interval for the mean commuting time? (Hint: First, find the Z value associated with a 90% confidence level.)*

*Answer:*

$$90\% \; CI = 7.5 \pm 1.65(0.07)$$
$$= 7.5 \pm 0.12$$
$$= 7.38 \text{ to } 7.62$$

# Chapter 8

(p. 190)

*Learning Check. Examine Table 8.2. Make sure that you can identify all the parts just described and that you understand how the numbers were obtained. Can you identify the independent and dependent variables in the table? You will need to know this to convert the frequencies to percentages.*

*Answer:*

*The independent variable is race, and home ownership is the dependent variable.*

(p. 203)

*Learning Check. The data we will use to practice calculating chi-square are also from Inman and Mayes's research. We will examine the relationship between age (independent variable) and first-generation college status (the dependent variable), as shown in the following bivariate table:*

| First-Generation Status | Years of Age | | Total |
| --- | --- | --- | --- |
| | 19 Years or Younger | 20 Years or Older | |
| Firsts | 916 (33.7%) | 1,018 (53.6%) | 1,934 (41.9%) |
| Nonfirsts | 1,802 (66.3%) | 881 (46.4%) | 2,683 (58.1%) |
| Total (N) | 2,718 (100.0%) | 1,899 (100.0%) | 4,617 (100.0%) |

*Source:* Adapted from W. Elliot Inman and Larry Mayes, "The Importance of Being First: Unique Characteristics of First Generation Community College Students," *Community College Review* 26, no. 3 (1999): 8.

*Construct a bivariate table (in percentages) showing no association between age and first-generation college status.*

*Answer:*

*Age and First-Generation College Status*

|  | 19 Years or Younger | 20 Years or Older |  |
|---|---|---|---|
| Firsts | 41.9% | 41.9% | 41.9% (1,934) |
| Nonfirsts | 58.1% | 58.1% | 58.1% (2,683) |
|  | 100.0% | 100.0% | 4617 |

(p. 204)

*Learning Check. Write out the research and the null hypotheses for your practice data.*

*Answer:*

*Null hypothesis: There is no association between age and first-generation college status.*

*Research hypothesis: Age and first-generation college status are statistically dependent.*

(p. 206)

*Learning Check. Calculate the expected frequencies for age and first-generation college status and construct a bivariate table. Are your column and row marginals the same as in the original table?*

*Answer:*

|  | $f_o$ | $f_e$ | $f_o - f_e$ | $(f_o - f_e)^2$ | $(f_o - f_e)^2 / fe$ |
|---|---|---|---|---|---|
| 19/Firsts | 916 | 1138.53 | −222.53 | 49519.60 | 43.49 |
| 19/Nonfirsts | 1802 | 1579.47 | 222.53 | 49519.60 | 31.35 |
| 20/Firsts | 1018 | 795.47 | 222.53 | 49519.60 | 62.25 |
| 20/Nonfirsts | −881 | 1103.53 | −222.53 | 49519.60 | 44.87 |

*Chi-square = 181.96, with Yates correction = 181.15.*

(p. 209)

*Learning Check. What decision can you make about the association between age and first-generation college status? Should you reject the null hypothesis at the .05 alpha level or at the .01 level?*

*Answer:*

*We would reject the null hypothesis of no difference. Our calculated chi-square is significant at the .05 and the .01 level. We have evidence that age is related to first-generation college status—53.6% of students 20 years or older are first-generation students versus 33.7% of students 19 years or younger.*

(p. 212)

*Learning Check.* *For the bivariate table with age and first-generation college status, the value of the obtained chi-square is 181.15 with 1 degree of freedom. Based on Appendix C, we determine that its probability is less than .001. This probability is less than our alpha level of .05. We reject the null hypothesis of no relationship between age and first-generation college status. If we reduce our sample size by half, the obtained chi-square is 90.58. Determine the P value for 90.58. What decision can you make about the null hypothesis?*

*Answer:*

*Even if we reduce the chi-square by half, we would still reject the null hypothesis.*

# Chapter 9

(p. 236)

*Learning Check.* *Use Figure 9.3 to predict the percentage of residents with a bachelor's degree in a state with a median household income of $47,500 and one with a median household income of $50,000.*

*Answer:*

*The percentage of residents with a bachelor's degree in a state with a median household income of $47,500 is about 28%. The comparable percentage in a state with a median household income of $50,000 is about 29.5%.*

(p. 239)

*Learning Check.* *For each of these four lines, as X goes up by 1 unit, what does Y do? Be sure you can answer this question using both the equation and the line.*

*Answer:*

*For the line Y = 1X, as X goes up by 1 unit, Y also goes up by 1 unit. In the second line, Y = 2 + 0.5X, Y increases by 0.5 units as a result of 1-unit increase in X. The line Y = 6 − 2X tells that every 1-unit increase in X results in 2-unit decrease in Y. Finally, in the fourth line, Y decreases by 0.33 units as a result of 1-unit increase in X.*

(p. 240)

*Learning Check.* *Use the linear equation describing the relationship between seniority and salary of teachers to obtain the predicted salary of a teacher with 12 years of seniority.*

*Answer:*

*The predicted salary of a teacher with 12 years of seniority is $36,000 (Y = 12,000 + 2,000(12)).*

(p. 244)

*Learning Check.* *Use the prediction equation to calculate the predicted values of Y for New York, Georgia, and Ohio. Verify that the regression line in Figure 9.6 passes through these points.*

*Answer:*

$$\text{New York: } \hat{Y} = 3.57 + 0.0005(51,384) = 29.3\%$$

$$\text{Georgia: } \hat{Y} = 3.57 + 0.0005(46,832) = 27.0\%$$

$$\text{Ohio: } \hat{Y} = 3.57 + 0.0005(44,532) = 25.8\%$$

# ANSWERS TO ODD-NUMBERED EXERCISES

## Chapter 1. The What and the Why of Statistics

1. Once our research question, the hypothesis, and the study variables have been selected, we move on to the next stage in the research process—measuring and collecting the data. The choice of a particular data collection method or instrument depends on our study objective. After our data have been collected, we have to find a systematic way to organize and analyze our data and set up some set of procedures to decide what they mean.

3. a. Interval ratio
   b. Nominal
   c. Interval ratio
   d. Ordinal
   e. Nominal
   f. Interval ratio
   g. Interval ratio
   h. Nominal

5. There are many possible variables from which to choose. Some of the most common selections by students will probably be type of occupation or industry, work experience, and educational training or expertise. Students should first address the relationship between these variables and gender: "Men have more years of work experience than women in the same occupation." Students may also consider measuring structural bias or discrimination.

7. In general, the difficulty with studying criminal acts (including hate crimes) is that the criminal act needs to be reported first. It is estimated that the majority of crimes are not reported to authorities. Data on reported crimes are routinely collected by the Federal Bureau of Investigation and the Bureau of Justice.

9. *Individual age:* This variable could be measured as an interval-ratio variable, with actual age in years reported. As discussed in the chapter, interval-ratio variables are the highest level of measurement and can also be measured at ordinal and nominal levels.

   *Annual income:* This variable could be measured as an interval-ratio variable, with actual dollar earnings reported.

   *Religiosity:* This variable could be measured in several ways. For example, as church attendance, the variable could be ordinal (number of times attended church in a month: every week, at least twice a month, less than two times a month, none at all).

   *Student performance:* This could be measured as an interval-ratio variable as GPA or test score.

*Social class:* This variable is an ordinal variable, with categories low, working, middle, and high.

*Attitude toward affirmative action:* This variable is an ordinal variable, with categories strongly disagree, disagree, neutral, agree, and strongly agree.

# Chapter 2. The Organization and Graphic Presentation of Data

1. a. Race is a nominal variable. Class is an ordinal variable, since the categories can be ordered from lower to higher status.
   b.

**Frequency Table for Race**

| Race | Frequency |
|------|-----------|
| White | 17 |
| Nonwhite | 13 |

**Frequency Table for Class**

| Class | Frequency |
|-------|-----------|
| Lower | 3 |
| Working | 15 |
| Middle | 11 |
| Upper | 1 |

   c. Proportion of nonwhite is 13/30 = 0.43. The percentage of white is (17/30) × 100 = 56.7%.
   d. The proportion of middle class is 11/30 = 0.37.

3. a.

**Frequency Table for Traumas Experienced**

| Number of Traumas | Frequency (f) |
|-------------------|---------------|
| 0 | 15 |
| 1 | 11 |
| 2 | 4 |
| Total ($N$) = 30 | 30 |

   Trauma is an interval-ratio variable, since it has a real zero point and a meaningful numeric scale.
   b. People in this survey are more likely to have experienced no traumas last year (50% of the group).
   c. The proportion who experienced one or more traumas is calculated by first adding 36.7% and 13.3% = 50%. Then divide that number by 100 to obtain 0.50, or half the group.

5. In this case, the independent variable is racial/ethnic background and the dependent variable is views on illegal immigration policies. For each racial/ethnic group, the majority of respondents (more than 50%) disagreed with current government efforts to deal with illegal immigration. There is no relationship between racial/ethnic background and views on illegal immigration policies because the percentages disagreeing from each group are nearly identical.

7. a.

| | f | % | Cf | C% |
|---|---|---|---|---|
| **Male** | | | | |
| Some high school | 100 | 15 | 100 | 15 |
| High school | 327 | 48 | 427 | 63 |
| Some college | 68 | 10 | 495 | 73 |
| College graduate | 190 | 28 | 685 | 101 |
| Total | 685 | 101 | | |
| **Female** | | | | |
| Some high school | 116 | 14 | 116 | 14 |
| High school | 407 | 50 | 523 | 64 |
| Some college | 76 | 9 | 599 | 73 |
| College graduate | 216 | 27 | 815 | 100 |
| Total | 815 | 100 | | |
| **White** | | | | |
| Some high school | 131 | 11 | 131 | 11 |
| High school | 567 | 49 | 698 | 60 |
| Some college | 113 | 10 | 811 | 70 |
| College graduate | 335 | 29 | 1,146 | 100 |
| Total | 1,146 | 100 | | |
| **Black** | | | | |
| Some high school | 50 | 24 | 50 | 24 |
| High school | 109 | 53 | 159 | 77 |
| Some college | 19 | 9 | 178 | 86 |
| College graduate | 28 | 14 | 206 | 100 |
| Total | 206 | 100 | | |
| **Hispanic** | | | | |
| Some high school | 21 | 38 | 21 | 38 |
| High school | 25 | 45 | 46 | 83 |
| Some college | 5 | 9 | 51 | 92 |
| College graduate | 5 | 9 | 56 | 100 |
| Total | 56 | 101 | | |

b. 10 + 28 = 38%; of males: 9 + 27 = 36% of females
c. 60% of whites; 77% of blacks; 83% of Hispanics
d. Cumulative percentages are more similar for men and women (almost 85% in group had at least a high school degree—85% for men and 86% for women) than the comparison of whites and blacks (about 87% of whites had at least a high school degree, while 76% of blacks reported the same, as did 62% of Hispanics). It appears that there is more educational inequality based on race than gender.

9. a. Interval ratio
b. Males outnumber females in these categories, more than double—4.2% of males; 1.6% of females.
c.

**Number of Moving Violation Tickets in the Last 12 Months × Sex Cross-Tabulation**

| | Sex | | |
|---|---|---|---|
| Count | 1 Male | 2 Female | Total |
| None | 445 | 562 | 1,007 |
| 1 | 107 | 91 | 198 |
| 2 | 46 | 35 | 81 |
| 3 | 13 | 5 | 18 |
| 4 or more | 13 | 6 | 19 |
| Total | 624 | 699 | 1,323 |

    d. Females and males are likely to report not having a ticket—though the percentage of females is higher (80%) than males (71%) in this category. In the 1, 2, 3, and 4 or more categories, there is a higher percentage of males receiving tickets than females.

11. Overall, a higher percentage of minority respondents indicated that the government should play a "major role" in improving the economic and social position of minorities. Sixty nine percent of blacks and 67% of Hispanics indicated this in their responses. On the other hand, 32% of whites agreed with this position (this is about 35–37% less than what was indicated by Hispanics and blacks). More whites, 51%, indicated that the government should play a "minor role."

13. Percentage reporting "Yes," homosexuality is an acceptable lifestyle.—*Sex:* More women than men; *Age:* younger respondents were more likely to respond yes (highest percentage is among 18- to 34-year-olds, 75%); *Political affiliation:* More Democrats (72%) and Independents (60%) responded yes than Republicans; and *Religious service attendance:* Yes responses are highest among those who do not attend church weekly (33% vs. 57% attend nearly weekly/monthly, and 74% attend less often or never).

# Chapter 3. Measures of Central Tendency

1. a. The mode can be found by two ways, by looking either for the highest frequency (749) or the highest percentage (51.5%). The mode is the category that corresponds to these values, "Strongly disapprove."
   b. The median can be found by two ways, by using either the frequencies column or the cumulative percentages.

| Using Frequencies | Using Cumulative Percentages |
|---|---|
| $$\frac{N+1}{2} = \frac{1,455+1}{2} = 728\text{th case}$$ | Notice that 48.5% of the observations fall in the "Disapprove" cumulative percentage category; 100% fall in the "Strongly disapprove" cumulative percentage category. |
| Starting with the frequency in the first category (288), add up the frequencies until you find where the 728th case falls. The 728th case corresponds to the category "Strongly disapprove," which is the median. | The 50% mark, or the median, is located somewhere within the "Strongly disapprove" category. So the median is "Strongly disapprove." |

    c. The mode is simply the category with the highest frequency (or percentage) in the distribution. The median divides the distribution into two equal parts so that half the cases are below it and half above it.
    d. Because this variable is an ordinal-level variable.

3. a. Interval ratio. The mode can be found by two ways, by looking either for the highest frequency (27) or the highest percentage (41.5%). The mode is the category that corresponds to the value, "52 weeks worked last year." The median can be found by two ways, by using either the frequencies column or the cumulative percentages.

| Using Frequencies | Using Cumulative Percentages |
|---|---|
| $$\frac{N+1}{2} = \frac{65+1}{2} = 33\text{th case}$$ | Note that 47.4% of the observations fall in the "45 weeks worked last year" category; 50.5% fall in the "47 weeks worked last year" category. |
| Starting with the frequency in the first category (10), add up the frequencies until you find where the 33rd case falls. The 33rd case corresponds to the category "47 weeks worked last year," which is the median. | The 50% mark, or the median, is located somewhere within the "47 weeks worked last year" category. So the median is "47 weeks worked last year." |

b. Since the median is merely a synonym for the 50th percentile, we already know that its value is 47 weeks worked last year.

25th percentile = $(65 \times 0.25) = 16.25$th case = 25.5 weeks worked last year

75th percentile = $(65 \times 0.75) = 48.75$th case = 52 weeks worked last year

5. a. The mode can be found by looking for the highest frequency in each column; the mode for each group is listed below:

18–29: Good
30–39: Good
40–49: Good
50–59: Good

The median can be found by two ways, by using either the frequencies column or the cumulative percentages. However, since the problem gives only the frequencies, we'll use those to solve the median.

| Age Group | | | |
|---|---|---|---|
| *18–29* | *30–39* | *40–49* | *50–59* |
| $\frac{N+1}{2} = \frac{188+1}{2} = 94.5$th case | $\frac{N+1}{2} = \frac{176+1}{2} = 88.5$th case | $\frac{N+1}{2} = \frac{185+1}{2} = 93$rd case | $\frac{N+1}{2} = \frac{182+1}{2} = 91.5$th case |
| Starting with the frequency in the first category (64), add up the frequencies until you find where the 94th and 95th cases fall. Both of these cases correspond to the category "Good," which is the median. | Starting with the frequency in the first category (57), add up the frequencies until you find where the 88th and 89th cases fall. Both of these cases correspond to the category "Good," which is the median. | Starting with the frequency in the first category (46), add up the frequencies until you find where the 93rd case falls. The 93rd case corresponds to the category "Good," which is the median. | Starting with the frequency in the first category (37), add up the frequencies until you find where the 91st and 92nd cases fall. Both of these cases correspond to the category "Good," which is the median. |

b. Since the mode and median for all four age groups is "Good," it appears that health status does not vary by age. However, we ought to question why this is the case. Perhaps it has to do with how respondents interpreted the question. For instance, it is possible that one's health status was assessed relative to his or her age. Neither the median nor the mode provides a better description of the data since they provide the same information.

7. We begin by multiplying each household size by its frequency.

| Household Size | Frequency | fY |
|---|---|---|
| 1 | 384 | 384 |
| 2 | 541 | 1,082 |
| 3 | 228 | 684 |
| 4 | 195 | 780 |
| 5 | 98 | 490 |
| 6 | 36 | 216 |
| 7 | 12 | 84 |
| 8 | 4 | 32 |
| 9 | 1 | 9 |
| 11 | 1 | 11 |
| Total | 1,500 | 3,772 |

$$\bar{Y} = \frac{\Sigma fY}{N} = \frac{3,772}{1,500} = 2.51$$

So the mean number of people per U.S. household is 2.51.

9. a. There appear to be a few outliers (i.e., extremely high values); this leads us to believe that the distribution is skewed in the positive direction.

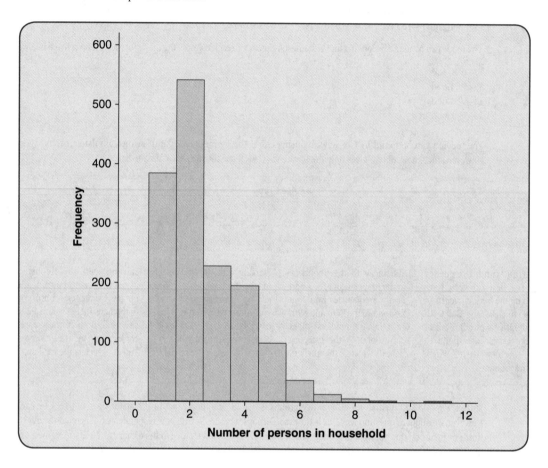

b. The median can be found by two ways, by using either the frequencies column or the cumulative percentages. The data are in frequencies, we'll use those to solve the median. Since the median (2) is less than the mean (2.51), we can conclude that the distribution is skewed in a positive direction. Our answer to Question 9a is further supported.

*Using Frequencies*

$$\frac{N + 1}{2} = \frac{1,500 + 1}{2} = 750.5\text{th case}$$

Starting with the frequency in the first category (384), add up the frequencies until you find where the 750th and 751st cases fall. Both of these cases correspond to the category "2," which is the median.

11. Yes, both of these politicians can be correct, at least in a technical sense. One politician can be referring to the mean; the other could be using the median. It would be unusual if these two statistics were exactly equal. The average or mean income of Americans can be greater than the median if the distribution of income is positively skewed, which is certainly true.

13. The mode can be found by looking for the highest frequency in each column; the mode for each group is listed below:

Males: "Don't Disapprove"
Females: "Don't Disapprove"

The median can be found two ways, either by using the frequencies column or by using the cumulative percentages. However, since the problem gives only the frequencies, we'll use those to solve for the median.

| Males | Females |
|---|---|
| $\dfrac{N+1}{2} = \dfrac{670+1}{2} = 335.5$th case | $\dfrac{N+1}{2} = \dfrac{719+1}{2} = 360$th case |
| Starting with the frequency in the first category (325), add up the frequencies until you find where the 335.5th case falls. The 335.5th case corresponds to the category "Disapprove," which is the median. | Starting with the frequency in the first category (302), add up the frequencies until you find where the 360th case falls. The 360th case corresponds to the category "Disapprove," which is the median. |

When using both the mode and the median to judge attitudes about marijuana among males and females, it appears that there are not substantial differences between boys and girls. This might be due to a lack of stigma associated with marijuana use that exists for both boys and girls. Perhaps the lack of stigma has more to do with age than sex/gender.

## Chapter 4. Measures of Variability

1. a. The range of convictions in 1990 is (583 − 79) = 504. The range of convictions in 2007 is (405 − 85) = 320. The range of convictions is larger in 1990 than in 2007.
   b. The mean number of convictions is 295.67 in 1990 and 255 in 2007.
   c. The standard deviation is 259.32 in 1990 and 160.93 in 2007.

**1990**

| Government Level | # of Convictions | $Y - \bar{Y}$ | $(Y - \bar{Y})^2$ |
|---|---|---|---|
| Federal | 583 | 287.33 | 82,558.53 |
| State | 79 | −216.67 | 46,945.89 |
| Local | 225 | −70.67 | 4,994.25 |
| Total | 887 | 0.01 | 134,498.67 |

$$\bar{Y} = 295.67$$

$$S_Y = \sqrt{S_Y^2} = \sqrt{\frac{\Sigma(Y - \bar{Y})^2}{N - 1}} = \sqrt{\frac{134,498.67}{2}} = 259.32$$

**2007**

| Government Level | # of Convictions | $Y - \bar{Y}$ | $(Y - \bar{Y})^2$ |
|---|---|---|---|
| Federal | 405 | 150 | 22,500 |
| State | 85 | −170 | 28,900 |
| Local | 275 | 20 | 400 |
| Total | 765 | 0.00 | 51,800 |

$$\bar{Y} = 255$$

$$S_Y = \sqrt{S_Y^2} = \sqrt{\frac{\Sigma(Y - \bar{Y})^2}{N - 1}} = \sqrt{\frac{51,800}{2}} = 160.93$$

d. The standard deviation is larger in 1990 than in 2007, thus indicating more variability in number of convictions in 1990 than in 2007. This supports our results from 1a.

3. a. The range is 3.6 (6.5 − 2.9). The 25th percentile, 3.05, means that 25% of cases fall below 3.05 divorce rate per 1,000 population. Likewise, the 75th percentile means that 75% of all cases fall below 4.6 divorce rate per 1,000 population.

| 25th percentile | 10(0.25) = 2.5th case | So (3.0 + 3.1)/2 = 3.05 |
|---|---|---|
| 75th percentile | 10(0.75) = 7.5th case | So (4.5 + 4.7)/2 = 4.6 |

b.

| State | Divorce Rate per 1,000 Population | $Y - \bar{Y}$ | $(Y - \bar{Y})^2$ |
|---|---|---|---|
| Alaska | 4.3 | 0.2 | 0.04 |
| Florida | 4.7 | 0.6 | 0.36 |
| Idaho | 4.9 | 0.8 | 0.64 |
| Maine | 4.5 | 0.4 | 0.16 |
| Maryland | 3.1 | −1 | 1 |
| Nevada | 6.5 | 2.4 | 5.76 |
| New Jersey | 3.0 | −1.1 | 1.21 |
| Texas | 3.3 | −0.8 | 0.64 |
| Vermont | 3.8 | −0.3 | 0.09 |
| Wisconsin | 2.9 | −1.2 | 1.44 |
| Total | 41 | 0.00 | 11.34 |

$$\bar{Y} = \frac{\Sigma Y}{N} = \frac{41}{10} = 4.1$$

$$S_Y = \sqrt{S_Y^2} = \sqrt{\frac{\Sigma(Y - \bar{Y})^2}{N - 1}} = \sqrt{\frac{11.34}{9}} = 1.12$$

The mean divorce rate is 4.1. On average, states deviated from the mean by 1.12 per 1,000 population.

c. Divorce rates may vary by state due to factors such as variation in religiosity, state policy (i.e., no-fault divorce laws), or employment opportunities.

5. a. The mean number of crimes is 3,066.8, and the mean amount of dollars (in millions) spent on police protection is $1,530.05.

| Number of Crimes | Police Protection Expenditures |
|---|---|
| $\bar{Y} = \dfrac{\Sigma Y}{N} = \dfrac{64,403}{21} = 3,066.8$ | $\bar{Y} = \dfrac{\Sigma Y}{N} = \dfrac{32,131}{21} = 1,530.05$ |

b. The standard deviation for the number of crimes is 725.58 and the standard deviation for the amount of dollars spent on police protection (in millions) is $1,754.80.

| Number of Crimes | $Y - \bar{Y}$ | $(Y - \bar{Y})^2$ | Police Protection Expenditures | $Y - \bar{Y}$ | $(Y - \bar{Y})^2$ |
|---|---|---|---|---|---|
| 2,635 | −431.8 | 186,451.24 | 221 | −1,309.05 | 1,713,611.90 |
| 2,013 | −1,053.8 | 1,110,494.44 | 274 | −1,256.05 | 1,577,661.60 |
| 2,442 | −624.8 | 390,375.04 | 136 | −1,394.05 | 1,943,375.40 |
| 2,838 | −228.8 | 52,349.44 | 1,673 | 142.95 | 20,434.70 |
| 2,815 | −251.8 | 63,403.24 | 286 | −1,244.05 | 1,547,660.40 |
| 2,785 | −281.8 | 79,411.24 | 905 | −625.05 | 390,687.50 |
| 2,488 | −578.8 | 335,009.44 | 7,585 | 6,054.95 | 36,662,419.5 |
| 2,644 | −422.8 | 178,759.84 | 3,010 | 1,479.95 | 2,190,252.00 |
| 2,883 | −183.8 | 33,782.44 | 2,535 | 1,004.95 | 1,009,924.50 |
| 4,029 | 962.2 | 925,828.84 | 2,689 | 1,158.95 | 1,343,165.10 |
| 3,817 | 750.2 | 562,800.04 | 1,039 | −491.05 | 241,130.10 |
| 3,562 | 495.2 | 245,223.04 | 3,761 | 2,230.95 | 4,977,137.90 |
| 3,775 | 708.2 | 501,547.24 | 2,333 | 802.95 | 644,728.70 |
| 3,102 | 35.2 | 1,239.04 | 1,434 | −96.05 | 9,225.60 |
| 3,398 | 331.2 | 109,693.44 | 1,294 | −236.05 | 55,719.60 |
| 3,087 | 20.2 | 408.04 | 570 | −960.05 | 921,696.00 |
| 4,373 | 1,306.2 | 1,706,158.44 | 1,179 | −351.05 | 123,236.10 |
| 2,128 | −938.8 | 881,345.44 | 106 | −1,424.05 | 2,027,918.40 |
| 1,791 | −1,275.8 | 1,627,665.64 | 132 | −1,398.05 | 1,954,543.80 |
| 3,623 | 556.2 | 309,358.44 | 334 | −1,196.05 | 1,430,535.60 |
| 4,175 | 1,108.2 | 1,228,107.24 | 635 | −895.05 | 801,114.50 |
| | | $\Sigma = 10,529,411.24$ | | | $\Sigma = 61,586,178.95$ |

$$S_Y = \sqrt{S_Y^2} = \sqrt{\frac{\Sigma(Y - \bar{Y})^2}{N-1}} = \sqrt{\frac{10,529,411.24}{20}} = 725.58 \qquad S_Y = \sqrt{S_Y^2} = \sqrt{\frac{\Sigma(Y - \bar{Y})^2}{N-1}} = \sqrt{\frac{61,586,178.95}{20}} = 1,754.80$$

c. Because the number of crimes and police protection expenditures is measured according to different scales, it isn't appropriate to directly compare the mean and standard deviation for one variable with the other. But we can talk about each distribution separately. We know from examining the mean and standard deviation for the number of crimes that the standard deviation is large, 725.58, indicating a wide dispersion of scores from the mean. For the number of crimes, states such as Missouri and South Dakota contribute more to its variability because they have values far from the mean (both above and below). With respect to police protection expenditures, we can see that there is a large dispersion from the mean of $1,530.05, as the standard deviation is $1,754.80. States such as North Dakota and New York contribute more to its variability because they have values far from the mean (both above and below).

d. Among other considerations, we need to consider the economic conditions in each state. A downturn in the local and state economy may play a part in the number of crimes and police expenditures per capita.

7. The data indicate that Chinese Americans, on average, have a slightly higher educational attainment (15.83–13.87 = 1.96) than Filipino Americans. Chinese Americans, on average, also desire fewer children than Filipino Americans (4.00–2.20 = 1.80). The standard deviation of years of education is slightly higher for Filipino Americans than Chinese Americans (3.335 vs. 2.691). Likewise, there is greater variability in the ideal number of children for Filipino Americans ($S_Y = 2.769$) than for Chinese Americans ($S_Y = 0.422$).

9. Looking at the means, we can see that it is likely young adults face driving offenses. We can see this by examining the means of each variable (.38; .28). However, without having access to the number of driving offenses for all drivers on the road, we cannot conclude that they are any less safe than other drivers. For instance, we might ask, what is the mean number of moving violations and automobile collisions for all licensed drivers? With respect to variability, we can see that there is more variability in moving violations (0.804) than in automobile collisions (0.610).

# Chapter 5. The Normal Distribution

1.  a. The $Z$ score for a person who watches more than 8 hr/day:

$$Z = \frac{8 - 2.95}{2.58} = 1.96$$

b. We first need to calculate the $Z$ score for a person who watches 5 hr/day:

$$Z = \frac{5 - 2.95}{2.58} = 0.79$$

The area between $Z$ and the mean is 0.2852. We then need to add 0.50 to 0.2852 to find the proportion of people who watch television less than 5 hr/day. Note, the question is asking for the proportion of people who watch television less than 5 hr/day, not the proportion of people who watch less than 5 hr of television a day but more than the mean. Thus, we conclude that the proportion of people who watch television less than 5 hr/day is 0.7852. This corresponds to 767.93 respondents ($0.7852 \times 978$).

c. 5.53 television hours per day corresponds to a $Z$ score of +1.

$$Y = Y + Z(S_Y) = 2.95 + 1(2.58) = 5.53 \, \text{hr/day}$$

d. The $Z$ score for a person who watches 1 hr of television per day is

$$Z = \frac{1 - 2.95}{2.58} = -0.76$$

The area between $Z$ and the mean is 0.2764.
The $Z$ score for a person who watches 6 hr of television per day is

$$Z = \frac{6 - 2.95}{2.58} = 1.18$$

The area between $Z$ and the mean is 0.3810.
Therefore, the percentage of people who watch between 1 and 6 hr of television per day is 65.74% (0.2764 + 0.3810 = 0.6574 × 100).

3.  a. For an individual with 13.44 years of education, his or her $Z$ score would be

$$Z = \frac{13.44 - 13.44}{3.1} = 0$$

b. Solve for the $Z$ score of your friend's years of education.

$$Z = \frac{14.22 - 13.44}{3.1} = .25$$

Looking to column b, we see that .0987 of the population fall between your friend's level of education and the mean (your level).

Since we already know that the proportion between our number of years of education (13.44) and your friend's number of years of education (14.22) is 0.0987, we can multiply $N$ (1,497) by this proportion. Thus, 147.8 people have between 13.44 and 14.22 years of education.

5.  a.  The mean and standard deviation are:

$$\bar{Y} = \frac{14{,}246}{51} = 279.33 \qquad\qquad S_Y = 350.73$$

b.  One standard deviation above the mean is $279.33 + 1.0(350.73) = 630.06$. Six states have unemployment numbers above this value. We would expect, from Appendix A, about 15.9% of the distribution to lie above a $Z$ score of 1.0. Then 15.9% of 51 is 8.1 states, which means that the number of states with scores greater than 1 standard deviation above the mean is less than what we would expect from a normal distribution. States with largest unemployment numbers are those that have very large urban areas.

c.  It is obvious from the histogram that the distribution of unemployment (in 1,000s) is positively skewed.

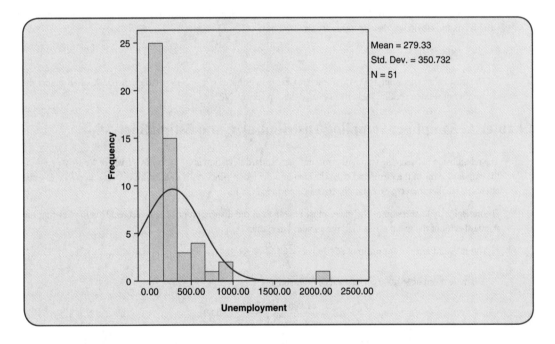

7.  a.  $Z = \dfrac{625 - 500}{100} = 1.25$

The area beyond the $Z$ is about 0.1056, so 10.56% of students should score above 625.

b.  $Z = \dfrac{400 - 500}{100} = -1.00$

The area between this score and the mean is 0.3413.

$$Z = \frac{600 - 500}{100} = 1.00$$

The area between this score and the mean is also 0.3413, so 68.26% of all students should score between 400 and 600 (or $\pm 1$ standard deviation from the mean of 500).

9.  a.  $Z = \dfrac{60 - 42.26}{14.34} = 1.24$

The area between the value and the upper tail of the distribution is 0.1075. So, the probability that someone will work more than 60 hours per week is 0.1075. This translates into approximately 96 ($894 \times 0.1075$) respondents in the sample.

b.  $Z = \dfrac{30 - 42.26}{14.34} = -0.85$

The area between the value and the lower tail of the distribution is 0.1977. So, the probability that someone will work less than 30 hours per week is 0.1977. This translates into approximately 177 ($894 \times 0.1977$) respondents in the sample.

11.  a.  The $Z$ score that is greater than 95% of the cases is 1.65. Translating this into a raw score for the number of women needing shelter yields

$$Y = \bar{Y} + Z(S_Y) = 250 + 1.65(75) = 373.75 \text{ women}$$

Unfortunately, a capacity of 350 is below this value, so there will not be enough space for all abused women on 95% of all nights. Obviously, the city needs at least 374 beds.

b.  $Z = \dfrac{220 - 250}{75} = -0.40$

The area below this value is 0.345, so the area exceeding this $Z$ is $1 - .345 = 0.655$. Or 65.5% of all nights the number of women seeking shelter will exceed the capacity of 220.

## Chapter 6. Sampling, Sampling Distributions, and Estimation

1.  The relationship between the standard error and the standard deviation is $\sigma_{\bar{Y}} = \sigma_Y / \sqrt{N}$ where $\sigma_{\bar{Y}} = \sigma_Y / \sqrt{N}$ is the standard error of the mean and $\sigma_Y$ is the standard deviation. Since $\sigma_Y$ is divided by $\sqrt{N}$, $\sigma_{\bar{Y}} = \sigma_Y / \sqrt{N}$ must always be smaller than $\sigma_Y$, except in the trivial case where $N = 1$.

Theoretically, the dispersion of the mean must be less than the dispersion of the raw scores. This implies that the standard error of the mean is less than the standard deviation.

3.  a.  The standard error of the mean of $64,750 is $\dfrac{\$59,750}{\sqrt{200}} = \$4,225.6$

A sample mean of $61,000 corresponds to a $Z$ score of

$$Z = \dfrac{61,000 - 64,750}{4,225.6} = -0.89$$

The area between the score and the mean (which is $64,750) is about .3133, which is also the probability of a mean between $61,000 and $64,750.

b.  For $75,000

$$Z = \dfrac{75,000 - 64,750}{4,225.6} = 2.43$$

The area beyond $75,000 is 0.0075. So, the probability that the sample mean exceeds $75,000 is 0.0075.

5.  a.  The standard error is calculated as follows

$$\sigma_{\bar{Y}} = \dfrac{\sigma_Y}{\sqrt{N}} = \dfrac{93,500}{\sqrt{7}} = 35,283.02$$

This value represents the average deviation of any sample mean from the mean of means. Accordingly, it may also be referred to as the standard deviation of the sampling distribution.

b. With the exception of cases where $N = 1$, the standard error will always be less in value than the standard deviation of the population. This is expressed by the formula

$$\sigma_{\bar{Y}} = \frac{\sigma_Y}{\sqrt{N}}$$

The shape of the sampling distribution is normal; thus, even when working with a skewed distribution, we know that the sampling distribution is normal. Suggestions for reducing the sampling error include increasing the sample size.

7. a.

$$S_p = \sqrt{\frac{(0.41)(1 - 0.41)}{1{,}501}} = 0.013$$

Confidence interval $= 0.41 \pm 1.96(0.013)$
$$= 0.41 \pm 0.025$$
$$= 0.385 \text{ to } 0.435$$

b.

Confidence interval $= 0.41 \pm 2.58(0.013)$
$$= 0.41 \pm 0.034$$
$$= 0.376 \text{ to } 0.444$$

c. There is very little difference between the 95% or 99% confidence intervals here because the sample size is reasonably large. The former interval is only 1/2 of a percentage point wide, the latter nearly 2/3 of a percentage point. Most large survey organizations use the 95% confidence interval routinely and that seems like the best choice here. Our conclusions about Americans' opinions about global warming will be the same in either case. The intent of this problem is to get students to recognize that they always have a choice as to what confidence interval they choose for a particular problem.

9. a.

$$S_P = \sqrt{\frac{(0.786)(1 - 0.786)}{1{,}303}} = 0.011$$

Confidence interval $= 0.786 \pm 1.96(0.011)$
$$= 0.786 \pm 0.022$$
$$= 0.764 \text{ to } 0.808 \text{ or } 76.4\% \text{ to } 80.8\%$$

b. Based on our answer in 9a, we know that a 90% confidence interval will be more precise than a 95% confidence interval that has a lower bound of 76.4% and an upper bound of 80.8%. Accordingly, a 90% confidence interval will have a lower bound that is greater than 76.4% and an upper bound that is less than 80.8%.

Additionally, we know that a 99% confidence interval will be less precise than what we calculated in 9a. Thus, the lower bound for a 99% confidence interval will be less than 76.4% and the upper bound will be greater than 80.8%.

11.

$$S_{\bar{Y}} = \frac{S_Y}{\sqrt{N}} = \frac{1.70}{\sqrt{2{,}020}} = 0.038$$

Confidence interval $= 1.94 \pm 1.65(0.038)$
$$= 1.94 \pm 0.063$$
$$= 1.877 \text{ to } 2.003$$

13.

$$\text{Standard error} = \sqrt{\frac{27(100 - 27)}{225}} = 2.96$$

$$\text{Confidence interval} = 27 \pm 1.96(2.96)$$
$$= 27 \pm 5.80$$
$$= 21.2\% \text{ to } 32.8\%$$

# Chapter 7. Testing Hypotheses

Please note that in this chapter, the calculations are often done by squaring very small numbers or by multiplying (dividing) a small number by another small number. This can lead to a loss of accuracy in calculations, so small differences in calculations may occur between student results and those listed below.

1.  a. $H_0: \mu_Y = 13.5$ years; $H_1: \mu_Y < 13.5$ years.
    b. The $Z$ value obtained is $-4.19$. The $P$ value for a $Z$ of $-4.19$ is less than .001 for a one-tailed test. This is less than the alpha level of .01, so we reject the null hypothesis and conclude that the doctors at the HMO do have less experience than the population of doctors at all HMOs.

3.  a. Two-tailed test, $\mu_1 \neq \$50,303$; null hypothesis, $\mu_1 = \$50,303$
    b. One-tailed test, $\mu_1 > 3.2$; null hypothesis, $\mu_1 = 3.2$
    c. One-tailed test, $\mu_1 < \mu_2$; null hypothesis, $\mu_1 = \mu_2$
    d. Two-tailed test, $\mu_1 \neq \mu_2$; null hypothesis, $\mu_1 = \mu_2$
    e. One-tailed test, $\mu_1 > \mu_2$; null hypothesis, $\mu_1 = \mu_2$
    f. One-tailed test, $\mu_1 < \mu_2$; null hypothesis, $\mu_1 = \mu_2$

5.  a. $H_0: \mu_1 = 37.7$; $H_1: \mu_1 \neq 37.7$.
    b. Based on a two-tailed test with significance level of .05, the $t$ is 25.67 and its $P$ level is <.001.

$$t = \frac{47.71 - 37.7}{\frac{17.35}{\sqrt{2,013}}} = \frac{10.01}{.39} = 25.67$$

    c. We can reject the null hypothesis in favor of the research hypothesis. There is a difference between the mean age of the GSS sample and the mean age of all American adults. We conclude that relative to age, the GSS sample is not representative of all American adults.

7.  a. When dealing with proportions, the appropriate test statistic is $Z$.
    b. $Z$ obtained is $-6$, $P < .0001$. Since $P(.0001) < \alpha(.05)$, we reject the null hypothesis. This indicates that there is a statistical difference between conservatives and liberals on their views toward affirmative action. Liberals are more likely to support affirmative action policies in the workplace than conservatives.

$$S_{P1-P2} = \sqrt{\frac{0.10(1 - 0.10)}{424} + \frac{0.28(1 - 0.28)}{336}} = 0.03$$

$$Z = \frac{0.10 - 0.28}{0.03} = -6$$

    c. For a two-tailed test, we would have to multiply $P$ by 2, $.0001 \times 2 = .0002$. Our decision would remain the same, we reject the null hypothesis.

9.  a. It would be a one-tailed test. $H_0: \pi_1 = \pi_2$; $H_1: \pi_1 < \pi_2$
    b. Since the probability of our obtained $Z$ is 0.0228 (Column C for $Z = 2.000$), we reject the null hypothesis. We can conclude that there is a significant relationship between one's gender and support for Hillary Clinton's presidential

candidacy. The proportion of women who supported Clinton is significantly greater than the proportion of men who supported her.

$$S_{p_1 p_2} = \sqrt{\frac{0.41(1 - 0.41)}{240} + \frac{0.49(1 - 0.49)}{381}} = 0.04$$

$$Z = \frac{0.41 - 0.49}{0.04} = -2.0$$

    c. If alpha were changed to .01, we would fail to reject the null hypothesis. The $P$ value of our obtained $Z$ is greater than 0.01.

11.  a. One-tailed test, $\alpha = .05$.

$$df = 149 + 306 - 2 = 453$$

$$S_{\bar{Y}_1 - \bar{Y}_2} = \sqrt{\frac{(149 - 1)1.52^2 + (306 - 1)0.79^2}{(149 + 306) - 2}} \sqrt{\frac{149 + 306}{149(306)}} = 0.11$$

$$t = \frac{1.76 - 0.73}{0.11} = 9.36$$

Based on the $t$ obtained of 9.36, we reject the null hypothesis. U.S. students use the Internet longer (more hours) than Indian students for their coursework.

    b. It should be noted that the variances are not equal, one is more than two times as large as the other. You should calculate the standard error and degrees of freedom using Formulas 7.8 and 7.9.

$$S_{\bar{Y}_1 - \bar{Y}_2} = \sqrt{\frac{1.91^2}{149} + \frac{0.78^2}{306}}$$
$$= \sqrt{.024483892 + .001988235}$$
$$= \sqrt{.026472127} = .16$$

$$df = \frac{(1.91^2/149 + 0.78^2/306)^2}{(1.91^2/149)/148 + (0.78^2/306)/305} = \frac{0.00070077356}{0.000171951} = 4.075$$

$$t = \frac{(2.08 - 0.87)}{0.16} = 7.56$$

Based on our $t$ obtained of 7.56, we can reject the null hypothesis. Based on the data, we conclude that there is a relationship between nationality and amount of personal Internet use. U.S. college students used the Internet more hours per week for personal use than did Indian college students.

13.

$$S_{p_1 p_2} = \sqrt{\frac{0.18(1 - 0.18)}{670} + \frac{0.14(1 - 0.14)}{723}} = 0.02$$

$$Z = \frac{0.18 - 0.14}{0.02} = 2$$

The probability of obtaining $Z(2)$ is $0.0228 \times 2 = 0.0456$. $P$ is greater than our $\alpha$ of .01, we cannot reject the null hypothesis. There is no significant relationship between respondent's sex and easy access to cocaine at $\alpha = .01$.

## Chapter 8. Relationships Between Two Variables: Cross-Tabulation

1.  a.  The independent variable is race; the dependent variable is fear of walking alone at night.

| | Race | |
|---|---|---|
| **Fear of Walking Alone at Night** | **Black** | **White** |
| Yes | 3 | 4 |
| No | 5 | 9 |

b.  Approximately 69% of whites (69.2%) are not afraid to walk alone in their neighborhoods at night, whereas approximately 63% of blacks (62.5%) are not afraid to walk alone. This amounts to about a 7% difference (69.2% − 62.5%) between whites and blacks who are not afraid to walk alone at night, indicating a weak relationship. Also, although we went ahead and compared percentage differences in this exercise; it is important to keep in mind that our sample size inhibits our ability to make any meaningful comparisons.

| | Race | |
|---|---|---|
| **Fear of Walking Alone at Night** | **Black** | **White** |
| Yes | 37.5% | 30.8% |
| No | 62.5% | 69.2% |

c.  There is some difference in fears between homeowners and renters. A total of 25.0% of homeowners and 38.5% of renters are afraid to walk in their neighborhood at night. The difference between the two groups is 13.5%. Thus, there is a weak to moderate relationship between homeownership and fear of walking in one's neighborhood at night.

| | Home Ownership | |
|---|---|---|
| **Fear of Walking Alone at Night** | **Yes** | **No** |
| Yes | 2 | 5 |
| Percentage | 25.0% | 38.5% |
| No | 6 | 8 |
| Percentage | 75.0% | 61.5% |

3.  a.  We will make 3,331 errors, because we predict that all victims fall in the modal category (white). E1 = 6,866 − 3,535 = 3,331.

b.  For white offenders, we could make 295 errors; for black offenders, 591 errors; and for other offenders, we would make 62 errors. E2 = 948.

c.  The proportional reduction in error is then (3,331 − 948)/3,331 = .7154. This indicates a very strong relationship between the two variables. We can reduce the error in predicting victim's race based upon race of offender by 71.54%.

5.  a.  Degrees of freedom = (3 − 1) (4 − 1) = 6

b.  Chi-square = 17.64. With 6 degrees of freedom, the probability of our obtained chi-square lies somewhere between .01 and .001. We reject the null hypothesis and conclude that social class and attitude toward welfare spending are dependent. Lower-class (48.9%) and upper-class (32.0%) respondents were more likely to indicate that we were spending "too little" on welfare.

| Social Class/Welfare Spending | $f_o$ | $f_e$ | $f_o - f_e$ | $(f_o - f_e)^2$ | $\dfrac{(f_o - f_e)^2}{f_e}$ |
|---|---|---|---|---|---|
| Lower/Too little | 23 | 13.1 | 9.9 | 98.01 | 7.48 |
| Lower/About right | 12 | 17.7 | −5.7 | 32.49 | 1.84 |
| Lower/Too much | 12 | 16.2 | −4.2 | 17.64 | 1.09 |
| Working/Too little | 92 | 92.8 | −.8 | .64 | .01 |
| Working/About right | 113 | 125.0 | −12 | 144 | 1.15 |
| Working/Too much | 127 | 114.2 | 12.8 | 163.84 | 1.43 |
| Middle/Too little | 76 | 86.1 | −10.1 | 102.01 | 1.18 |
| Middle/About right | 133 | 115.9 | 17.1 | 292.41 | 2.52 |
| Middle/Too much | 99 | 106 | −7 | 49 | .46 |
| Upper/Too little | 8 | 7 | 1 | 1 | .14 |
| Upper/About right | 10 | 9.4 | .6 | .36 | .04 |
| Upper/Too much | 7 | 8.6 | −1.6 | 2.56 | .30 |
| | | $\chi^2 = 17.64$ | | | |

7. a. For females, the percentage of those indicating that they "definitely will" attend college increased from 33.6 to 62.4. This is an increase of 28.8% in 21 years! The other dramatic change in reported percentages is among girls who indicated that they definitely or probably won't attend college. From a high of 45% in 1980, only 17.5% of girls in 2001 reported that they won't attend college—a decrease of 27.5%!

   b. The pattern is also true for males. The largest shifts were in the categories for "definitely will" and "definitely/probably won't" attend college. As with the girls, the highest percentage of boys who reported that they would definitely attend college were increasing from 35.6% in 1980 to 51.5% in 2001. Boys had smaller percentage changes than girls for the "definitely will" and "definitely/probably not" percentages reported from 1990 to 2001. For example, the difference in the percentage of boys who reported that they would not be attending college in 1990 and 2001 is $30.2 − 24.1 = 6.1\%$. For the same time period, the difference is $28.8 − 17.5 = 11.3\%$ for girls. Finally, in 2001, 24.1% of boys reported that they would definitely or probably not attend college. A smaller percentage of girls, 17.5%, reported the same.

9. Bible and PREMARSX: The calculated lambda = .20. E1 = 1859 − 1017 = 842; E2 = 590 + 82 = 672.

   Bible and HOMOSEX: The calculated lambda = .12. E1 = 1838 − 1243 = 595; E2 = 396 + 127 = 523.

   Bible and PORNLAW: The calculated lambda = .0. E1 = 1897 − 1831 = 66; E2 = 48 + 18 = 66.

   The strongest association with belief about the Bible is attitudes about premarital sex. Twenty percent of the error in predicting responses to the statement can be reduced based on information about belief about the Bible. Only 12% of the error is reduced in predicting attitudes toward homosexuality based on one's belief about the Bible and 0% of the error is reduced in predicting attitudes toward pornography laws based on one's belief about the Bible.

11.

| GPA/Trying cocaine | $f_o$ | $f_e$ | $f_o - f_e$ | $(f_o - f_e)^2$ | $\dfrac{(f_o - f_e)^2}{f_e}$ |
|---|---|---|---|---|---|
| <= 2.0/don't disapprove | 28 | 14.1 | 13.9 | 193.21 | 13.70 |
| <= 2.0/disapprove | 27 | 20.3 | 6.7 | 44.89 | 2.21 |
| <= 2.0/strongly disapprove | 76 | 96.6 | −20.6 | 424.36 | 4.39 |
| <= 3.0/don't disapprove | 63 | 58.1 | 4.9 | 24.01 | 0.41 |
| <= 3.0/disapprove | 89 | 83.7 | 5.3 | 28.09 | 0.34 |
| <= 3.0/strongly disapprove | 388 | 398.2 | −10.2 | 104.04 | 0.26 |
| <= 4.0/don't disapprove | 59 | 77.8 | −18.8 | 353.44 | 4.54 |
| <= 4.0/disapprove | 100 | 112 | −12 | 144 | 1.29 |
| <= 4.0/strongly disapprove | 564 | 533.2 | 30.8 | 948.64 | 1.78 |
| | | $\chi^2 = 28.92$ | | | |

Chi-square: 28.92. With four degrees of freedom, the probability of our obtained chi-square is less than .001. We reject the null hypothesis and conclude that there is a relationship between GPA and attitude toward trying cocaine.

# Chapter 9. Regression and Correlation

1. a. On the scatterplot below, the regression line has been plotted to make it easier to see the relationship between the two variables.

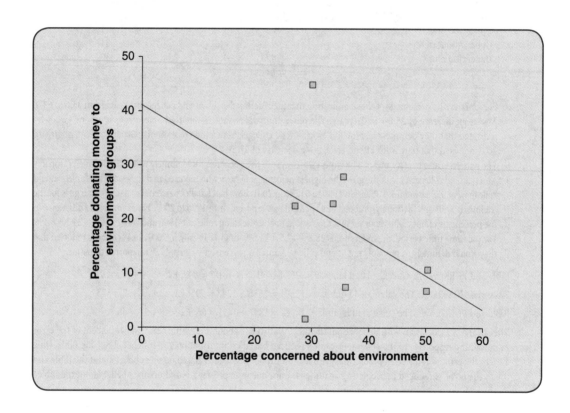

b. The scatterplot shows that there is a general linear relationship between the two variables. There is not a lot of scatter about the straight line describing the relationship. As the percentage of respondents concerned about the environment increases, the percentage of respondents donating money to environmental groups decreases.

c. The Pearson correlation coefficient between the two variables is −0.40. This seems reasonable because it is in line with the fact that the scatterplot indicated a negative relationship. A correlation coefficient of −0.40 indicates a moderate, negative relationship between being concerned about the environment and actually donating money to environmental groups.

| | (1) | (2) | (3) | (4) | (5) | (6) | (7) |
|---|---|---|---|---|---|---|---|
| | *Percentage Concerned* | *Percentage Donating* | | | | | |
| *Country* | X | Y | $(X - \bar{X})$ | $(X - \bar{X})^2$ | $(Y - \bar{Y})$ | $(Y - \bar{Y})^2$ | $(X - \bar{X})(Y - \bar{Y})$ |
| United States | 33.8 | 22.8 | −2.69 | 7.22 | 4.77 | 22.80 | −12.83 |
| Austria | 35.5 | 27.8 | −0.99 | 0.98 | 9.77 | 95.55 | −9.65 |
| Netherlands | 30.1 | 44.8 | −6.39 | 40.80 | 26.77 | 716.90 | −171.03 |
| Slovenia | 50.3 | 10.7 | 13.81 | 190.79 | −7.32 | 53.66 | −101.18 |
| Russia | 29.0 | 1.6 | −7.49 | 56.06 | −16.42 | 269.78 | 122.98 |
| Philippines | 50.1 | 6.8 | 13.61 | 185.30 | −11.22 | 126.00 | −152.80 |
| Spain | 35.9 | 7.4 | −0.59 | 0.35 | −10.62 | 112.89 | 6.24 |
| Denmark | 27.2 | 22.3 | −9.29 | 86.26 | 4.28 | 18.28 | −39.70 |
| | $\Sigma X = 91.9$ | $\Sigma Y = 144.2$ | 0.0[a] | 567.76 | 0.0[a] | 1,415.85 | −357.97 |

$$\text{Mean } X = \bar{X} = \frac{\Sigma X}{N} = \frac{291.9}{8} = 36.49$$

$$\text{Mean } Y = \bar{Y} = \frac{\Sigma Y}{N} = \frac{144.2}{8} = 18.03$$

$$\text{Variance } (Y) = S_Y^2 = \frac{\Sigma(Y - \bar{Y})^2}{N - 1} = \frac{1,415.9}{7} = 202.3$$

$$\text{Standard deviation } (Y) = S_Y = \sqrt{202.3} = 14.22$$

$$\text{Variance } (X) = S_X^2 = \frac{\Sigma(X - \bar{X})^2}{N - 1} = \frac{567.8}{7} = 81.11$$

$$\text{Standard deviation } (X) = S_X = \sqrt{81.11} = 9.01$$

$$\text{Covariance } (X, Y) = S_{YX} = \frac{\Sigma(X - \bar{X})(Y - \bar{Y})}{N - 1} = \frac{-357.97}{7} = -51.14$$

$$r = \frac{S_{YX}}{S_X S_Y} = \frac{-51.14}{(9.01)(14.22)} = -0.40^a$$

a. Answers may differ slightly due to rounding.

3.  a. The correlation coefficient is −0.45.

| *GNP per Capita* | *Percentage Willing to Pay* | $(X - \bar{X})$ | $(X - \bar{X})^2$ | $(Y - \bar{Y})$ | $(Y - \bar{Y})^2$ | $(X - \bar{X})(Y - \bar{Y})$ |
|---|---|---|---|---|---|---|
| 29.24 | 44.9 | 2.72 | 7.40 | −1.64 | 2.69 | −4.46 |
| 18.71 | 53.3 | −7.81 | 61.00 | 6.76 | 45.70 | −52.80 |
| 24.78 | 61.2 | −1.74 | 3.03 | 14.66 | 214.92 | −25.51 |
| 34.31 | 40.7 | 7.79 | 60.68 | −5.84 | 34.11 | −45.49 |
| 25.58 | 32.6 | −0.94 | 0.88 | −13.94 | 194.32 | 13.10 |
| $\Sigma X = 132.62$ | $\Sigma Y = 232.7$ | 0.0[a] | 132.99 | 0.0[a] | 491.74 | −115.16 |

*(Continued)*

(Continued)

$$\text{Mean } X = \bar{X} = \frac{\Sigma X}{N} = \frac{132.62}{5} = 26.52$$

$$\text{Mean } Y = \bar{Y} = \frac{\Sigma Y}{N} = \frac{232.7}{5} = 46.54$$

$$\text{Variance } (Y) = S_Y^2 = \frac{\Sigma(Y - \bar{Y})^2}{N - 1} = \frac{491.74}{4} = 122.94$$

$$\text{Standard deviation } (Y) = S_Y = \sqrt{122.94} = 11.09$$

$$\text{Variance } (X) = S_X^2 = \frac{\Sigma(X - \bar{X})^2}{N - 1} = \frac{132.99}{4} = 33.25$$

$$\text{Standard deviation } (X) = S_X = \sqrt{33.25} = 5.77$$

$$\text{Covariance } (X, Y) = S_{YX} = \frac{\Sigma(X - \bar{X})(Y - \bar{Y})}{N - 1} = \frac{-115.16}{4} = -28.79$$

$$r = \frac{S_{YX}}{S_X S_Y} = \frac{-28.79}{(11.09)(5.77)} = -0.45^{a}$$

a. Answers may differ slightly due to rounding.

b. A correlation coefficient of −0.45 means that relatively high values of GNP are moderately negatively associated with low values of percentage of residents willing to pay higher prices to protect the environment.

5. a. Yes, as indicated by the negative $b$ of −0.196, the relationship between the variables is negative.
   b. Based on the coefficient output, we can predict that an individual with 16 years of education will watch television 2.439 hr/week. In contrast, someone with 12 years of education (high school degree) is predicted to watch 3.223 hr of television per week. This is 0.784 hr more than someone with a college degree. We used the regression equation $\hat{Y} = 5.575 + (-0.196)X$ to make these predictions.
   c. Looking at the scatterplot, a straight line does approximate the data; however, it does so rather poorly, leaving quite a bit of scatter on either side of the regression line.

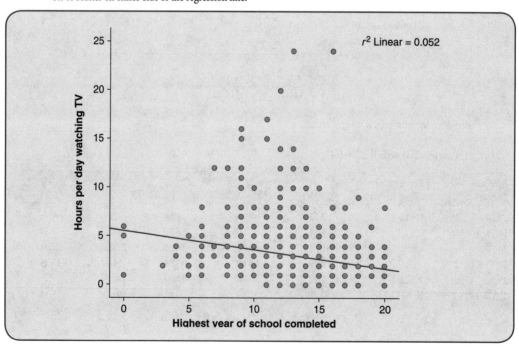

7.  a.  The scatterplot indicates a weak-to-moderate, positive relationship between a respondent's mother's education and a respondent's education.

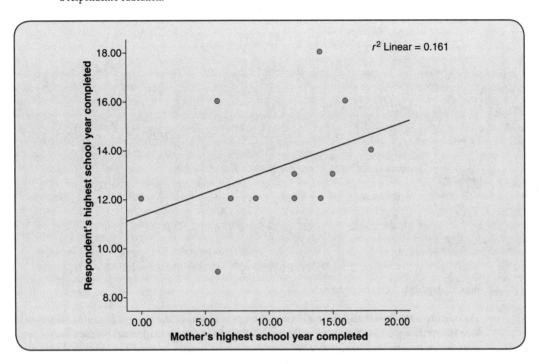

b.  See the regression line on the scatterplot above. The slope of the regression equation is 0.186. The intercept is 11.25. A straight line fits the data reasonably well. In general, the higher a respondent's mother's education, the higher a respondent's education.

| MAEDUC | EDUC | $(X - \bar{X})$ | $(X - \bar{X})^2$ | $(Y - \bar{Y})$ | $(Y - \bar{Y})^2$ | $(X - \bar{X})(Y - \bar{Y})$ |
|---|---|---|---|---|---|---|
| 0 | 12 | −10.75 | 115.56 | −1.25 | 1.56 | 13.44 |
| 15 | 13 | 4.25 | 18.06 | −0.25 | 0.06 | −1.06 |
| 6 | 9 | −4.75 | 22.56 | −4.25 | 18.06 | 20.19 |
| 9 | 12 | −1.75 | 3.06 | −1.25 | 1.56 | 2.19 |
| 16 | 16 | 5.25 | 27.56 | 2.75 | 7.56 | 14.44 |
| 12 | 12 | 1.25 | 1.56 | −1.25 | 1.56 | −1.56 |
| 6 | 16 | −4.75 | 22.56 | 2.75 | 7.56 | −13.06 |
| 18 | 14 | 7.25 | 52.56 | 0.75 | 0.56 | 5.44 |
| 12 | 13 | 1.25 | 1.56 | −0.25 | 0.06 | −0.31 |
| 14 | 12 | 3.25 | 10.56 | −1.25 | 1.56 | −4.06 |
| 14 | 18 | 3.25 | 10.56 | 4.75 | 22.56 | 15.44 |
| 7 | 12 | −3.75 | 14.06 | −1.25 | 1.56 | 4.69 |
| $\Sigma X = 129$ | $\Sigma Y = 159$ | 0.0[a] | 300.22 | 0.0[a] | 64.22 | 55.78 |

*(Continued)*

(Continued)

$$\text{Mean } X = \bar{X} = \frac{\Sigma X}{N} = \frac{129}{12} = 10.75$$

$$\text{Mean } Y = \bar{Y} = \frac{\Sigma Y}{N} = \frac{159}{12} = 13.25$$

$$\text{Variance } (Y) = S_Y^2 = \frac{\Sigma(Y - \bar{Y})^2}{N - 1} = \frac{64.22}{11} = 5.84$$

$$\text{Standard deviation } (Y) = S_Y = \sqrt{5.84} = 2.42$$

$$\text{Variance}(X) = S_X^2 = \frac{\Sigma(X - \bar{X})^2}{N - 1} = \frac{300.22}{11} = 27.29$$

$$\text{Standard deviation } (X) = S_X = \sqrt{27.29} = 5.22$$

$$\text{Covariance } (X, Y) = S_{YX} = \frac{\Sigma(X - \bar{X})(Y - \bar{Y})}{N - 1} = \frac{55.78}{11} = 5.07$$

$$b = \frac{S_{YX}}{S_X^2} = \frac{5.07}{27.29} = 0.186^a$$

$$a = \bar{Y} - b\bar{X} = 13.25 - (0.186)(10.75) = 11.25^a$$

a. Answers may differ slightly due to rounding.

c. The error of prediction for the second case is about 1.04 years of education. That is, we predicted that he or she would have 14.04 years of education $\hat{Y} = 11.25 + 0.186(15)$, but this person actually has 13 years of education. The error of prediction for a person whose mother had 18 years of education and who himself or herself had 14 years of education is −0.598 years of education $\hat{Y} = 11.25 + 0.186(18)$. Here, we predicted a value of 14.598 but the person has less than that amount.

d. For someone whose mother received 4 years of education, we predict about 11.99 years of education $\hat{Y} = 11.25 + 0.186(4)$. For someone whose mother received 12 years of education, we predict about 13.48 years of education $\hat{Y} = 11.25 + 0.186(12)$.

e. On the least squares line. Because both the means are used to calculate the least squares line, it makes sense that the point should fall on the line.

9.  a. A straight line does seem to fit the data, as shown in the scatterplot.

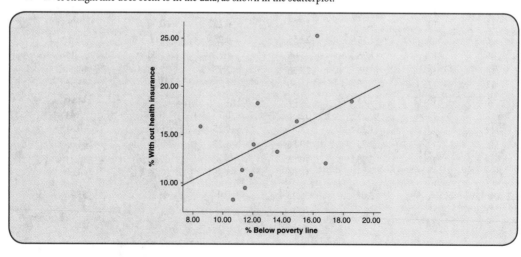

b. The equation, $\hat{Y} = 3.796 + 0.8X$, supports the assertion that a straight line best fits these data. In fact, $b$ is 0.8, which indicates that for a 1% increase in those living below the poverty line, there is a corresponding 0.8% increase in the percentage of people without health insurance. With 25.2% without health insurance, Texas falls especially far from the regression line.

| Percentage Below Poverty | Percentage Without Health Insurance | $(X - \bar{X})$ | $(X - \bar{X})^2$ | $(Y - \bar{Y})$ | $(Y - \bar{Y})^2$ | $(X - \bar{X})(Y - \bar{Y})$ |
|---|---|---|---|---|---|---|
| 16.9 | 12 | 3.62 | 13.10 | -2.42 | 5.86 | -8.76 |
| 12.4 | 18.2 | -0.88 | 0.77 | 3.78 | 14.29 | -3.33 |
| 12.1 | 13.9 | -1.18 | 1.39 | -0.52 | 0.27 | 0.62 |
| 18.6 | 18.5 | 5.32 | 28.30 | 4.08 | 16.65 | 21.71 |
| 8.6 | 15.8 | -4.68 | 21.90 | 1.38 | 1.90 | -6.46 |
| 13.7 | 13.2 | 0.42 | 0.18 | -1.22 | 1.49 | -0.51 |
| 11.6 | 9.5 | -1.68 | 2.82 | -4.92 | 24.21 | 8.27 |
| 12 | 10.8 | -1.28 | 1.64 | -3.62 | 13.10 | 4.64 |
| 15 | 16.4 | 1.72 | 2.96 | 1.98 | 3.92 | 3.41 |
| 16.3 | 25.2 | 3.02 | 9.12 | 10.78 | 116.21 | 32.56 |
| 11.4 | 11.3 | -1.88 | 3.53 | -3.12 | 9.73 | 5.87 |
| 10.8 | 8.2 | -2.48 | 6.15 | -6.22 | 38.69 | 15.43 |
| $\Sigma X = 159.4$ | $\Sigma Y = 173$ | $0.0^a$ | 91.86 | $0.0^a$ | 246.32 | 73.45 |

$$\text{Mean } X = \bar{X} = \frac{\Sigma X}{N} = \frac{159.4}{12} = 13.28$$

$$\text{Mean } Y = \bar{Y} = \frac{\Sigma Y}{N} = \frac{173}{12} = 14.42$$

$$\text{Variance } (Y) = S_Y^2 = \frac{\Sigma(Y - \bar{Y})^2}{N - 1} = \frac{246.32}{11} = 22.39$$

$$\text{Standard deviation } (Y) = S_Y = \sqrt{22.39} = 4.73$$

$$\text{Variance } (X) = S_X^2 = \frac{\Sigma(X - \bar{X})^2}{N - 1} = \frac{91.86}{11} = 8.35$$

$$\text{Standard deviation } (X) = S_X = \sqrt{8.35} = 2.89$$

$$\text{Covariance } (X, Y) = S_{YX} = \frac{\Sigma(X - \bar{X})(Y - \bar{Y})}{N - 1} = \frac{73.45}{11} = 6.68$$

$$b = \frac{S_{YX}}{S_X^2} = \frac{6.68}{8.35} = 0.8$$

$$a = \bar{Y} - b\bar{X} = 14.42 - 0.8(13.28) = 3.796$$

$$\hat{Y} = 3.796 + 0.8X$$

a. Answers may differ slightly due to rounding.

c. $5 = 3.796 + 0.8X$

$X = 1.51\%$. So about 1.51% need to be living below the poverty level to have only 5% without health insurance.

d. You cannot go outside the scope of your data. That said, it is interesting to see how closely poverty and lack of health insurance are related. Although we cannot generalize statistically, this example does help us see just how closely the two are related. Thus, as we go about our research, this may be a consideration to keep in mind for future studies.

11. a. False, both $b$ and $r$ will always have the same sign because both tell us the direction of the relationship.
    b. Both $a$ and $b$ refer to changes in the dependent variable.
    c. The coefficient of determination, $r^2$, is PRE measure. PRE stands for proportional reduction of error. This means that $r^2$ indicates the extent to which prediction error is reduced when we take into account the independent variable in our predictions.
    d. False. The regression equation serves many functions, most notably to make predictions. Whether a regression equation models a causal relationship is a matter of meeting the causal requirements as they were discussed in Chapter 1.
       Regression equations are commonly used to model relationships wherein researchers look for how changes in one or more variables (referred to as the independent variables) correspond with changes in another variable (referred to as the dependent variable).

# Chapter 10. Analysis of Variance

1. a. Males = 25.59; females = 22.57. On average, men are slightly (3.02 years) older than women at the birth of their first child.
   b. Based on $F$ obtained of 55.933, we would reject the null hypothesis of no difference. We know from the SPSS output that probability of the $F$ obtained is 0.000, less than our alpha of .05. The output indicates that men are older at the birth of their first child than women.

3. a.

| $\bar{Y}_1 = 2.875$ | $\bar{Y}_2 = 2.250$ | $\bar{Y}_3 = 2.00$ | $\bar{Y}_4 = 1.375$ |
|---|---|---|---|
| $\Sigma Y_1 = 23$ | $\Sigma Y_2 = 18$ | $\Sigma Y_3 = 16$ | $\Sigma Y_4 = 11$ |
| $\Sigma Y_1^2 = 71$ | $\Sigma Y_2^2 = 44$ | $\Sigma Y_3^2 = 38$ | $\Sigma Y_4^2 = 17$ |
| $n_1 = 8$ | $n_2 = 8$ | $n_3 = 8$ | $n_4 = 8$ |

$$\bar{Y} = 2.125$$
$$N = 32$$

$$SSB = 8(2.875 - 2.125)^2 + 8(2.250 - 2.125)^2 + 8(2.00 - 2.125)^2 + 8(1.375 - 2.125)^2$$
$$= 8(0.5625) + 8(.015625) + 8(.015625) + 8(.5625)$$
$$= 4.5 + .125 + .125 + 4.5$$
$$SSB = 9.25$$
$$df_b = 4 - 1$$
$$df_b = 3$$

**Mean square between = 9.25/3 = 3.08**

$$SSW = (71 + 44 + 38 + 17) - [(23^2/8) + (18^2/8) + (16^2/8) + (11^2/8)]$$
$$= 170 - (66.125 + 40.5 + 32 + 15.125)$$
$$= 170 - 153.75$$
$$SSW = 16.25$$
$$df_w = 32 - 4$$
$$df_w = 28$$

**Mean square within = 16.25/28 = 0.58**

$$F = 3.08/0.58$$
$$F = 5.31$$

*Decision.* If we set alpha at 0.05, $F$ critical would be 2.95 ($df_1 = 3$ and $df_2 = 28$). Based on our $F$ obtained of 5.31, we would reject the null hypothesis and conclude that at least one of the means is significantly different than the others. Upper class respondents rate their health the highest (1.375), followed by middle and working class respondents (2.00 and 2.25 respectively) and lower class respondents (2.875) on a scale where 1 = excellent, 4 = poor.

5.

| $\bar{Y}_1 = 16$ | $\bar{Y}_2 = 14$ | $\bar{Y}_3 = 60$ |
|---|---|---|
| $\Sigma Y_1 = 16$ | $\Sigma Y_2 = 14$ | $\Sigma Y_3 = 6$ |
| $\Sigma Y_1^2 = 30$ | $\Sigma Y_2^2 = 24$ | $\Sigma Y_3^2 = 8$ |
| $n_1 = 10$ | $n_2 = 10$ | $n_3 = 10$ |
| | $\bar{Y} = 1.2$ | |
| | $N = 30$ | |

$$SSB = 10(1.6 - 1.2)^2 + 10(1.4 - 1.2)^2 + 10(0.6 - 1.2)^2$$
$$= 10(0.16) + 10(0.04) + 10(0.36)$$
$$= 1.6 + 0.4 + 3.6$$
$$SSB = 5.6$$
$$df_b = 3 - 1$$
$$df_b = 2$$

Mean square between $= 5.6/2 = 2.8$

$$SSW = (30 + 24 + 8) - [(16^2/10) + (14^2/10) + (6^2/10)]$$
$$= 62 - (25.6 + 19.6 + 3.6)$$
$$= 62 - 48.8$$
$$SSW = 13.2$$
$$df_w = 30 - 3$$
$$df_w = 27$$

Mean square within $= 13.2/27 = 0.488889$

$$F = 2.8/0.488889$$
$$F = 5.727273$$

*Decision.* If we set alpha at 0.01, $F$ critical would be 5.49 ($df_1 = 2$ and $df_2 = 27$). Based on our $F$ obtained of 5.727, we would reject the null hypothesis and conclude that at least one of the means is significantly different than the others. Respondents with less than a high school degree rate their church attendance highest (1.6), followed by respondents with a high school degree (1.4) and then respondents with some college degree (0.6).

7.

| $\bar{Y}_1 = 0.8$ | $\bar{Y}_2 = 1.75$ | $\bar{Y}_3 = 3.20$ |
|---|---|---|
| $\Sigma Y_1 = 4$ | $\Sigma Y_2 = 7$ | $\Sigma Y_3 = 16$ |
| $\Sigma Y_1^2 = 6$ | $\Sigma Y_2^2 = 15$ | $\Sigma Y_3^2 = 8$ |
| $n_1 = 5$ | $n_2 = 4$ | $n_3 = 5$ |
| | $\bar{Y} = 1.93$ | |
| | $N = 14$ | |

$$SSB = 5(.8 - 1.93)^2 + 4(1.75 - 1.93)^2 + 5(3.20 - 1.93)^2$$
$$= 5(1.2769) + 4(.0324) + 5(1.6129)$$
$$= 6.3845 + 0.1296 + 8.0645$$
$$SSB = 14.58$$
$$df_b = 3 - 1$$
$$df_b = 2$$

Mean square between $= 14.58/2 = 7.29$

$$SSW = (6 + 15 + 54) - [(4^2/5) + (7^2/4) + (16^2/5)]$$
$$= 75 - (3.2 + 12.25 + 51.2)$$
$$= 75 - 66.65$$
$$SSW = 8.35$$
$$df_w = 14 - 3$$
$$df_w = 11$$

Mean square within $= 8.35/11 = 0.76$
$$F = 7.29/0.76$$
$$F = 9.59$$

*Decision.* If we set alpha at .05, $F$ critical would be 3.98 ($df_1 = 2$ and $df_2 = 11$). Based on our $F$ obtained of 9.59, we would reject the null hypothesis and conclude that at least one of the means is significantly different from the others. The average number of moving violations is the highest for large-city respondents (3.2); medium-sized-city residents are next (1.75), followed last by small-town respondents (0.8).

# GLOSSARY

**Alpha ($\alpha$)** The level of probability at which the null hypothesis is rejected. It is customary to set alpha at the .05, .01, or .001 level

**Analysis of variance (ANOVA)** An inferential statistics technique designed to test for significant relationship between two variables in two or more sample effects of authority structures and gender

**Asymmetrical measure of association** A measure whose value may vary depending on which variable is considered the independent variable and which the dependent variable

**Bar graph** A graph showing the differences in frequencies or percentages among categories of a nominal or an ordinal variable. The categories are displayed as rectangles of equal width with their height proportional to the frequency or percentage of the category

**Between-group sum of squares ($SSB$)** The sum of squared deviations between each sample mean to the overall mean score

**Bivariate analysis** A statistical method designed to detect and describe the relationship between two variables

**Bivariate table** A table that displays the distribution of one variable across the categories of another variable

**Cell** The intersection of a row and a column in a bivariate table

**Central limit theorem** If all possible random samples of size $N$ are drawn from a population with a mean $\mu_Y$ and a standard deviation $\sigma_Y$, then as $N$ becomes larger, the sampling distribution of sample means becomes approximately normal, with mean $\mu_{\bar{Y}}$ and standard deviation, $\sigma_{\bar{Y}} = \sigma_Y / \sqrt{N}$

**Chi-square (obtained)** The test statistic that summarizes the differences between the observed ($f_o$) and the expected ($f_e$) frequencies in a bivariate table

**Chi-square test** An inferential statistics technique designed to test for significant relationships between two variables organized in a bivariate table

**Coefficient of determination ($r^2$)** A PRE measure reflecting the proportional reduction of error that results from using the linear regression model. It reflects the proportion of the total variation in the dependent variable, $Y$, explained by the independent variable, $X$

**Column variable** A variable whose categories are the columns of a bivariate table

**Confidence interval (CI)** A range of values defined by the confidence level within

**Confidence level** The likelihood, expressed as a percentage or a probability, that a specified interval will contain the population parameter

**Control variable** An additional variable considered in a bivariate relationship. The variable is controlled for when we take into account its effect on the variables in the bivariate relationship

**Cramer's $V$** A measure of association for nominal variables based on the value of chi-square and ranges between 0.0 to 1.0

**Cross-tabulation** A technique for analyzing the relationship between two variables that have been organized in a table

**Cumulative frequency distribution** A distribution showing the frequency at or below each category (class interval or score) of the variable

**Cumulative percentage distribution** A distribution showing the percentage at or below each category (class interval or score) of the variable

**Data** Information represented by numbers, which can be the subject of statistical analysis

**Degrees of freedom ($df$)** The number of scores that are free to vary in calculating a statistic

**Dependent variable** Variable to be explained (the "effect")

**Descriptive statistics** Procedures that help us organize and describe data collected from either a sample or a population

**Deterministic (perfect) linear relationship** A relationship between two interval-ratio variables in which all the observations (the dots) fall along a straight line. The line provides a predicted value of $Y$ (the vertical axis) for any value of $X$ (the horizontal axis)

**Dichotomous variable** A variable that has only two values

**Direct causal relationship** A bivariate relationship that cannot be accounted for by other theoretically relevant variables

**Elaboration** A process designed to further explore a bivariate relationship; it involves the introduction of control variables

**Empirical research** Research based on evidence that can be verified by using our direct experience

**Estimation** A process whereby we select a random sample from a population and use a sample statistic to estimate a population parameter

**Expected frequencies ($f_e$)** The cell frequencies that would be expected in a bivariate table if the two variables were statistically independent

**F critical** $F$-test statistic that corresponds to the alpha level, $df_w$, and $df_b$

**F obtained** The $F$-test statistic that is calculated

**F ratio or F statistic** The test statistic for ANOVA, calculated by the ratio of mean square to mean square within

**Frequency distribution** A table reporting the number of observations falling into each category of the variable

**Gamma** A symmetrical measure of association suitable for use with ordinal variables or with dichotomous nominal variables. It can vary from 0.0 to ±1.0 and provides us with an indication of the strength and direction of the association between the variables

**Histogram** A graph showing the differences in frequencies or percentages among categories of an interval-ratio variable. The categories are displayed as contiguous bars, with width proportional to the width of the category and height proportional to the frequency or percentage of that category

**Hypothesis** A tentative answer to a research problem

**Independent variable** The variable expected to account for (the "cause" of) the dependent variable

**Inferential statistics** The logic and procedures concerned with making predictions or inferences about a population from observations and analyses of a sample

**Interquartile range (IQR)** The width of the middle 50% of the distribution. It is defined as the difference between the lower and upper quartiles ($Q_1$ and $Q_3$)

**Interval-ratio measurement** Measurements for all cases are expressed in the same units

**Kendall's tau-*b*** A symmetrical measure of association suitable for use with ordinal variables. Unlike gamma, it accounts for pairs tied on the independent and dependent variable. It can vary from 0.0 to ±1.0. It provides an indication of the strength and direction of the association between the variables

**Lambda** An asymmetrical measure of association, lambda is suitable for use with nominal variables and may range from 0.0 to 1.0. It provides us with an indication of the strength of an association between the independent and dependent variables

**Least squares line (best-fitting line)** A line where the residual sum of squares, or $\Sigma e^2$, is at a minimum

**Least squares method** The technique that produces the least squares line

**Left-tailed test** A one-tailed test in which the sample outcome is hypothesized to be at the left tail of the sampling distribution

**Linear relationship** A relationship between two interval-ratio variables in which the observations displayed in a scatter diagram can be approximated with a straight line

**Line graph** A graph showing the differences in frequencies or percentages among categories of an interval-ratio variable. Points representing the frequencies of each category are placed above the midpoint of the category and are joined by a straight line

**Marginals** The row and column totals in a bivariate table

**Margin of error** The radius of a confidence interval

**Mean** The arithmetic average obtained by adding up all the scores and dividing by the total number of scores

**Mean square between** Sum of squares between divided by its corresponding degrees of freedom

**Mean squares regression** An average computed by dividing the regression sum of squares ($SSR$) by its corresponding degrees of freedom

**Mean squares residual** An average computed by dividing the residual sum of squares ($SSE$) by its corresponding degrees of freedom

**Mean square within** Sum of squares within divided by its corresponding degrees of freedom

**Measure of association** A single summarizing number that reflects the strength of a relationship, indicates the usefulness of predicting the dependent variable from the independent variable, and often shows the direction of the relationship

**Measures of central tendency** Numbers that describe what is average or typical of the distribution

**Measures of variability** Numbers that describe diversity or variability in the distribution

**Median** The score that divides the distribution into two equal parts so that half the cases are above and half below

**Mode** The category or score with the highest frequency (or percentage) in the distribution of main points

**Negatively skewed distribution** A distribution with a few extremely low values

**Negative relationship** A bivariate relationship between two variables measured at the ordinal level or higher in which the variables vary in opposite directions

**Nominal measurement** Numbers or other symbols are assigned to a set of categories for the purpose of naming, labeling, or classifying the observations

**Normal distribution** A bell-shaped and symmetrical theoretical distribution with the mean, the median, and the mode all coinciding at its peak and with the frequencies gradually decreasing at both ends of the curve

**Null hypothesis** $(H_0)$ A statement of "no difference," which contradicts the research hypothesis and is always expressed in terms of population parameters

**Observed frequencies** $(f_o)$ The cell frequencies actually observed in a bivariate table

**One-tailed test** A type of hypothesis test that involves a directional hypothesis. It specifies that the values of one group are either larger or smaller than some specified population value

**One-way ANOVA** Analysis of variance application with one dependent variable and one independent variable

**Ordinal measurement** Numbers are assigned to rank-ordered categories ranging from low to high

**Parameter** A measure (e.g., mean or standard deviation) used to describe the population distribution

**Pearson's correlation coefficient** $(r)$ The square root of $r^2$; it is a measure of association for interval-ratio variables, reflecting the strength of the linear association between two interval-ratio variables. It can be positive or negative in sign

**Percentage** A relative frequency obtained by dividing the frequency in each category by the total number of cases and multiplying by 100

**Percentage distribution** A table showing the percentage of observations falling into each category of the variable

**Percentile** A score below which a specific percentage of the distribution falls

**Pie chart** A graph showing the differences in frequencies or percentages among categories of a nominal or an ordinal variable. The categories are displayed as segments of a circle whose pieces add up to 100% of the total frequencies

**Point estimate** A sample statistic used to estimate the exact value of a population parameter

**Population** The total set of individuals, objects, groups, or events in which the researcher is interested

**Positively skewed distribution** A distribution with a few extremely high values

**Positive relationship** A bivariate relationship between two variables measured at the ordinal level or higher in which the variables vary in the same direction

**Probability sampling** A method of sampling that enables the researcher to specify for each case in the population the probability of its inclusion in the sample

**Proportion** A relative frequency obtained by dividing the frequency in each category by the total number of cases

**Proportional reduction of error** (PRE) The concept that underlies the definition and interpretation of several measures of association. PRE measures are derived by comparing the errors made in predicting the dependent variable while ignoring the independent variable with errors made when making predictions that use information about the independent variable

**P value** The probability associated with the obtained value of $Z$

**Range** A measure of variation in interval-ratio variables. It is the difference between the highest (maximum) and the lowest (minimum) scores in the distribution

**Rate** A number obtained by dividing the number of actual occurrences in a given time period by the number of possible occurrences

**Regression sum of squares** (SSR) Reflects the improvement in the prediction error resulting from using the linear prediction equation, $SST - SSE$

**Research hypothesis** $(H_1)$ A statement reflecting the substantive hypothesis. It is always expressed in terms of population parameters, but its specific form varies from test to test

**Research process** A set of activities in which social scientists engage to answer questions, examine ideas, or test theories

**Residual sum of squares** (SSE) Sum of squared differences between observed and predicted $Y$

**Right-tailed test** A one-tailed test in which the sample outcome is hypothesized to be at the right tail of the sampling distribution

**Row variable** A variable whose categories are the rows of a bivariate table

**Sample** A relatively small subset selected from a population

**Sampling distribution** The sampling distribution is a theoretical probability distribution of all possible sample values for the statistics in which we are interested

**Sampling distribution of the difference between means** A theoretical probability distribution that would be obtained by calculating all the possible mean differences $(\bar{Y}_1 - \bar{Y}_2)$ that would be obtained by drawing all the possible independent random samples of size $N_1$ and $N_2$ from two populations where $N_1$ and $N_2$ are each greater than 50

**Sampling distribution of the mean** A theoretical probability distribution of sample means that would be obtained by drawing from the population all possible samples of the same size

**Sampling error** The discrepancy between a sample estimate of a population parameter and the real population parameter

**Scatter diagram (scatterplot)** A visual method used to display a relationship between two interval-ratio variables

**Simple random sample** A sample designed in such a way as to ensure that (1) every member of the population has an equal chance of being chosen and (2) every combination of $N$ members has an equal chance of being chosen

**Skewed distribution** A distribution with a few extreme values on one side of the distribution

**Slope** (*b*) The amount of change in a dependent variable per unit change in an independent variable

**Standard deviation** A measure of variation for interval-ratio variables; it is equal to the square root of the variance

**Standard error of the mean** The standard deviation of the sampling distribution of the mean. It describes how much dispersion there is in the sampling distribution of the mean

**Standard normal distribution** A normal distribution represented in standard ($Z$) scores

**Standard normal table** A table showing the area (as a proportion, which can be translated into a percentage) under the standard normal curve corresponding to any $Z$ score or its fraction

**Standard ($Z$) score** The number of standard deviations that a given raw score is above or below the mean

**Statistic** A measure (e.g., mean or standard deviation) used to describe the sample distribution

**Statistical hypothesis testing** A procedure that allows us to evaluate hypotheses about population parameters based on sample statistics

**Statistical independence** The absence of association between two cross-tabulated variables. The percentage distributions of the dependent variable within each category of the independent variable are identical

**Statistics** A set of procedures used by social scientists to organize, summarize, and communicate information

**Symmetrical distribution** The frequencies at the right and left tails of the distribution are identical; each half of the distribution is the mirror image of the other

**Symmetrical measure of association** A measure whose value will be the same when either variable is considered the independent variable or the dependent variable

**Systematic random sampling** A method of sampling in which every *K*th member (*K* is a ratio obtained by dividing the population size by the desired sample size) in the total population is chosen for inclusion in the sample after the first member of the sample is selected at random from among the first *K* members in the population

***t* distribution** A family of curves, each determined by its degrees of freedom (*df*). It is used when the population standard deviation is unknown and the standard error is estimated from the sample standard deviation

**Theory** An elaborate explanation of the relationship between two or more observable attributes of individuals or groups

**Time-series chart** A graph displaying changes in a variable at different points in time. It shows time (measured in units such as years or months) on the horizontal axis and the frequencies (percentages or rates) of another variable on the vertical axis

**Total sum of squares (*SST*)** The total variation in scores, calculated by adding *SSB* and *SSW*

***t* statistic (obtained)** The test statistic computed to test the null hypothesis about a population mean when the population standard deviation is unknown and is estimated using the sample standard deviation

**Two-tailed test** A type of hypothesis test that involves a nondirectional research hypothesis. We are equally interested in whether the values are less than or greater than one another. The sample outcome may be located at both the low and high ends of the sampling distribution

**Type I error** The probability associated with rejecting a null hypothesis when it is true

**Type II error** The probability associated with failing to reject a null hypothesis when it is false

**Unit of analysis** The level of social life on which social scientists focus. Examples of different levels are individuals and groups

**Variable** A property of people or objects that takes on two or more values

**Variance** A measure of variation for interval-ratio variables; it is the average of the squared deviations from the mean

**Within-group sum of squares** (*SSW*) Sum of squared deviations within each group, calculated between each individual score and the sample mean

**Y-intercept** (*a*) The point where the regression line crosses the *Y*-axis and where $X = 0$

**Z statistic (obtained)** The test statistic computed by converting a sample statistic (such as the mean) to a *Z* score. The formula for obtaining *Z* varies from test to test

# NOTES

## Chapter 1

1. U.S. Census Bureau, *American Community Survey*, 2008.
2. U.S. Census Bureau, *Statistical Abstract of the United States*, 2010, Table 603.
3. Rampell, Catherine. "Women Now a Majority in American Workplaces," *The New York Times,* February 5, 2010. Retrieved February 5, 2010, from www.nytimes.com/2010/02/06/business/economy/06women.html.
4. U.S. Census Bureau, *Statistical Abstract of the United States*, 2010, Table 607.
5. Chava Frankfort-Nachmias and David Nachmias, *Research Methods in the Social Sciences* (New York: Worth Publishers, 2000), p. 56.
6. Barbara Reskin and Irene Padavic, *Women and Men at Work* (Thousand Oaks, CA: SAGE, 2002), p. 65, 2002, p. 144.
7. Frankfort-Nachmias and Nachmias, 2000, p. 50.
8. Ibid., p. 52.

## Chapter 2

1. Gary Hytrek and Kristine Zentgraf, *America Transformed: Globalization, Inequality and Power* (New York: Oxford University Press, 2007).
2. U.S. Census Bureau, *American Community Survey*, 2008.
3. Ibid., 2008.
4. U.S. Census Bureau, *Statistical Abstract for the United States*, 2010, Table 43.
5. Ibid., 2010, Table 69.
6. Ibid., 2010, Table 56.
7. David Knoke and George W. Bohrnstedt, *Basic Social Statistics* (New York: Peacock, 1991), p. 25.
8. Ibid., p. 41.
9. The idea of "Reading the Research Literature" sections that appear in most chapters was inspired by Joseph F. Healey, *Statistics: A Tool for Social Research*, 5th ed. (Belmont, CA: Wadsworth, 1999).
10. Eric Fong and Kumiko Shibuya, "Urbanization and Home Ownership: The Spatial Assimilation Process in U.S. Metropolitan Areas," *Sociological Perspectives* 43, no. 1 (2000): 137–157. Used with permission.

11. The U.S. Census Bureau notes that persons of Hispanic origin may be of any race.
12. U.S. Census Bureau, *Marital Status and Living Arrangements: March 1996,* Current Population Reports, P20–496, 1998, p. 5.
13. U.S. Census Bureau, *65+ in America,* Current Population Reports, Special Studies, P23–190, 1996, pp. 6-1, 6-8.
14. U.S. Census Bureau, *65+ in America,* Current Population Reports, Special Studies, P23–190, 1996, pp. 2–3.
15. Ibid., p. 6-2.
16. Edward R. Tufte, *The Visual Display of Quantitative Information* (Cheshire, CI: Graphics Press, 1983), p. 53.
17. U.S. Census Bureau, *Statistical Abstract for the United States*, 2010, Table 224.
18. Catherine Freeman, *Trend in Educational Equity of Girls and Women: 2004* (Washington, DC: U.S. Department of Education, National Center for Education Statistics, 2004), pp. 9–10.

## Chapter 3

1. Bureau of Labor Statistics, Current Population Survey 2009, *Household Data Annual Averages*, Table 37.
2. This rule was adapted from David Knoke and George W. Bohrnstedt, *Basic Statistics* (New York: Peacock Publishers, 1991), pp. 56–57.
3. The rates presented in Table 3.4 are computed for aggregate units (cities) of different sizes. The mean of 13.95 is therefore called an unweighted mean. It is not the same as the murder rate for the population in the combined cities.
4. Three variables, TVHOURS, SIBS, and EDUC, were taken from a GSS sample; EDUC was then recoded into another variable including only the respondents without a high school diploma.

## Chapter 4

1. Johnneta B. Cole, "Commonalities and Differences," in *Race, Class, and Gender*, ed. Margaret L. Andersen and Patricia Hill Collins (Belmont, CA: Wadsworth, 1998), pp. 128–129.
2. Ibid., pp. 129–130.
3. Recent Census data reveal that the recession of 2008–2009 has halted this dominant migration trend.

4. The percentage increase in the population 65 years and above for each state and region was obtained by the following formula:

$$\text{Percentage increase} = [(2015 \text{ population} - 2008 \text{ population})/ \\ 2008 \text{ population}] \times 100$$

5. $N - 1$ is used in the formula for computing variance because usually we are computing from a sample with the intention of generalizing to a larger population. $N - 1$ in the formula gives a better estimate and is also the formula used in SPSS.

6. Herman J. Loether and Donald G. McTavish, *Descriptive and Inferential Statistics: An Introduction* (Boston: Allyn and Bacon, 1980), pp. 160–161.

7. Stephanie A. Bohon, Monica Kirkpatrick Johnson, and Bridget K. Gorman, "College Aspirations and Expectations among Latino Adolescents in the United States," *Social Problems* 53, no. 2 (2006): 207–225.

8. Ibid., p. 210.

9. Ibid., p. 213.

10. Ibid.

# Chapter 6

1. This discussion has benefited from a more extensive presentation on the aims of sampling in Richard Maisel and Caroline Hodges Persell, *How Sampling Works* (Thousand Oaks, CA: SAGE, 1996).

2. The discussion in these sections is based on Chava Frankfort-Nachmias and David Nachmias, *Research Methods in the Social Sciences* (New York: Worth Publishers, 2007), pp. 167–177.

3. The population of the 20 individuals presented in Table 6.2 is considered a finite population. A finite population consists of a finite (countable) number of elements (observations). Other examples of finite populations include all women in the labor force in 2008 and all public hospitals in New York City. A population is considered infinite when there is no limit to the number of elements it can include. Examples of infinite populations include all women in the labor force, in the past or the future. Most samples studied by social scientists come from finite populations. However, it is also possible to form a sample from an infinite population.

4. Here we are using an idealized example in which the sampling distribution is actually computed. However, please bear in mind that in practice one never computes a sampling distribution because it is also infinite.

5. The relationship between sample size and interval width when estimating means also holds true for sample proportions. When the sample size increases, the standard error of the proportion decreases, and therefore, the width of the confidence interval decreases as well.

6. "More Americans Favor than Oppose Arizona Immigration Law," Gallup Poll, April 27–28, 2010.

7. Frank Newport, "Americans' Views on Healthcare Law Remain Stable," *Gallup*, April 15, 2010.

8. Michael Luo and Megan Thee-Brenan, "Poll reveals Trauma of Joblessness in U.S." *The New York Times*, December 15, 2009.

9. Andrew Kohut, "Getting It Wrong," *The New York Times*, January 10, 2008.

10. Data sources: http://stats.org/faq_margin.htm; www.charneyresearch.com; www.pollingreport.com/sampling.htm; www.robertniles.com/stats/margin.shtml; http://us-elections.suite101.com/article.cfm/new_hampshire_primary_poll_wrong.

11. Data from "Little Consensus on Global Warming," Pew Research Center, July 12, 2006.

12. Data from "Internet's Broader Role in Campaign 2008," Pew Research Center, January 11, 2008.

13. Data from "The Millenials: Confident. Connected. Open to Change," Pew Research Center, February 24, 2010.

# Chapter 7

1. Steve Hargreaves, *Gas prices hit working class,* 2007. Retrieved February 11, 2008, from http://money.cnn.com/2007/11/13/news/economy/gas_burden/index.htm.

2. American Automobile Association, 2010. *Daily Fuel Gauge Report,* April 27. Retrieved April 27, 2010, from www.fuelgaugereport.com/.

3. To compute the sample variance for any particular sample, we must first compute the sample mean. Since the sum of the deviations about the mean must equal 0, only $N - 1$ of the deviation scores are free to vary with each variance estimate.

4. Current Population Survey, PINC-02. People 18 Years Old and Over, by Total Money Income in 2008.

5. Degrees of freedom formula based on Dennis Hinkle, William Wiersma, and Stephen Jurs, *Applied Statistics for the Behavioral Sciences* (Boston: Houghton Mifflin, 1998), p. 268.

6. The sample proportions are unbiased estimates of the corresponding population proportions. Therefore, we can use the Z statistic, although our standard error is estimated from the sample proportions.

7. Paula Y. Goodwin, William D. Mosher, and Anjani Chandra, "Marriage and cohabitation in the United States: A statistical portrait based on Cycle 6 (2002) of the National Survey of Family Growth," *Vital Health Statistics* 23, no. 28 (2010): 1–45.

8. Robert E. Jones and Shirley A. Rainey, "Examining Linkages Between Race, Environmental Concern, Health and Justice in a Highly Polluted Community of Color," *Journal of Black Studies* 36, no. 4 (2006): 473–496.

9. Marcelline Fusilier, Subhash Durlabhji, Alain Cucchi, and Michael Collins, "A Four Country Investigation of Factors Facilitating Student Internet Use," *CyberPsychology and Behavior* 8, no. 5 (2005): 454–464.

# Chapter 8

1. Damien Cave, "A Generation Gap Over Immigration," *The New York Times*, May 17, 2010. Retrieved May 2010, from www .nytimes.com/2010/05/18/us/18divide.html.

2. *USA Today*, October 9, 1992.

3. National Center for Health Statistics, *Summary Health Statistics for U.S. Population*, U.S. National Health Interview Survey, 2002.

4. Full consideration of the question of detecting the presence of a bivariate relationship requires the use of inferential statistics. Inferential statistics is discussed in Chapters 7 through 10.

5. Note that this group is but a small sample taken from the GSS national sample. The relationship between home ownership and race noted here may not necessarily hold true in other (larger) samples.

6. Another way in which percentages are sometimes expressed is with the total number of cases ($N$) used as the base. These overall percentages express the proportion of the sample who share two properties. For example, 7 of 89 respondents (7.9%) support abortion and have job security. Overall percentages do not have as much research utility as row and column percentages and are used less frequently.

7. The same three properties are also discussed by Joseph P. Healey in *Statistics: A Tool for Social Research*, 5th ed. (Belmont, CA: Wadsworth, 1999), pp. 314–320.

8. Church attendance has been recoded into three categories.

9. For purposes of illustration, only selected categories of educational level and attendance of religious services are shown.

10. Because statistical independence is a symmetrical property, the distribution of the independent variable within each category of the dependent variable will also be identical. That is, if gender and first-generation college status were statistically independent, we would also expect to see the distribution of gender identical in each category of the variable first generation.

11. Paul Mazerolle, Alex Piquero, and Robert Brame, "Violent Onset Offenders: Do Initial Experiences Shape Criminal Career Dimensions?" *International Criminal Justice Review* 20, no. 2 (2010): 132–146.

12. Ibid., p. 136.

13. Although this general formula provides a framework for all PRE measures of association, only lambda is illustrated with this formula. Gamma, which is discussed in the next section, is calculated with a different formula. Both are interpreted as PRE measures.

# Chapter 9

1. Refer to Paul Allison's *Multiple Regression: A Primer* (Thousand Oaks, CA: SAGE, 1999) for a complete discussion of multiple regression—statistical methods and techniques that consider the relationship between one dependent variable and one or more independent variables.

2. If you obtain $r$ simply by taking the square root of $r^2$, make sure not to lose the sign of $r$ ($r^2$ is always positive, but $r$ can also be negative), which can be ascertained by looking at the sign of $S_{yx}$.

3. Guttmacher Institute, *U.S. Teenage Pregnancies, Births and Abortions: National and State Trends and Trends by Race and Ethnicity*, January 2010.

4. J. M. Greene and C. L. Ringwalt, "Pregnancy Among Three National Samples of Runaway Homeless Youth," *Journal of Adolescent Health* 23 (1998): 370–377.

5. William J. Wilson, *The Truly Disadvantaged: The Inner City, the Underclass, and Public Policy* (Chicago: University of Chicago Press, 1987).

6. Stephanie Coontz, "The Welfare Discussion We Really Need," *Christian Science Monitor* (December 29, 1994): 19.

7. The District of Columbia was removed from the analysis due to its extremely high teen pregnancy rate relative to other states.

8. Analysis is limited to women 40 years and older.

9. Michael L. Benson, John Wooldredge, Amy B. Thistlethwaite, and Greer Litton Fox, "The Correlation Between Race and Domestic Violence Is Confounded With Community Context," *Social Problems* 51, no. 3 (2004): 326–342.

10. Michael L. Benson and Greer L. Fox, *Economic Distress, Community Context and Intimate Violence: An Application and Extension of Social Disorganization Theory*, U.S. Department of Justice, Document No. 193433, 2002.

11. Ibid.

# Chapter 10

1. Sarah Crissey, *Educational Attainment in the United States: 2007*, Current Population Report, P20-560 (Washington, DC: U.S. Census Bureau, 2009).

2. Since the $N$ in our computational example is small ($N = 21$), the assumptions of normality and homogeneity of variance are required. We've selected a small $N$ to demonstrate the calculations for $F$ and have proceeded with Assumptions 3 and 4. If a researcher is not comfortable with making these assumptions for a small sample, she or he can increase the size of $N$. In general, the $F$ test is known to be robust with respect to moderate violations of these assumptions. A larger $N$ increases the $F$ test's robustness to severe departures from the normality and homogeneity of variance assumptions.

3. Carol Musil, Camille Warner, Jaclene Zauszniewski, May Wykle, and Theresa Standing, "Grandmother Caregiving, Family Stress and Strain, and Depressive Symptoms," *Western Journal of Nursing Research* 31, no. 3 (2009): 389–408.

4. Ibid., p. 395.

5. Ibid., p. 391.

# Index

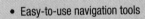